Intelligent Solutions for Optimizing Agriculture and Tackling Climate Change:

Current and Future Dimensions

Ahmed Karmaoui
Moulay Ismail University, Meknès, Morocco & Moroccan Center for Culture and Science, Morocco

A volume in the Advances in Environmental Engineering and Green Technologies (AEEGT) Book Series

IGI Global
PUBLISHER of TIMELY KNOWLEDGE

Published in the United States of America by
IGI Global
Engineering Science Reference (an imprint of IGI Global)
701 E. Chocolate Avenue
Hershey PA, USA 17033
Tel: 717-533-8845
Fax: 717-533-8661
E-mail: cust@igi-global.com
Web site: http://www.igi-global.com

Library of Congress Cataloging-in-Publication Data

Names: Karmaoui, Ahmed, 1982- editor.
Title: Intelligent solutions for optimizing agriculture and tackling climate change : current and future dimensions / Ahmed Karmaoui, editor
Description: Hershey, PA : Engineering Science Reference, [2022] | Includes bibliographical references and index. | Summary: “This book will research the impacts of climate change on agriculture including water shortage, water salinity, soil degradation, infectious disease, thermal and saline stress, and agricultural yield decreasing in an attempt to highlight the importance of artificial intelligence and how it will change the future of agriculture”-- Provided by publisher.
Identifiers: LCCN 2022013987 (print) | LCCN 2022013988 (ebook) | ISBN 9781668446492 (h/c) | ISBN 9781668446508 (s/c) | ISBN 9781668446515 (eISBN)
Subjects: LCSH: Crops and climate. | Climatic changes. | Meteorology, Agricultural--Methodology. | Agricultural ecology. | Agricultural innovations.
Classification: LCC S600.4 .I58 2022 (print) | LCC S600.4 (ebook) | DDC 630.2/515--dc23/eng/20220401
LC record available at https://lccn.loc.gov/2022013987
LC ebook record available at https://lccn.loc.gov/2022013988

This book is published in the IGI Global book series Advances in Environmental Engineering and Green Technologies (AEEGT) (ISSN: 2326-9162; eISSN: 2326-9170)

British Cataloguing in Publication Data
A Cataloguing in Publication record for this book is available from the British Library.

Advances in Environmental Engineering and Green Technologies (AEEGT) Book Series

ISSN:2326-9162
EISSN:2326-9170

MISSION

Growing awareness and an increased focus on environmental issues such as climate change, energy use, and loss of non-renewable resources have brought about a greater need for research that provides potential solutions to these problems. Research in environmental science and engineering continues to play a vital role in uncovering new opportunities for a "green" future.

The **Advances in Environmental Engineering and Green Technologies (AEEGT)** book series is a mouthpiece for research in all aspects of environmental science, earth science, and green initiatives. This series supports the ongoing research in this field through publishing books that discuss topics within environmental engineering or that deal with the interdisciplinary field of green technologies.

COVERAGE

- Sustainable Communities
- Alternative Power Sources
- Water Supply and Treatment
- Electric Vehicles
- Air Quality
- Policies Involving Green Technologies and Environmental Engineering
- Industrial Waste Management and Minimization
- Cleantech
- Radioactive Waste Treatment
- Green Transportation

IGI Global is currently accepting manuscripts for publication within this series. To submit a proposal for a volume in this series, please contact our Acquisition Editors at Acquisitions@igi-global.com or visit: http://www.igi-global.com/publish/.

The Advances in Environmental Engineering and Green Technologies (AEEGT) Book Series (ISSN 2326-9162) is published by IGI Global, 701 E. Chocolate Avenue, Hershey, PA 17033-1240, USA, www.igi-global.com. This series is composed of titles available for purchase individually; each title is edited to be contextually exclusive from any other title within the series. For pricing and ordering information please visit http://www.igi-global.com/book-series/advances-environmental-engineering-green-technologies/73679. Postmaster: Send all address changes to above address.

Titles in this Series

For a list of additional titles in this series, please visit: http://www.igi-global.com/book-series/

Recent Advances in Global Meliponiculture
Shamsul Bahri Abd Razak (Universiti Malaysia Terengganu, Malaysia) Tuan Zainazor Tuan Chilek (Universiti Malaysia Terengganu, Malaysia) and Jumadil Saputra (Universiti Malaysia Terengganu, Malaysia)
Engineering Science Reference • © 2023 • 250pp • H/C (ISBN: 9781668462652) • US $215.00

Implications of Nanoecotoxicology on Environmental Sustainability
Rafiq Lone (Central University of Kashmir, Ganderbal, Jammu and Kashmir, India) and Javid Ahmad Malik (Guru Ghasidas Vishwavidyalaya, Bilaspur Chhattisgarh, India)
Engineering Science Reference • © 2023 • 300pp • H/C (ISBN: 9781668455333) • US $250.00

Food Sustainability, Environmental Awareness, and Adaptation and Mitigation Strategies for Developing Countries
Ahmad Ni'matullah Al-Baarri (Diponegoro University, Indonesia) and Diana Nur Afifah (Diponegoro University, Indonesia)
Engineering Science Reference • © 2023 • 315pp • H/C (ISBN: 9781668456293) • US $110.00

Biomass and Bioenergy Solutions for Climate Change Mitigation and Sustainability
Ashok Kumar Rathoure (M/s Akone Services, India) and Shankar Mukundrao Khade (Ajeenkya D.Y. Patil University, India)
Engineering Science Reference • © 2023 • 412pp • H/C (ISBN: 9781668452691) • US $250.00

Transcending Humanitarian Engineering Strategies for Sustainable Futures
Yiannis Koumpouros (University of West Attica, Greece) Angelos Georgoulas (University of West Attica, Greece) and Georgia Kremmyda (University of Warwick, UK)
Engineering Science Reference • © 2023 • 315pp • H/C (ISBN: 9781668456194) • US $270.00

IGI Global
PUBLISHER of TIMELY KNOWLEDGE

701 East Chocolate Avenue, Hershey, PA 17033, USA
Tel: 717-533-8845 x100 • Fax: 717-533-8661
E-Mail: cust@igi-global.com • www.igi-global.com

Table of Contents

Section 2
Ecological and Socio-Economic Techniques to Optimize Agriculture in Context of Climate Change

Detailed Table of Contents

Anthropogenic activities intensify greenhouse gases (GHG) emission. The emissions lead to air pollution, public health risks, and climate change vagaries. Global deaths due to air pollution amounted to 15 deaths/minute in 2016. Moreover, floods, storms, and droughts accounted for 59%, 26%, and 15% respectively of crop losses between 2003 and 2013. Carbon recycling is among efforts to curb GHG which form 75% of GHG. The recycling methods include carbon capture and storage (CCS), carbon capture and utilization (CCU) and carbon capture, storage, and utilization (CCUS). However, these methods are too expensive for developing countries like Tanzania. Agroforestry is a cost-effective carbon recycler compared to other solutions. Besides, the Neem tree has a higher capacity of sequestering carbon at an average of 161% compared to other tree species in the tropics. Application of artificial intelligence can intensify Neem tree-based farming to hasten carbon sequestration.

Climate change is expected to greatly alter and modify all life forms on Earth. Therefore, maintaining natural and agroecosystems sustainability is one of the major scientific concerns in the current era. Reducing greenhouse gas (GHG) emissions represents an urgent need to mitigate the effects and consequences of climate change. The present chapter analyzes carbon footprint and sequestration in different olive plantations, from extensively managed to super-intensively managed plantations, in three geographic districts of Marrakech-Safi region in Morocco. Olive tree plantations showed an important capacity to store CO2, especially in high-density plantations, however, the carbon footprint is still high compared with extensively managed plantations. The chapter contributes to an innovative solution to reduce GHG and, consequently, to maintain the sustainability of the agricultural sector in Morocco under the current context of global changes.

Climate change has negatively affected crop productivity and induced intensive degradation of various ecosystems components. Eco-friendly management practices, such as the use of arbuscular mycorrhizal fungi (AMF), are becoming a crucial tool to improve crop yield and productivity under changing climate. AMF are generally known as biofertilizers, which can improve plant growth and development and provide protection to plants against environmental constraints such as soil poverty, drought, salinity, as well as metal toxicity. AMF are natural root symbionts that provide essential mineral nutrients to plants in order to improve their growth under normal and stressed conditions. AMF's role as bio-fertilizers can potentially fortify plants development against a changing environment. The present chapter will provide current comprehensive knowledge on AMF and their influence on plants growth and yield through improving nutrient uptake as well as soil characteristics, which can contribute to the maintenance of ecosystem health and sustainability under harsh environmental conditions.

Worldwide, the arid lands occupy one of the largest areas on Earth, covering more than 30% of the total area. In these lands, scarcity of resources has emerged as a consequence of climate change (CC) and anthropogenic pressure. The reduced resources require sustainable management, including the use of efficient methods and techniques to adapt to CC. The purpose of this chapter is to present a collection of affordable techniques and practices to optimize farming productivity in the context of CC. This chapter also presents the strengths and weaknesses and highlights the replication of these methods where water supply and food security are far from being achieved. Each technic must be studied and adapted with caution in context of the defined arid area, taking into account the local socio-economic and ecological characteristics. The proposed techniques may be used in governmental planning to support the adaptation and mitigation strategies.

Foreword

Intelligent Solutions for Optimizing Agriculture and Tackling Climate Change: Current and Future Dimensions is another outstanding book edited by Prof. Ahmed Karmaoui and released in a critical moment. In a fascinating, comprehensible, and delightful manner, the editor put together several pieces to solve the "puzzle" of the human future under climate change.

The relationship of humans with their environment has changed over time and the development of agriculture. Until the beginning of the Holocene, humans were hunters and gatherers. Then, with an increasing population, they started growing plants and animals. Thus, domestication has greatly contributed to improving the quality and quantity of agricultural products. However, this in turn created a profound transformation of earth's ecosystems that is now apparent and often referred to as "global change". The overexploitation of natural resources has undoubtedly reshaped the functioning of the Earth system, including the effects of climate change. Current evidence suggested that human-driven changes to the global environment are in many cases unprecedented. Therefore, and more than ever, looking for innovative and intelligent solutions to mitigate the effects of global changes is an urgent need to define the human future on Earth.

This book focuses on "Intelligent" solutions that can be used or at least considered to mitigate the impact of climate change on agriculture. It addresses several important but difficult questions. How could artificial intelligence be used to analyze the effects of climate change on agriculture? How can intelligent agricultural management practices mitigate climate change? What are the implications of global change for the future and the sustainability of agriculture? Can smart solutions be used in optimizing agriculture and tackling climate change? By raising and attempting to address these questions in this book, the editor hope to give some direction to the future of agroecosystems and to put forward critical perspectives to find answers to these questions.

This book could not have been possible without the active involvement of a large number of researchers. The book's production has truly been a community effort, where Prof. Karmaoui led us on an incredible journey where the reader could discover a wide range of topics through carefully selected chapters to solve the "puzzle" of the Human future.

Finally, *Intelligent Solutions for Optimizing Agriculture and Tackling Climate Change: Current and Future Dimensions* stands as one contribution of the many required to build the knowledge base to support the sustainability of agriculture in the current Era of Human history.

Jalal Kassout
National Institute of Agronomic Research, CRRA-Marrakech, Morocco

Preface

The interactions between climate change (CC), agriculture, and technology are of increasing concern to academics, educators, and planners. After the publication of the first report of the Intergovernmental Panel on Climate Change (IPCC), studies of the impact of climate change on productive systems such as agriculture have multiplied. In recent years, there has been a huge amount of scientific work on climate change and its effects on agriculture. Climate change, water overuse, droughts, soil salinity, groundwater reduction, salinity, land degradation, and desertification are the principal causes of the degradation of agroecosystems. These risks push for better management of agricultural systems and natural resources using sustainable solutions. The best solutions become often from new technologies and artificial intelligence (AI) offers promising tools. This smart technology was used in a wide range of fields, such as biology geosciences, parasitology, and natural resource management. It can make decisions in real-time and with low effort, which changes the agriculture sectors using sensors and cameras. Through the collection and analysis of data, the use of AI can intensify farming to increase carbon sequestration, help in protecting crops make decisions about irrigation, plant nutrition, weed protection, phyto-health, yield, and harvesting, and forecast weather. Ai is powerful and useful through many techniques such as Agriculture robots, the use of soil moisture and raindrop sensors, weed control automated systems, Unmanned Aeronautical Vehicles (UAVs) to map and monitor the farms, and machine learning (ML). Intelligent solutions include artificial intelligence using sensors, images, and robots, recent methods, and eco-friendly management techniques such as the use of arbuscular mycorrhizal fungi, and many traditional techniques based on ancestral knowledge.

Recent techniques such as the conversion of agricultural land to fallow land, the use of rainwater harvesting to cope with water scarcity, the dew collection for harvesting water, and buried diffusers can be used to both save irrigation water and maintain soil quality, the magnetic water irrigation to fight against the salinity of water, the fight against desertification using physical and biologic stabilizing systems. The use of renewable energy techniques includes solar and eolian energy systems and pumped hydroelectric. Traditional technologies are also important alternatives that have proven to be effective in several areas such as Khattaras, composting, waste recycling, etc. These practices can be used to increase productivity, diversify crops, and control pests. History records many examples of the use of tools and techniques to improve and maintain agricultural production. It started by harnessing the physical strength of the group, exploiting animal power, then using steam power, and currently renewable energies after using the machine. Agriculture was mechanized with the introduction of the mechanized plow that was drawn by animals. Later the generalization of the use of tractors, trucks, harvesters, and planes for spraying pesticides expanded, and then the use of computers, satellite images, and the Global Positioning System. Currently, artificial intelligence is being developed through the development of software, methods, and algorithms. This development allowed the optimal use of natural resources, soil, water, human forces, and environmental factors, which led to an increase in their preservation, an increase in the quantity of production, and an improvement in quality. These methods are becoming essential to support soil, water, and biodiversity as the primary pillars to improve animal and crop productivity in the context of climate change. The adoption of these techniques is associated with weaknesses. The application and the use of any method or practice should be done in light of its reliability, feasibility, energy consumption, cost, security, utility and productivity, transferability, and ease of use.

This edited book groups the contributions of many researchers and scholars in the field of climate change and agriculture exploring the importance of intelligent techniques, methods, and practices to optimize farm productivity. It represents innovative approaches and reviews of the literature that provides useful and updated information about the topic. The relevance of this topic is evident since academics are increasingly aware of its importance in the context of CC.

The impacts of climate change on agriculture are numerous and diversified including water shortage, water salinity, soil degradation, infectious disease, thermal and saline stress, and agricultural yield decreasing … These results are exacerbated due by anthropogenic pressure such as air pollution, over-use of water, soil and biodiversity resources, urbanization, and industrialization, etc.

In this edited book, different aspects of CC associated with agriculture and artificial intelligence will be explored. Relevant theoretical Approaches and current empirical research findings were investigated. Two main axes were explored including 1) The importance of artificial intelligence, in the agriculture sector, current and future dimensions, and 2) Recent and traditional techniques in agriculture, the opportunities, and challenges.

The current edited book provides an essential resource for academics, scholars, practitioners, and for professionals in this field of agriculture. It was a forum to share updated, original, and relevant knowledge, empirical research, and experiences in tackling climate change and optimizing agriculture production. Intelligent solutions for optimizing agriculture and tackling CC refer to the summation of all traditional and modern techniques and methods to increase farm productivity sustainably. The edited book includes a collection of chapters about artificial intelligence and some practices such as soil and water improvement. This edited book is divided into two sections. Section 1 is about AI techniques to optimize agriculture in the context of climate change and is divided into three chapters:

Chapter 1 explores in detail the use of AI in agroforestry as economic solution for carbon recycling in Tanzania. Anthropogenic activities intensify greenhouse gases (GHG) emission. The emissions lead to air pollution, public health risks and climate change vagaries. Global deaths due to air pollution amounted to 15 deaths/minute in 2016. Moreover, floods, storms and droughts accounted for 59%, 26% and 15% respectively of crop losses between 2003 and 2013. Carbon recycling is among efforts to curb GHG which form 75% of GHG. The recycling methods include carbon capture and storage (CCS), carbon capture and utilization (CCU) and carbon capture, storage and Utilization (CCUS). However, these methods are too expensive for developing countries like Tanzania. Agroforestry is the cost-effective carbon recycler compared to other solutions. Besides, Neem tree has higher capacity of sequestering carbon at an average of 161% compared to other tree species in the tropics. Application of Artificial Intelligence can intensify neem tree-based farming to hasten carbon sequestration.

Chapter 2 shifts focus on machine learning tools for ensuring optimized agribusiness recovery strategies. This chapter presents machine learning approaches for business continuity and recovery optimization in agribusiness. Firstly, a mathematical method, entitled Business Continuity Points (BCPTs) is tested with domain data for its potential to predict process recovery results, namely recovery time and criticality ranking of key operations. A 72.22% accuracy has been estimated. Then, decision tree prediction with 10-fold cross validation and random forest, has been 92.31% accurate in classifying business functions as critical or not. Additionally, a new multi-approach and multi-class decision tree classifier with some of the BCPTs input variables is presented, with 55.36% accuracy and 70.37% and 88.89% accuracy rates when boosted with the 10 folds and the random forest. Finally, regression analysis techniques are used to improve the initial recovery time BCPTs formula. Exponential regression has been more precise compared to the quadratic model (R2exp=0.954, R2quad=0.85). Despite current data limitations, the inferred prediction patterns are robust and highly accurate in the given field.

Chapter 3 reports the future of artificial intelligence in agricultural field using a bibliometric analysis. Artificial intelligence (AI) is well known for diverse applications in many fields including agriculture. The use of AI in the agricultural field has allowed for optimization in real-time and with less cost and effort. This chapter presents a bibliometric analysis of the future of AI in agriculture. This analysis was carried out using the online Scopus tool and VOSviewer software. The outputs show an increasing trend of documents from 1979 to 2021 and 2020 is the most productive year in terms of quantity. Computer science, engineering, Agricultural and Biological sciences, and environmental science are the main subjects while the United States, China, and India are the most influential in the field. United States, China, and India are the most influential in the field, while article and conference papers are the top two doctypes. For the trend, research in this field is turning toward machine learning, the Internet of things, deep learning, agricultural robots, learning algorithms, big data, data analysis, image processing, support vector machines, and automation.

Section 2 explored the ecological and socio-economic techniques to optimize agriculture in context of climate change and included five chapters:

Chapter 4 assesses the carbon balance in olive cultivation systems in Morocco: presenting an intelligent solution in climate change mitigation. Climate change is expected to greatly alter and modify all life forms on Earth. Therefore, maintaining natural and agroecosystems sustainability is one of the major scientific concerns in the current era. Reducing greenhouse gas (GHG) emissions represents an urgent need to mitigate the effects and consequences of climate change. The present chapter analyzes carbon footprint and sequestration in different olive plantations from extensively managed to super-intensively managed plantations in three geographic districts of Marrakech-Safi region in Morocco. Olive tree plantations showed an important capacity to store CO2, especially in high-density plantations, however, the carbon footprint is still high compared with extensively managed plantations. The chapter contributes to an innovative solution to reduce GHG and, consequently, to maintain the sustainability of the agricultural sector in Morocco under the current context of global changes.

Chapter 5 presented the arbuscular mycorrhizal fungi technique to enhance the production and crop yield under changing climate. Climate change has negatively affected crop productivity and induced intensive degradation of various ecosystems components. Eco-friendly management practices, such as the use of arbuscular mycorrhizal fungi (AMF), are becoming a crucial tool to improve crop yield and productivity under changing climate. AMF are generally known as biofertilizers, which can improve plant growth and development and provide protection to plants against environmental constraints such as soil poverty, drought, salinity, as well as metals toxicity. AMF are natural root symbionts that provide essential mineral nutrients to plants in order to improve their growth under normal and stressed conditions. AMF role as bio-fertilizers can potentially fortify plants development against a changing environment. The present chapter will provide current comprehensive knowledge on AMF and their influence on plants growth and yield through improving nutrient uptake as well as soil characteristics, which can contribute to the maintenance of ecosystem health and sustainability under harsh environmental conditions.

Chapter 6 reviewed some affordable methods and techniques used to optimize farming productivity in a context of climate change in arid area. Worldwide, the arid lands occupy one of the largest areas on Earth covering more than 30% of the total area. In these lands, scarcity of resources has emerged as a consequence of climate change (CC) and anthropogenic pressure. The reduced resources require sustainable management including the use of efficient methods and techniques to adapt to CC. The purpose of this chapter is to present a collection of affordable techniques and practices to optimize farming productivity in the context of CC. This chapter also presents the strengths and weaknesses and highlights the replication of these methods where water supply and food security are far from being achieved. Each technic must be studied and adapted with caution in context of the defined arid area taking into account the local socio-economic and ecological characteristics. The proposed techniques may be used in governmental planning to support the adaptation and mitigation strategies.

Chapter 7 analyzed the population projections, food consumption and agricultural production under climatic constraints. Climate change is expected to have a significant impact on agricultural production both locally and globally. High temperatures and irregular precipitation projected by the IPCC could lead to a decline in agricultural production, which would require significant changes in agricultural practices. A concomitant factor to climate change will be the increase in population and its distribution and level of consumption which also influence agricultural production strategies, the conversion of agricultural land, the type of irrigation. Determining the impacts of these forcings on land use will depend on our understanding of past changes in the agricultural sector and how future climate change will occur. It is proposed to develop an innovative methodology that will integrate agricultural production, socio-economic statistics economics and demographic trends and patterns of agricultural land use caused by climate change and human development.

Chapter 8 reviews the global warming reassessing the origin of the problem, and proposing an effective human-scale solution. In this chapter, the authors presented a bibliometric analysis and a short discussion of some solutions to global warming. The current chapter attempts to reassess the origin of Global Warming, and proposes an effective human-scale solution. The bibliometric review reveals an increasing trend of indexed publication from 1936 to 2020. The trend analysis shows research turning towards topics such as performance, life-cycle assessment, electric vehicles, construction, thermal comfort, mechanical properties, and quality. With regard to the origin of Global Warming, doubt is cast on the view that increased heat in the atmosphere is chiefly caused by increased carbon. From the point of view of chemistry and physics, CO2 is not a cause of heat, but a result of what creates heat (combustion). Moreover, the low-carbon strategies suggested to control the increase of heat seem slow, difficult to measure and control, and so far, ineffective. In industry, it is axiomatic that good technologies have measurable results, and always work in the same way if conditions are as foreseen.

Ahmed Karmaoui
Moulay Ismail University, Morocco

Acknowledgment

It is widely recognized that climate change in the form of decreasing precipitation and increasing temperatures is affecting many productive sectors, including agriculture. The challenges are multiplying, like the increase in extreme events frequency, pest invasion, soil and water degradation, and decrease in animal and crop productivity. This negative trend required substantial efforts to combat climate change and achieve sustainable development goals, such as food security support and poverty alleviation. Intelligent solutions, namely artificial intelligence using sensors, robots, eco-friendly management techniques, such as the use of arbuscular mycorrhizal fungi, and many traditional techniques based on ancestral knowledge. These methods are becoming essential to support soil, water, and biodiversity as the primary pillars to improve animal and crop productivity in context of climate change.

This edited book groups the contribution of many researchers and scholars in the field of climate change and agriculture exploring the importance of intelligent techniques, methods, and practices to optimize farm productivity.

I would like to thank the authors who contributed to this project, and the reviewers who assisted us in reading and reviewing the chapters. I would like to thank everyone who supported us to finalize this project. Special thanks to the IGI Global team for the editorial work, assistance, and patience to produce this collective book.

Editorial Advisory Board

Ahmed Karmaoui
Moulay Ismail University, Morocco & Moroccan Center for Culture and Science, Morocco

Section 1

Artificial Intelligence Techniques to Optimize Agriculture in Context of Climate Change

Chapter 1
Using Artificial Intelligence in Agroforestry as an Economic Solution for Carbon Recycling in Tanzania

Proscovia Paschal Kamugisha
Mbeya Campus College, Mzumbe University, Tanzania

Sebastian Faustin Mhanga
https://orcid.org/0000-0002-9991-1394
Expand (T) Limited, Tanzania

ABSTRACT

Anthropogenic activities intensify greenhouse gases (GHG) emission. The emissions lead to air pollution, public health risks, and climate change vagaries. Global deaths due to air pollution amounted to 15 deaths/minute in 2016. Moreover, floods, storms, and droughts accounted for 59%, 26%, and 15% respectively of crop losses between 2003 and 2013. Carbon recycling is among efforts to curb GHG which form 75% of GHG. The recycling methods include carbon capture and storage (CCS), carbon capture and utilization (CCU) and carbon capture, storage, and utilization (CCUS). However, these methods are too expensive for developing countries like Tanzania. Agroforestry is a cost-effective carbon recycler compared to other solutions. Besides, the Neem tree has a higher capacity of sequestering carbon at an average of 161% compared to other tree species in the tropics. Application of artificial intelligence can intensify Neem tree-based farming to hasten carbon sequestration.

DOI: 10.4018/978-1-6684-4649-2.ch001

INTRODUCTION

Background

The sustainable development goal 13 of the United Nations intend to achieve net-zero greenhouse gas (GHG) emissions by 2050. To attain this objective, transformation of value chains across the economy is the prerequisite (Rizos et al, 2018). Transformation encompasses new products and business models to look for means of converting what a normal linear economy considered as wastes into a new product circle. Energy-intensive industries normally emit Carbon dioxide (CO_2) and hydrocarbons as wastes that in turn pollute the environment, impair food security and pose health risks to the public. Carbon dioxide (CO_2) and methane (CH_4) accounted for 73% and 19% of greenhouse gases (GHGs) emitted in 2019 (Olivier & Peters, 2020). The concentration of CO_2 in the atmosphere has increased by 31% since the commencement of second half of nineteenth century during the industrial era (UNCTAD, 2019).

This situation has resulted into increased global air temperatures, melting of ice and snow that in turn lead to increased risen sea levels (UNCTAD, 2019). The rise of air temperature is supported by evidence from the five warmest years ever occurred since 1880 as from 2015 that reached the maximum in 2019 with global land and ocean temperature of 0.95^0C above the average (Olivier & Peters, 2020). Public health concerns and substantial agricultural losses have also been documented. According to (Landrigan et al., 2018) deaths attributable to air pollution accounted for 16% leading to the tune of 9 million premature deaths in 2015 worldwide. Likewise, WHO (2016) revealed that in 2016, almost a quarter (23%) of all global deaths were attributable to air pollution, amounting to 8.2 million deaths annually; which is equivalent to 936 deaths per minute. The ambient air pollution was responsible for 4.2 million premature deaths in 2016; whose 7.11% (300,000) cases were children under the age of 5 years (WHO, 2018). The global welfare losses associated with air pollution stood at US$ 4.6 trillion which is equivalent to 6.2% of the world economic output, 1.7% losses of GDP in low & middle income countries and 2% losses of GDP in high income countries per annum (Landrigan et al., 2018). However, the burden of disease attributable to air pollution is heaviest in low and middle-income countries (LMICs) (Figure 1) (WHO, 2018; WHO, 2016). It is estimated that deaths associated with outdoor air pollution in Africa has increased from 164,000 cases in 1990 to 258,000 cases in 2017 (Rees et al., 2019) that is equivalent to a growth of 60% in a 17 years period.

Figure 1.

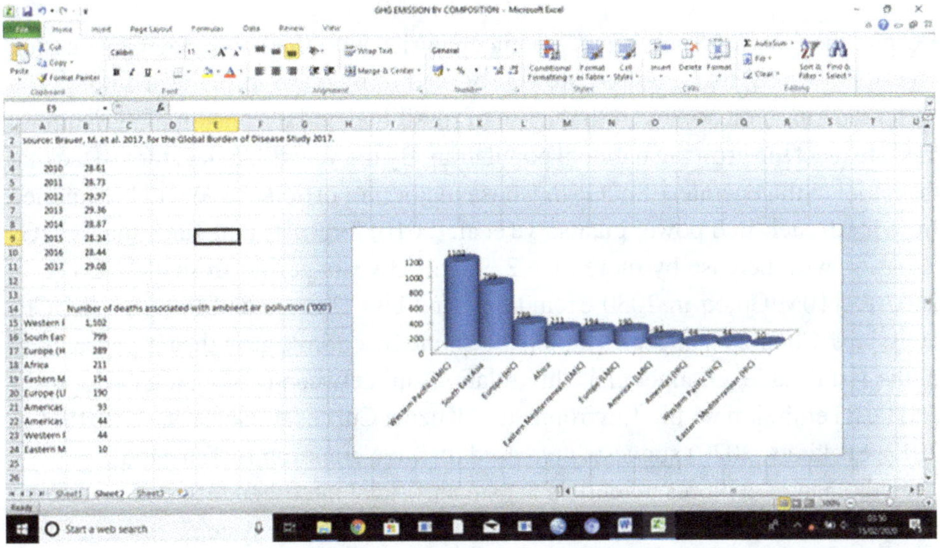

In Tanzania, almost 19% of annual deaths is associated with impacts of air pollution related diseases that amounted to 74,170 cases in 2016 (URT, 2019b). ; At the same time, the proportion of Death and Disability Adjusted Life Years (DALYs) attributable to communicable and non-communicable diseases accounted for 62% and 30.8% respectively in 2016 (URT, 2019a). Moreover, the report estimated that diseases attributable to air pollution have costed the country up to US$ 540 and US$ 665 million in 2015 due to lost productivity and welfare damages (URT, 2019a).

Episodes of adverse weather conditions rank the second negative impact of air pollution. The climate change accounted for 65% of vagaries of weather conditions in more than half of last decade worldwide (Herring et al., 2018). Floods, storms and droughts accounted for 59%, 26% and 15% respectively of global crop losses and damages to agricultural infrastructure between 2003 and 2013 (FAO, 2018). In monetary terms, natural disasters were responsible for $96 billion worth of agricultural produce loss in developing countries between 2005 and 2015 (FAO, 2018). Meanwhile the increased temperature of 1^0C in developing countries is associated with a decline of 2.66% agricultural output; and decreased export growth by 2.0–5.6% (Jones & Olken, 2010).

Status of GHGs in Tanzania

Despite the fact that Tanzania is among countries with the lowest GHGs emission rate in the world, its emission rate is projected to increase in the near future. The increased emission comes from the use of fossil fuel and expansion of agricultural lands to meet the growing national population. By the year 2025, it is forecasted that the nation will generate 1,800GWh whose major proportion (75%) will be supplied from fossil fuel fired power plants Pye et al. (2010) forecasts that Tanzania's GHGs emission will increase by more than 229% in 23 years period from 48MtCo$_2$eq in 2007 to 110MtCo$_2$eq in 2030 excluding Land Use Change and Forestry (LUCF) emissions. Coupled with recent national sustained economic growth of 7% per annum in the last decade, Granger causality relationship between economic development and GHG emission within Environmental Kuznet Curve (Adzawla et al., 2019; Lu, 2017; Seekings, 2020) suggests increased massive emission in Tanzania.

This chapter explores potentials of using artificial intelligence to optimize neem tree plantations and neem tree-based agroforestry farming systems to recycle the increasing levels of carbon. In this chapter, AI means employment of automation devices in agroforestry & forestry production; and farm management. It entails the use of Unmanned Aerial Vehicles (UAVs), Internet of Things (IoT) machine learning (ML), fuzzy logic and Robotic technologies in planting, weeding, identifying diseases; and spraying fertilizers, herbicides, pesticides & fungicides (Bhagat et al., 2022). The use of Artificial Intelligence (AI) has a potential to lessen GHG emission, raise agroforestry farming and increase farm yield through precision agriculture (Debangshi, 2021; Giacomo & David, 2018) . Similarly, Internet of Things (IoT) gather data on farm elements including weather, soil quality attributes, temperature, water levels and invasion of weeds in farms (Betteena et al., 2020). The drones are capable to plant 40,000 seeds/day and can accomplish the man-day planting within ten minutes. It reduces the planting costs by 85% and saves costs of pesticides by 30% (Debangshi, 2021). Soylu & Çarman (2021) revealed that the use of fuzzy logic reduced tractor slippage and fuel consumption by 42% and 44% respectively.

Essentially, CO_2 that has been long regarded as threatening waste will be turned to business potential as carbon stock in neem trees to meet rural and urban dwellers as a feasible low-cost investment in forestry and agroforestry farming. This will in turn contribute to close the national supply gap of wood related products. Tanzanian annual per capita demand for wood in 2015 stood at 1.39m^3 which is far more above the allowable cut of 0.95m^3 per annum (URT, 2015a).

The changing role of CO_2 is linked to introduction of circular economy principles that change industrial waste into valuable resource (Akash et al., 2016). Neem tree has proved to be an effective means of sequestrating carbon dioxide into stocks of woods that can be drawn to save and expand the national timber sub-sector (Bohre

& Chaubey, 2016; Saka, 2020). The neem trees portray significant higher carbon sequestration potentials up to 243% higher compared to other tree cultivars including *Ficus benghalesis, Tamarindius indica, Casurina equisetiflolia and Conocarps eretus* (Arya et al., 2017; Moussa et al., 2018; Noumi et al., 2018); and develop huge canopy at young age that hasten micro climate amelioration on the ground. Bohre & Chaubey (2016) found that at the age of ten years neem tree sequestered carbon amounting to 144tons/ha and 529ton/ha of carbon dioxide (CO_2) in Indian Coal Mined Lands.

A number of studies on options to capture carbon and store for the purpose of transforming carbon from harmful waste to other products in the country exists (Hansson et al., 2020; Jones & Olken, 2010; Lupala et al., 2014). However, these studies focused on financing mechanisms of low carbon investment potentials (Pye et al., 2010), feasibility of biomass production to meet bioenergy needs with carbon capture and storage (Hansson et al., 2020); and carbon sequestration in natural forests (Lupala et al., 2014; Malunguja et al., 2020). Hence, this book chapter, bridges this gap of knowledge by shedding light on investment potentials of neem tree (*Azadrachta Indica*) plantations and neem tree-based agroforestry using AI for investors to make informed decisions in their undertakings.

The rest of this chapter is organised as follows: Section 1.3 presents reviews of prevailing carbon recycling technologies. Section 2 explores sources of GHG emission in Tanzania that include deforestation and agriculture while section 3 presents current forest plantations and agroforestry practices in the country with the projected demand for the wood - based products. Section 4 portrays legal and regulatory framework for forest plantations and agroforestry; and section 5 describes neem tree carbon recycling potential and its possibility to meet national growing need of timber-based products. Moreover, section 6 discusses application of artificial intelligence in agriculture and forest sector management while section 7 provides conclusion and recommendations.

Efforts addressing Carbon Recycling

Conventional carbon recycling through sequestration is thought from three streams: carbon capture and storage (CCS), carbon capture and utilization (CCU) and carbon capture utilization and storage (CCUS) (Tcvetkov et al., 2019). CCS entails capturing and disposing CO_2 mainly by offshore storage (van der Zwaan & Gerlagh, 2016). CCU entails technologies which involve capturing of carbon dioxide and use it in the manufacturing industries as a constituent of various products such as carbonated beverages, mineralization of refrigerants and plastics. Meanwhile CCUS technologies encompasses the use of CO_2 to enhance efficiency of resource extraction like oil, geothermal energy and natural gas to mention a few (Nogia et al., 2016).

Numerous authors have documented significant reduction of CO_2 emission using CCS technologies (Akash et al., 2016; Odenberger et al., 2013; Singh et al., 2011). However, the technologies involve high costs such that it might face competitive disadvantage from other low carbon generation options. The economic assessment of CCUS technologies were found to have economic returns in different parts of the globe (Kim et al., 2016; Oladosu et al., 2016; Viebahn et al., 2015) such that CCUS enhanced coalbed Methane can be a significant competitor of natural gas. However, the technology development is still at its infant stage such that further investigations to validate its economic value are prerequisites. The economic assessment of various CCU based technologies indicate that these options are competitive with a win-win situation in the energy markets (Chauvy et al., 2019; Muthuraj & Mekonnen, 2018). However, these technologies are still at their early stage which can be used in processing of plastic wastes in future. Generally, CCU technological options have proved to reduce more quantities of CO_2 compared to CCS and CCUS based technologies (Tcvetkov et al., 2019).

The Integrated Assessment Models (IAMs) used by Intergovernmental Panel on Climate Change (IPCC) showed that Bioenergy with Carbon Capture and Storage (BECCS) was proved to be the most efficient Negative Emission Technologies (Beck & Mahony, 2018). Fuss et al. (2014) estimated the amounts of carbon removed from atmosphere to range between 2 and 10GtCO_2eq annually by 2050. Despite the fact that aforementioned carbon sequestration technologies reduce significant amount of CO_2 emission, their application is limited to developed countries since they involve huge amount of costs. Hansson et al. (2020) foresee barrier in adoption of BECCS technologies in Tanzania as they require massive land use that might end up to land use conflicts. Forest plantation has proved to be the best option to reduce carbon footprints as they act as carbon sinks due to its cost effectiveness compared to other engineering solutions (Engel et al., 2008; van Minnen et al., 2008).

Neem tree-based agroforestry practices are thought to have the best option with fast growing cultivars that lead to carbon sequestration, harness microclimate amelioration, leads to increased land productivity and improve the wood industry. Hence, Neem trees-based agroforestry and plantations also contribute to close the national supply gap in the wood industry. Bohre & Chaubey (2016) found that at the age of ten years neem tree sequestered carbon amounting to 144tons/ha and 529ton/ha of carbon dioxide (CO_2) in Indian Coal Mined Lands. Meanwhile, the study found increased content of carbons and CO_2 as trees increased in Diameter at Breast Height (DBH) and total height.

A number of emerging initiatives to address GHG emission using Artificial Intelligence in Agriculture and forestry exists in Tanzania. The *Ushaurikilimo* which means agricultural advice project is jointly implemented by Kilosa Local Government, Ministry of Agriculture, Food & Cooperatives, Kilosa Rural Services and Electronic Communication Center, Uyole Agricultural Research Institutes and Kilosa Community Radio (Shapa et al., 2021). Other initiatives include: Peresi Mgorozi who designed the AI to detect diseases in maize plants and detection of banana diseases using the plantae leaf (SAHARA-VENTURES, 2019). Meanwhile, Agrobot which enhances provision of advice regarding fertilization and usage of pesticides in Morogoro, Iringa and Dar-es-Salaam.

SOURCES OF GHGS EMISSION IN TANZANIA

Tanzania is acknowledged to be GHG sink whose forests absorbs more than what it emits in the atmosphere. WRI CAIT (2021) estimated that Tanzania's GHG emission stood at 315 $MtCO_2e$ in 2018; with annual per capita emission of 0.2 tCO2e (URT, 2015b). Almost three quarters (73%) of the emission was from Land Use Changes (LUC) sector (Figure 2). This, followed by agriculture (27%), energy (18%) and other fuel combustion (11%) (WRI-CAIT, 2021). Pye et al. (2010) projected the increase of national GHGs emission to 115Mt CO_2 by 2030; mainly from agricultural sector. Likewise, IPCC (2014) noted that agriculture, forestry, and other land use (AFOLU) accounted for 24% of global GHG emission in 2010; whose major components are Forest and other land uses (FOLU) that ranks second emitter after fossil fuel combustion.

Figure 2.

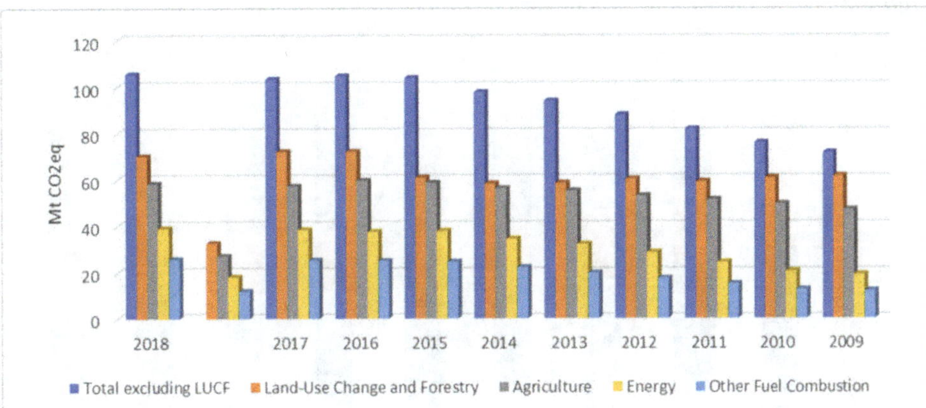

Deforestation

The change of forest into non-forest land uses leads to emission of CO_2 from the biomass that would have been stored in the biomass and deadwood. In Tanzania the area of forest degraded land is 500,000ha per annum (FAO, 2020; Kideghesho, 2015; URT, 2014) mainly due wood extraction for cooking energy, expansion of agricultural land and timber related activities (FAO, 2020). URT (2014) estimated the average standing wood volume to have declined from 180m^3 in 1961 to 50m^3 in 2012. Moreover, Lupala et al. (2014) noted that Tanzanian forest cover decline more than 50% in 1961 to about 36% in 1990s. Meanwhile, Lukumbuzya & Sianga (2017) noted a decreased forest growing stock of 1.1ha/capita during 2017 compared to that growing stock of 3ha/capita during 1980s. FAO (2020) depicted the declining trend of forest cover in Tanzania as from 1990s to 2020 (Figure 3). The declining trend of forest cover is associated with increased national GHGs emission from 10 Mt CO_2eq in 1990 to 40 Mt CO_2eq in 2030 (Pye et al., 2010).

Moreover, the national per capita wood demand of 1.39m^3/year is greater than allowable cut of 0.95m^3/year/capita (URT, 2015a) such that efforts to achieve sustainable production and efficient harvest of wood are pre-requisite. Msuya et al. (2011) projected that about 4,000,000ha will be cleared between 2010 and 2030 to meet national cooking needs. This is equivalent to 200,000ha forest loss annually that account for two fifth of the national deforestation rate of 500,000ha per annum (FAO, 2020).

Figure 3.

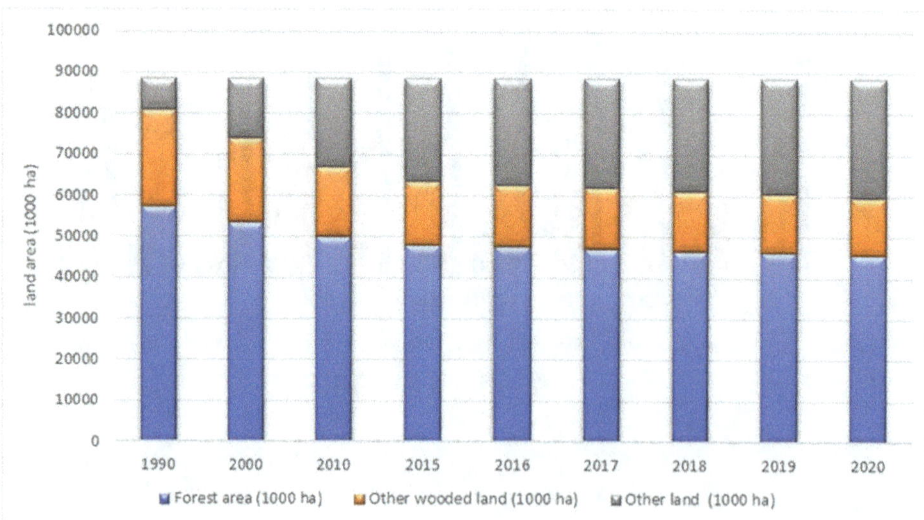

Agriculture

Agriculture is the second sector contributing to national GHG emission after LUCF accounting for 18.8% of the total national emission (WRI CAIT, 2021) mainly from enteric fermentation, manure management and burning of Savannah forests and agricultural residues (URT, 2020). Beghin (2015) projected the increase of per capita food demand from 138kg of grains equivalent per annum in 2013 to 187kg in 2023 that requires expansion of farmland. URT (2021) projected national population to increase from 46,356,279 in 2013 to 84,380,103 in 2030. Looking at minimum per capita food demand requirements of 138kg of grains equivalent/year, the national food demand will increase from 7,953,992 tons/year of grains equivalent in 2020 to 11,988,273tons/year grains equivalent in 2035. Based on (WB, 2018) estimates for grains yield of 1544kg/ha, Tanzania necessitates to expand area for food production from 5,151,549ha in 2020 to 7,764,425ha in 2035 with increase of agricultural land expansion of 171,191ha per annum.

Expansion of agricultural land reduces capacity to sequester CO_2 since forests are more effective in sequestrating the gas than agricultural related activities that normally replace forests (Pedersen, 2017). FAO (2017) asserted the compounding increase of global agricultural emission at annual rate of 1.2% between 1961 and 2016. The increasing national population of 2.7% growth rate coupled with increased per capita food demand, intensify potentials of encroaching more forest land that impair the GHG emission mitigation efforts. Therefore, determined efforts to meet national food demand coupled with hastened GHG emission mitigation is a crucial for our survival.

CURRENT FORESTRY AND AGROFORESTRY PRACTICES IN TANZANIA

Forestry Plantation in Tanzania

Tanzania's major portion of forest plantation is owned by the government (Ngaga, 2011; URT, 2015a). Dominance of government forest plantation can be traced back to 1970s when the government established large scale tree plantation in the southern highlands (Rojas-Downing et al., 2017) to cater for environmental concerns (Pye et al., 2010). Gradually, involvement of private sector took off such that until 2000s there were five private companies involved in forest plantations. (Mwaseba et al., 2020). Likewise, during 2000s a number of individuals joined forestry tree growing initiatives (PFP, 2016). Currently, it is estimated that 174,000 hectares (54%) of forest plantation in southern highlands are owned by small and medium scale tree growers, while 100,000 hectares (31%) by the government and 51,000 hectares (15%) by large private plantations (15%) (Held et al., 2017a). Lusambo et al. (2021) asserted that there are 16 plantations owned by the government throughout the country to date dominated by Sao hill (Table 1). The current 500,000ha woodlots for both central government and private investors is meant for timber & timber products; and paper industries (Chamuya & Khalid, 2018); and more than three quarters of forest plantations are located in the Southern Highlands (URT, 2015a).

Table 1a. Tanzania Governmental owned Forest Plantations. Source: (Lusambo et al., 2021; Ngaga, 2011)

S/n	Name of plantation	Location of the plantation (region)	Species within the plantation	Area of the plantation covered (ha)	Proportion of the total governmental forest plantations (%)
1	Sao Hill	Iringa	Cupressus lusitanica	104	
			Pinus spp	41,123	
			Eucalyptus spp	3,773	
		Sub total		**45,000**	**52.78**
2	Meru/USA	Arusha	Pines	950	
			Cypress	2,080	
			Grevillea	1,285	
			Acacia	199	
			Casuarina spp	205	
			Eucalyptus spp	492	
			Other tree spp	352	
		Sub – total		**5,563**	**6.53**
3	North Kilimanjaro	Kilimanjaro	Pinus patula	3,304	
			Cupressus Lusitanica	781	
			Mixed softwood	585	
			Grevillea robusta	232	
			Eucalyptus spp	131	
			Natural forest	193	
			Others	656	
		Sub total		**5,883**	**6.90**

continues on following page

Table 1. Continued

S/n	Name of plantation	Location of the plantation (region)	Species within the plantation	Area of the plantation covered (ha)	Proportion of the total governmental forest plantations (%)
4	West Kilimanjaro	Kilimanjaro	Pinus patula	38	
			Cupressus Lusitanica	954	
			Grevillea robusta	833	
			Eucalyptus spp	28	
			Others	632	
		Sub total		**2,485**	**2.91**
5	Buhindi	Mwanza	Pines	3,180	
			Cedrella odorata	38	
		Sub total		**3,218**	**3.77**
6	Kiwira	Mbeya	Pines	2,585	
			Other tree spp	199	
		Sub total		**2,784**	**3.27**
7	Rondo	Lindi	Pines and Cypress	1,557	
			Tectonal grandis	191	
			Milicia excelsa	297	
			Pinus caribea	1,142	
			Other hardwoods	38	
		Sub total		**2,083**	**2.44**
8	Kawetire	Mbeya	Pinus patula	1,465	
			Eucalyptus maidenii	136	
		Sub total		**1,601**	**1.90**
9	Rubya	Mwanza	Pines	1,524	
			Pines & cadrela	59	
		Sub total		**1,584**	**1.90**
10	Shume/Magamba	Tanga	Pines and Cypress	3,408	
			Eucalyptus spp	101	
			Juniperus procera	19	
			Acacia spp	241	
			Grevillea robusta	138	
			Cinamom camphora	89	
			Seed stands (TTSA)	16	
			Trial plots under TAFORI	66	
			Natural forests	77	
			Other species	719	
		Sub total		**4,558**	**5.35**

continues on following page

Table 1. Continued

S/n	Name of plantation	Location of the plantation (region)	Species within the plantation	Area of the plantation covered (ha)	Proportion of the total governmental forest plantations (%)
11	Longuza	Tanga	Tectonal grandis	1,346	
			Terminalia spp	12	
			Eucalyptus spp	3	
			Natural forests	739	
			C. odorata, C.tona & T. grandis	348	
			Chorophora regia	1	
		Sub total		**2 449**	**2.90**
12	Ukaguru	Morogoro	Pines spp	905	
			Cupressus lusitanica	36	
		Sub total		**941**	**1.10**
13	Mtibwa	Morogoro	Tectona grandis	1,354	
			Cadrella odorata	88	
		Sub total		**1,442**	**1.70**
14	Matorogoro	Songea	Pinus patula	402	
			Grevillea robusta	15	
			Indigenous spp	385	
			Natural forests	3,723	
			Other spp	66	
		Sub total		**4,592**	**5.39**
15	Ruvu wood fuel	Coast	-	633	
		Sub total		**633**	**0.74**
16	Rubare	Kagera	Pinus caribea	432	
		Sub total		**432**	**0.51**
		Grand total		**85, 248**	

The Tanzanian non-industrial private forestry (NIPF) are owned and managed by individuals or corporations whose objectives are not to meet timber production nor wood processing (Harrison et al., 2002). Recently, Tanzania has experienced steady growth in NIPF sub sector to meet market demand. Small holder forest plantation is becoming a significant part of the forest industry in the country as a number of individuals are joining this lucrative venture. The NIPFs supply about 250,000m^3 (rwe) annually from southern highlands of Tanzania (URT, 2011).

Tanzanian private forest plantation is dominated by pine trees (65%) and Eucalyptus (20%). The rest is made up with teak, grevillea and other tree species (Held et al., 2017).There is also an increasing trend of shifting from wood emanating from natural forests to plantation grown woods. The increasing patterns of private sector in tree plantations has been observed in Southern highlands whereby smallholder farmers' woodlots account for 67% of wood and Timber supply (Arvola et al., 2019).

Agroforestry in Tanzania

Conceptually, Agroforestry encompasses integration of trees with crops and/or animal production within the same piece of land. This application can be an involvement of a tree/shrub component and livestock in the same land (agro-silvopasture) or the raising of forest trees and food crops on the same piece of land (agri-silviculture) (Pye-Smith, 2008). In Tanzania, a number of agroforestry systems exist. The systems include Chagga home gardens in Kilimanjaro region, the Obohochere (Mara region home gardens), Ngitili (the Wasukuma silvopastoral system) in Shinyanga region (Mkonda & He, 2016) and the Gliricidia Agroforestry farming system in Dodoma region (Swamila et al., 2020). The Chagga home gardens integrate coffee, bananas & maize. Meanwhile, Ngitili system involves conserving of animal feeds in rain season and uses it during the dry season when pastures are scarce in the common grazed lands (Swamila et al., 2020). Generally, Agroforestry in Tanzania is practiced as a live fence and provision of sheds in the farmland. However, prevailing local agroforestry systems excludes neem trees despite its huge potentials of sequestering carbon relatively better than other tree cultivars.

Supply Gap of Wood - Based Products in Tanzania

URT (2015a) estimated that 48 million ha of Tanzania mainland is covered by forests with wood volume of 3.3m^3 billion whose 97% is natural forests. Computation from that base establishes that Tanzanian average volume of wood stands at 37.9m^3ha^{-1} with allowable cut of 0.95 $m^3capita^{-1}$. Nevertheless, this forest stock is insufficient to meet the national demand for wood-based products amounting to 1.39$m^3capita^{-1}$ per annum (Lukumbuzya & Sianga, 2017). Figure 4 portrays enlargement of supply gap for wood-based products as time passes with the business-as-usual situation (Held et al., 2017b). Essentially, demand for timber is projected to increase from 2.3million m^3(rwe) in 2013 to 5.2 million m^3(rwe) in 2035 (Held et al., 2017b) due to development of construction sub-sector, paper production and population growth. Lack of efforts to meet wood products supply gap lead to wood imports or encroachment to the reserved forests (Lusambo et al., 2021).

Figure 4.

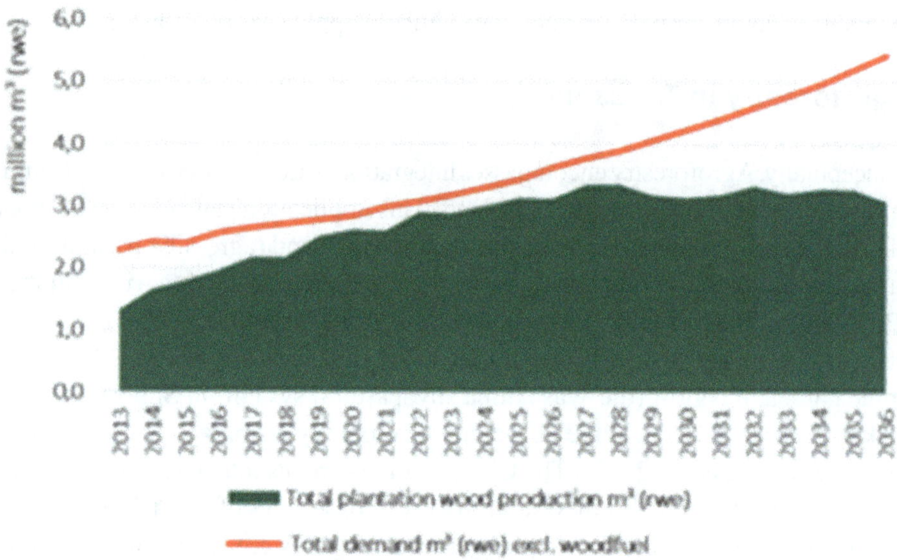

LEGAL AND REGULATORY FRAMEWORK ON CARBON SEQUESTRATION

Regulatory Framework in the Forest Plantations

The National Environmental Policy (2021) addresses two issues of environmental pollution and deforestation that are connected to GHG emission and climate change. In its caption 28, the policy postulates the need for technology used in production be environmentally sound such that it generates low or no waste to prevent pollution. Specifically, in caption 54 the policy stipulates minimization of air pollution mainly from mining sector; while caption 56 subsection (iv) the policy stipulates requirement of environmental audit/inventory to control pollution. Moreover, caption 76 puts the responsibility of a polluter to bear the costs of pollution caused where deemed appropriate. Meanwhile, the Environmental Management Act. 2004 caption 106 - 113 specify pollution control mechanisms. Caption 106 prosecutes any polluter as guilty. Moreover, the National Environmental Policy (2021) emphasizes collaboration of the government and private sector to address challenges related to environmental pollution, deforestation and climate change vagaries. In its section 3.5 – 3.7 the policy specifies the commitment of the government to partner with private sector in addressing climate change initiatives.

The National Forest Policy 1998 addresses climate change adaptation issues related to management of forest land resources and; conservation of forest ecosystems & its biodiversity that are endangered from anthropogenic activities. Moreover, National Forest policy and the Forest Act 2002 encourage involvement of private sector in expansion of national forest cover through management of existing plantations and establishment of private woodlots. Generally, Tanzanian policy and legislation favours involvement of private sector in national efforts for addressing GHG mitigation, climate change and carbon sequestration in particular.

Agroforestry Regulatory Framework

Tanzania has no specific policy to guide agroforestry practices. The only guideline is the National Agroforestry Strategy (NAS) of 2004 that envisaged to upscale agroforestry practices and benefit 4,000,000 farming households by 2025 (Msuya & Kidegesho, 2015). However, agroforestry related issues are embedded in other sectoral policies and legislations. These include: National Forest Policy 1998, National Forest Act. 2002, National Agricultural Policy 2013, The National Livestock Policy 2006, the National Beekeeping Policy 1998, the National Fisheries Policy of 1998, The National Fisheries Act 2003, the National Environmental Policy 1998 and National Environmental Act. 2004.

The National Forest Policy (1998) and the Forest Act. 2002 promote agroforestry practices by recognising the community and individually owned forests as a means to secure cooking energy needs, carbon sequestration and climate amelioration. However, the NAS lacks institutional framework as it was housed within the Ministry of Agriculture and Food Security (MAFS) and Ministry of Natural Resources and Tourism (MNRT) in 1993. It is suggested that a review of the National Forest Policy put on board the issue of agroforestry to establish its institutional set up. Therefore, authors are in the opinion that it is essential to develop agroforestry policy which will mainstream agroforestry issues into agricultural sector and establish its institutional framework.

Artificial Intelligence Regulatory Framework

The use of Artificial Intelligence (AI) in Tanzania is relatively new such that there is no specific policy and/or regulation guiding the use of AI perse (Masanja & Mkumbo, 2020). The policy, legal and guiding principles governing AI are embedded in the Information Technology and Communication (ITC) sectoral policies and regulatory framework.

The National ICT Policy 2016 supports the use of artificial intelligence. In its section 3.8 the policy necessitates the government to collaborate with other partners in promotion of research & development, innovation and encourage deployment of ICT software and hardware to support Tanzanian social economic development. Moreover, in its section 3.9 the policy recognizes the importance of developing productive sectors including agriculture, tourism & mineral resources by utilizing ICT innovations to hasten development in the productive sectors. Furthermore, artificial intelligence is promoted through the Community Development Policy (1996) whereby one of its objectives is to promote discovery and development of technology that boosts community development. In this regard AI is meant for developing communities through efficient and timely discovering of agricultural challenges. Furthermore, the use of AI in agricultural sector aligns to the National Science and Technology Policy for Tanzania (1996) which stipulates the need for generation of scientific and technological knowledge, inculcation of science and Technology culture and Promotion of emerging technologies that accelerates the National economy.

NEEM TREE- BASED AGROFORESTRY AS BIO-ECONOMIC SOLUTION TO CARBON SEQUESTRATION

Neem tree plantation and agroforestry farming would contribute to massive carbon recycling and meet climate change mitigation efforts. This is due to the fact that neem tree is a fast-growing cultivar known for its carbon sequestration potential, climate amelioration and conservation of biodiversity. The Neem Tree grows fast at a rate of 3cm girth increment per annum; records height of 5.7meters, girth of 0.29m and flowering percentage of 73% at nine years of age (Nanang et al., 1997). Moreover, Neem tree contains significant higher carbon sequestration potentials compared to other tree species (Arya et al., 2017; Moussa et al., 2018) . It also develops huge canopy at young age leading to micro-climate amelioration. Bohre & Chaubey, (2016) found that at the age of ten years neem tree sequestered carbon amounting to 144tons/ha and 529ton/ha of carbon dioxide (CO_2) in Indian Coal Mined Lands. Indeed, authors further found annual incremental stocks of carbon content to be 0.29, 0.42, 0.51, 5.64, 8.50, 14.39, 24.45; and 34.65 tons $ha^{-1}yr^{1}$ in 2, 5, 6, 7, 8, 10, 14; and 18 years old plantations, respectively. Likewise, Saka et al. (2020) revealed that the carbon stock was recorded to be higher in trunk (10.483 kg/ha) compared to 1.192 kg/ha and 0.004 kg/ha in tree branch and leaf respectively.

Table 2 portrays Neem tree potential of carbon sequestering from different parts of the globe. At the age of 10 years Noor et al. (2020) noted that, neem tree is able to sequester 31.82kg C/tree (88.15 tons/ha) above the ground mass and 17.97kg C/tree below ground mass (49.78 tons/ha). Moreover, Ajani & Shams (2016) revealed that neem tree has higher carbon sequestration potential (662.32kg C/tree compared to that of Conocarpus erectus 192.70 kg of carbon per tree at DBH of [3] 5cm. Suryawanshi et al. (2014) found that Neem tree ranked the second tree in terms of sequestering carbon after Moringa Olifera with 12.272 tons of carbon per tree in Niamey, Niger. Furthermore, Tak & Kakde (2020) found that, neem tree had the highest carbon sequestration to 3.3 tons of C/tree and 12 tons of CO_2/tree in Thane City, India compared to Ficus benghalensis (2.4 tonsC/tree), Tamarindius indica (1.7 tons C/ tree), Casurina equisetifolia (1.5 tonsC/tree) and Pongamia pinnata (0.8 tonsgC/tree).

Table 2. Carbon Sequestration Potentials of Neem Tree at DBH

S/n	Author	Location	Age of trees (years)	Carbon stock (tha^{-1})	Carbon dioxide (tha^{-1})	Carbon stock ($ttree^{-1}$)	Carbon dioxide ($ttree^{-1}$)
1	Suryawanshi et al. (2014)	Niamey, Niger				12.2	
2	Tak & Kakde, (2020)	Thane City, India				3.3	12
3	Ajani & Shams (2016)	Pakistan	10			0.66	
4	Noor et al. (2020)		10	88.15	-	31.82	-
5	Bohre & Chaubey (2016)	Madha Pradesh, India		144.48	529.78		
6	Arya et al. (2017)	Mahesan town, India				2.55	
7	Arya et al. (2017)	Gandhinagar, India				2.32	

In fact, Neem tree portrays more carbon sequestration potential than other types of trees in tropical regions. Table 3 shows carbon sequestration potentials of neem tree compared to other types of trees. It can be concluded that, Neem tree is able to sequester carbon by 243% higher than other known tree cultivars.

Table 3. Neem Tree Carbon Sequestration Capacity compared to other Trees

S/n	Author	Neem tree	Amount Carbon sequestered by Neem tree (tons/tree)	Other tree types	Amount of Carbon sequestered by Neem tree (tons/tree)	Determination for carbon sequestration capacity compared to other trees	The extent to which neem tree exceeds other tree species in terms of carbon sequestration (%)
1	Tak & Kakde (2020)	Neem tree	3.3	Ficus benghalensis	2.4	(3.3/2.4)*100 = 137	37
2	(Tak & Kakde, 2020)	Neem tree	3.3	Tamarindius Indica	1.7	(3.3/1.7)*100 = 194	94
3	(Tak & Kakde, 2020)	Neem tree	3.3	Casurina equisetifolia	1.5	(3.3/1.5)*100 = 220	120
4	(Tak & Kakde, 2020)	Neem tree	3.3	Pongamia Pinnata	0.8	(3.3/0.8)*100 = 412	312
5	Ajani & Shams (2016)	Neem tree	0.662	Conocarpus Erectus	0.192	(0.662/0.192)*100 = 345	245
Average excess carbon sequestration capacity of neem tree over other tree types						**[(37+94+120+312+245)/5]**	**161**

Tanzania's nationally determined contribution (NDCs) to United Nations in 2015 committed to reduce climate related disasters from 70% in 2015 to 50% in 2030 and lessen GHG emission up to 20% by 2030 (REIT, 2016). Adoption of neem tree-based agroforestry practices and forest plantation will lead to improved carbon recycling and mitigation of the GHGs emission. This will complement the current efforts of expanding tree plantation targeted by Tanzania Forest Services Agency (TFS) Five-year Strategic Plan, 2014-2019 annual tree planting of 10,000ha. However, the survival rate of planted trees is acknowledged to account for only 70% (United Republic of Tanzania (URT), 2021b). Moreover, if the TFS targets are to be implemented Tanzania requires more than five decades realizing the AFR100 commitment of restoring 5.2 million hectors of degraded forest land. For that case, diverse efforts are inevitable to attain the commitment Tanzania made for restoring 5.2 million ha and carbon sequestration to contribute to the global initiatives to recycle carbon which is the major component of GHG emission worldwide.

This can be done by mobilizing smallholder farmers into groups to have joint efforts to supply seedlings for neem plantations and neem tree-based agroforestry at individual farms. Furthermore, smallholder farmers can be considered for out grower schemes for neem tree-based plantations to assure markets of their products serving as an inclusive business model. The Private Public Partnership approach in Neem tree plantation coupled with sustainable harvesting would yield robust results in carbon recycling targets. Drones and other artificial intelligence facilities can also be applied to large plantations in tree planting, weeding and spraying of fertilizers &herbicides.

The neem tree-based agroforestry will increase agricultural productivity due to ameliorated climate and beneficial interactions between the trees and crops grown. Integration of bee keeping in the tree farms can increase the income of investors and improve agricultural productivity to nearby farmers through effective pollination. The increased agricultural productivity reduces the necessity to expand agricultural land. Technically investment in neem tree-based plantations and agroforestry can serve as the tool for carbon recycling. Similarly, the GHG emission and climate change mitigation will in turn reduce air pollution that will lessen public health concerns associated with air pollution country wide. Availability of relatively cheaper labour force amounting to annual per capita wage rate of U$1,200 (Haji, 2015; URT, 2018) attracts investors. Moreover, 28% of Tanzanian labour force is youth aged between 15 and 35 years (Morisset et al., 2014) who can be productively be employed in the tree plantations.

Essentially, reforestation mitigates GHG emission and climate change vagaries as evidenced in Brazil where per capita emission was reduced by 30% from 10.5 tCO_{2e} in 2005 to 6.6 tCO2e in 2014 (UNCTAD, 2019); amounting to 750 million tCO_2e annually due to reforestation efforts (Boucher et al., 2014). Likewise, India reversed its deforestation trend from 1990s at 145,000 hectares per annum (FAO, 2010). Consequently, the country sequestered 177 million tCO_2e in 2007 (Boucher et al., 2014).

APPLICATIONS OF ARTIFICIAL INTELLIGENCE IN AGRICULTURE AND FOREST MANAGEMENT.

The use of AI in farm management has been widely employed in developed countries to irrigation activities for optimizing water usage (Arvind et al., 2017; Cruz et al., 2017; García et al., 2022); weed control (Ngo et al., 2019; Partel et al., 2019; Rosle et al., 2021) and spraying of pesticides (Rahman et al., 2021; Zheng et al., 2016). In forestry management, the AI determined diameter and tree height (Hu et al., 2020; Montano et al., 2017); biomass volume and basal area (Martins Silva et al., 2019) and counting of trees in urban areas (Liu et al., 2021). The drones which are equipped with LiDAR sensors are capacitated to estimate biomass change of trees periodically (Debangshi, 2021). Actually, application of weed controller has reduced the use of herbicides by 90% in Brazilian soyabean & corn farms (Debangshi, 2021). The drones have also been used to inspect post timber harvest and mapping of site preparation and tree planting in West Virginia's Division of Forestry in USA. It can be concluded that the use of AI in crops farming and forestry management indicate the potentials of using the technology (AI) in agroforestry and forest plantations.

In Tanzania AI is widely used in integrated financial management systems, government payroll systems, aviation industry and higher learning institutions. However, the use of AI in Tanzanian agriculture is skewed towards solicitation of extension services to farmers through the use of mobile phone support tools. In this case, the mobile App has been integrated with Chatbot assistant in Android smartphones to provide solutions for challenges facing smallholder farmers (Shapa et al., 2021). A chatbot is a human computer interaction technology which provides information and facilitate informed decision making in different sectors.

One of the chatbot widely used in Tanzania's agricultural sector is '*Ushaurikilimo*'. The Ushaurikilimo is the web and mobile based platform that provides advisory services to farmers. The mostly used group of farmers in *Ushaurikilimo* is a hub of more than 700 farmers and 20 extensionists (Fue et al., 2016). The *Ushaurikilimo* is jointly implemented by Kilosa Local Government, Ministry of Agriculture, Food & Cooperatives, Kilosa Rural Services and Electronic Communication Center, Uyole Agricultural Research Institutes and Kilosa Community Radio (Shapa et al., 2021). Other initiatives include: Peresi Mgorozi who designed the AI to detect diseases in maize plants and detection of banana diseases using the plantae leaf (SAHARA-VENTURES, 2019). Others are Agrobot which enhances provision of advice regarding fertilization and usage of pesticides in Morogoro, Iringa and Dar-es-Salaam (Masanja & Mkumbo, 2020). Given the rising national population that requires more agricultural expansion and huge deforestation rate which lead to environmental hazards and climate change vagaries, it is imperative to employ artificial intelligence in intensifying the neem tree-based agroforestry and forestry management in Tanzania.

CONCLUSION AND RECOMMENDATIONS

The trend of annual deforestation rates indicates potentials of turning the country into a bare land and impairment of efforts to mitigate GHG, carbon recycling and food security initiatives. The increased GHG emission and climate change calamities will turn to air pollution that amplifies public health concerns. To address the above-mentioned tragedies, deliberate efforts are necessary to come up with sustainable solutions.

Neem forest plantation and neem tree-based agroforestry practices can be used as the massive carbon recycling, bringing economic benefits in meeting national timber supply gap and increase food productivity through immense biomass accumulation and climate amelioration. However, its promotion and adoption as forest plantation and agroforestry farming needs government policy formulation to provide the framework and incentives for the private sector and other interested stakeholders to commit resources for investment.

Currently, forest plantation for timber in Tanzania is dominated by Pines and Eucalyptus trees. Meanwhile, agroforestry systems practiced in the country that include Chagga home gardens, the Sukuma fodder enclosure 'Ngitili', the Mara and Kagera regions home gardens and Gliricidia Agroforestry system practiced in central corridor have not considered neem tree as a component of the systems. Otherwise, several types of trees ranging from local to exotic species are just raised to meet shading objectives, live fence in the homesteads, or just as boundaries in their farmland. Neem tree is hardly found in government managed plantations. Occasionally, neem trees are found along human settlements in urban areas to provide shade.

Neem trees suit in different agro-ecological conditions and even in marginal lands such as arid and semi-arid areas where other crops cannot grow. Given Tanzania's vast marginal land Neem-tree plantations and/or agroforestry can be adopted. Some studies have identified the rate of 160 trees per hector (5m x 12m) as the viable investment without compromising agricultural yields in tropical regions (Chaturvedi & Khurana, 2017; van der Zwaan & Gerlagh, 2016). Adoption of neem tree-based agroforestry can reduce the encroachment of communal natural forests. The neem tree-based agroforestry farms and/or plantations can also be used for carbon trading, processing of its seeds into medicinal products and cosmetics.

Intensification of neem tree plantations and/agroforestry can be enhanced by the use of artificial intelligence to hasten investments at small scale and/or large-scale farms. The small-scale farms can employ simple facilities for detection of diseases and/or pests' infestations and soil fertility analysis. Farmers could hire services of AI facilities from service providers available in Tanzania including Agrinfo, Tanzania Flying Labs and We Robotics just to mention a few. Meanwhile, for large scale farming, AI facilities can be used for site mapping, soil quality assessment, planting, irrigation, and spraying agrochemicals to fast-track the farming process which will ultimately lead to establishing high capacity of carbon sequestration resulting from the planted trees. Moreover, investors in large scale agroforestry and/or plantations can accommodate small scale farmers by establishing out growers' schemes.

REFERENCES

Adzawla, W., Sawaneh, M., & Yusuf, A. M. (2019). Greenhouse gasses emission and economic growth nexus of sub-Saharan Africa. *Scientific American, 3*, e00065. doi:10.1016/j.sciaf.2019.e00065

Ajani, A., & Shams, Z. I. (2016). Comparative Status of Sequestered Carbon Stock of Azadirachta indica and Conocarpus erectus at the University of Karachi Campus, Pakistan. *International Journal of Environment, 5*(2), 89–97. doi:10.3126/ije.v5i2.15009

Akash, A. R., Rao, A. B., & Chandel, M. K. (2016). Prospects of Implementing CO2 Capture and Sequestration (CCS) in the Proposed Supercritical Coal Power Plants in India. *Energy Procedia, 90*, 604–612. doi:10.1016/j.egypro.2016.11.229

Arvind, G., Anthra, V. G., Haripriya, H., Akshaya, R. R., & Aravind, S. (2017). Automated irrigation with advanced seed germination and pest control. *IEEE International Conference on Technological Innovations in ICT For Agriculture and Rural Development (TIAR 2017), Tiar*, (pp. 64–67). IEEE. 10.1109/TIAR.2017.8273687

Arvola, A., Malkamäki, A., Penttilä, J., & Toppinen, A. (2019). Mapping the Future Market Potential of Timber from Small-Scale Tree Farmers: Perspectives from the Southern Highlands in Tanzania. *Small-scale Forestry, 18*(2), 189–212. doi:10.100711842-019-09414-8

Arya, A., Shalini Negi, S., Kathota, J. C., Patel, A. N., Kalubarme, M. H., & Garg, J. K. (2017). Carbon Sequestration Analysis of dominant tree species using Geo-informatics Technology in Gujarat State (INDIA). *International Journal of Environment and Geoinformatics*, *4*(2), 79–93. doi:10.30897/ijegeo.312517

Beck, S., & Mahony, M. (2018). The IPCC and the new map of science and politics. *Wiley Interdisciplinary Reviews: Climate Change*, *9*(6), 1–16. doi:10.1002/wcc.547

Beghin. (2015). Assessing Food Security in Tanzania in the Next Decade. *Agricultural Policy Review, 201*(3), 3.

Betteena, D., Fernando, S., Sabarishwaran, S., Priya, R. R., & Santhoshini, S. (2020, August). Smart agriculture monitoring system using IoT. [UAE]. *IACSIT International Journal of Engineering and Technology*, *7*(4), 308–311. doi:10.35940/ijeat.D7916.049420

Bhagat, P. R., Naz, F., & Magda, R. (2022). Artificial intelligence solutions enabling sustainable agriculture : A bibliometric analysis. *PLoS One*, *2022*(June), 1–19. doi:10.1371/journal.pone.0268989 PMID:35679287

Bohre, P., & Chaubey, O. P. (2016). Biomass production and carbon sequestration by Azadirachta indica in coal mined lands. *International Journal of Bio-Science and Bio-Technology*, *8*(2), 111–120. doi:10.14257/ijbsbt.2016.8.2.10

Boucher, D., Elias, P., Faires, J., & Smith, S. (2014). Deforestation Success Stories. In *Union of Concerned Scientists* (June Issue). http://search.ebscohost.com/login.aspx?direct=true&db=8gh&AN=109565479&site=ehost-live

Chaturvedi, S., & Khurana, S. M. P. (2017). *Biotechnological Approaches for Urban Solid Waste Management*. 860–866. doi:10.21276/sajb.2017.5.12.7

Chauvy, R., Meunier, N., Thomas, D., & De Weireld, G. (2019). Selecting emerging CO2 utilization products for short- to mid-term deployment. *Applied Energy*, *236*(February), 662–680. doi:10.1016/j.apenergy.2018.11.096

Cruz, J. R. D., Baldovino, R. G., Bandala, A. A., & Dadios, E. P. (2017). Water usage optimization of Smart Farm Automated Irrigation System using artificial neural network. *2017 5th International Conference on Information and Communication Technology*. IEEE. 10.1109/ICoICT.2017.8074668

Debangshi, U. (2021). Drones - Applications in Agriculture. *Chronicle of Bioresource Management*, *5*(3), 115–120. doi:10.5281/zenodo.5554734

Engel, S., Pagiola, S., Wunder, S., & CIFOR. (2008). Moving ahead with REDD. In Angelson (Ed.), *Issues, options and implications. Bogor,* (*Vol. 65*, Issue 4). CIROR. http://scholar.google.com/scholar?hl=en&btnG=Search&q=intitle:Moving+Ahead+with+REDD#1

FAO. (2010). *Global Forest Resource Assessment.* [Data set]. FAO.

FAO. (2017). Sustainable Woodfuel For Food Security, a Smart Choice: gree, renewable and affordable. In FAO working paper.

FAO. (2018). *The state of world ' s forests: forest pathways to sustainable development.* FAO.

FAO. (2020). *Global Forest Resources Assessment 2020; Report United Republic of Tanzania.* FAO.

Fue, K., Geofrey, A., & Mlozi, P. M. R. S. (2016). Analyzing Usage of Crowdsourcing Platform "Ushaurikilimo" by Pastoral and Agro-Pastoral Communities in Tanzania. *International Journal of Instructional Technology and Distance Learning, 13*(January), 3–23.

Fuss, S., Canadell, J. G., Peters, G. P., Tavoni, M., Andrew, R. M., Ciais, P., Jackson, R. B., Jones, C. D., Kraxner, F., Nakicenovic, N., Le Quéré, C., Raupach, M. R., Sharifi, A., Smith, P., & Yamagata, Y. (2014). Commentary: Betting on negative emissions. *Nature Climate Change, 4*(10), 850–853. doi:10.1038/nclimate2392

García, L., Parra, L., Jimenez, J. M., Lloret, J., Lorenz, P., García, L., Parra, L., Jimenez, J. M., & Lloret, J., & Irri-, P. L. I. S. (2022). IoT-Based Smart Irrigation Systems : An Overview on the Recent Trends on Sensors and IoT Systems for Irrigation in Precision Agriculture To cite this version : IoT-Based Smart Irrigation Systems : An Overview on. *Sensors (Basel), 2020*(20).

Giacomo, R., & David, G. (2018). E-Argriculture In Action: Drones For Agriculture. In E-Agriculture in Action: Drones for Agriculture.

Haji, M. (2015). Youth employment in Tanzania: Taking stock of the evidence and knowledge gaps. International Development Research Centre, 1–30.

Hansson, A., Fridahl, M., Haikola, S., Yanda, P., Pauline, N., & Mabhuye, E. (2020). Preconditions for bioenergy with carbon capture and storage (BECCS) in sub-Saharan Africa: The case of Tanzania. *Environment, Development and Sustainability, 22*(7), 6851–6875. doi:10.100710668-019-00517-y

Harrison, S., Herbohn, J., & Niskanen, A. (2002). Non-industrial, smallholder, small-scale and family forestry: What's in a name? *Small-Scale Forest Economics. Management and Policy*, *1*(1), 1–11. doi:10.100711842-002-0001-3

Held, C., Jacovelli, P., Techel, G., Nutto, L., Wathum, G., & Wittmann, N. (2017a). *Tanzanian Wood Product Market Study*. Issue November.

Held, C., Jacovelli, P., Techel, G., Nutto, L., Wathum, G., & Wittmann, N. (2017b). *Tanzanian Wood Product Market Study Report for the Forestry Development Trust*. 139.

Herring, S. C., Christidis, N., Hoell, A., Kossin, J. P., Schreck, C. J. III, & Stott, P. A. (2018). Explaining Extreme Events of 2016 From A Climate Perspective. *Bulletin of the American Meteorological Society*, *99*(1), S1–S157. https://journals.ametsoc.org/view/journals/bams/99/1/bams-explainingextremeevents2016.1.xml. doi:10.1175/BAMS-ExplainingExtremeEvents2016.1

Hu, Y., Xu, X., Wu, F., Sun, Z., Xia, H., Meng, Q., Huang, W., Zhou, H., Gao, J., Li, W., Peng, D., & Xiao, X. (2020). Estimating forest stock volume in Hunan Province, China, by integrating in situ plot data, Sentinel-2 images, and linear and machine learning regression models. *Remote Sensing*, *12*(1), 186. Advance online publication. doi:10.3390/rs12010186

IPCC. (2014). Climate Change 2014 Part A: Global and Sectoral Aspects. In Climate Change 2014: Impacts, Adaptation, and Vulnerability. Part A: Global and Sectoral Aspects. Contribution of Working Group II to the Fifth Assessment Report of the Intergovernmental Panel on Climate Change. IEEE. papers2://publication/uuid/B8BF5043-C873-4AFD-97F9-A630782E590D

Jones, B. F., & Olken, B. A. (2010). Climate shocks and exports. *The American Economic Review*, *100*(2), 454–459. doi:10.1257/aer.100.2.454

Khalid, N. A. C. & S. (2018). Reshaping the Terrain Landscape Restoration in Tanzania. *Global Landscape Forum,* 1–4.

Kideghesho. (2015). Realities on Deforestation in Tanzania — Trends, Drivers, Implications and the Way Forward. In *Intech* (*Vol. 32*, pp. 137–144). http://www.trafficj.org/publication/17_Timber-trade-East-Southern-Africa.pdf

Kim, D. G., Thomas, A. D., Pelster, D., Rosenstock, T. S., & Sanz-Cobena, A. (2016). Greenhouse gas emissions from natural ecosystems and agricultural lands in sub-Saharan Africa: Synthesis of available data and suggestions for further research. *Biogeosciences*, *13*(16), 4789–4809. doi:10.5194/bg-13-4789-2016

Landrigan, P. J., Fuller, R., Acosta, N. J. R., Adeyi, O., Arnold, R., Basu, N., Baldé, A. B., Bertollini, R., Bose-O'Reilly, S., Boufford, J. I., Breysse, P. N., Chiles, T., Mahidol, C., Coll-Seck, A. M., Cropper, M. L., Fobil, J., Fuster, V., Greenstone, M., Haines, A., & Zhong, M. (2018). The Lancet Commission on pollution and health. *Lancet*, *391*(10119), 462–512. doi:10.1016/S0140-6736(17)32345-0 PMID:29056410

Liu, X., Ghazali, K. H., Han, F., & Mohamed, I. I. (2021). Automatic Detection of Oil Palm Tree from UAV Images Based on the Deep Learning Method. *Applied Artificial Intelligence*, *35*(1), 13–24. doi:10.1080/08839514.2020.1831226

Lu, W. C. (2017). Greenhouse gas emissions, energy consumption and economic growth: A panel cointegration analysis for 16 Asian countries. *International Journal of Environmental Research and Public Health*, *14*(11), 15. doi:10.3390/ijerph14111436 PMID:29165399

Lukumbuzya, K., & Sianga, C. (2017). Overview of the Timber Trade in East and Southern Africa: National Perspectives and Regional Trade Linkage. *Traffic Joint Report*. http://www.trafficj.org/publication/17_Timber-trade-East-Southern-Africa.pdf

Lupala, Z. J., Lusambo, L. P., & Ngaga, Y. M. (2014). Management, Growth, and Carbon Storage in Miombo Woodlands of Tanzania. *International Journal of Forestry Research*, *2014*, 1–11. doi:10.1155/2014/629317

Lusambo, L. P., Nyanda, S. S., & Mhando, D. G. (2021). Profitability Analysis of Tree Growing in the Southern Highlands of Tanzania. *International Journal of Forestry Research*, *2021*, 1–10. doi:10.1155/2021/8872211

Malunguja, G. K., Devi, A., Kilonzo, M., & Rubanza, C. D. K. (2020). Climate change mitigation through carbon dioxide (CO2) sequestration in community reserved forests of northwest Tanzania. *Archives of Agriculture and Environmental Science*, *5*(3), 231–240. doi:10.26832/24566632.2020.050301

Martins Silva, J. P., Marques da Silva, M. L., Ferreira da Silva, E., Fernandes da Silva, G., Ribeiro de Mendonça, A., Cabacinha, C. D., Araújo, E. F., Santos, J. S., Vieira, G. C., Felix de Almeida, M. N., & Fernandes, M. R. (2019). Computational techniques applied to volume and biomass estimation of trees in Brazilian savanna. *Journal of Environmental Management*, *249*(August), 109368. doi:10.1016/j.jenvman.2019.109368 PMID:31421480

Masanja, N. M., & Mkumbo, H. (2020). The Application of Open Source Artificial Intelligence as an Approach to Frugal Innovation in Tanzania. [IJRIAS]. *International Journal of Research and Innovation in Applied Science*, *5*(3), 2454–6194. www.rsisinternational.org

Mkonda, M. Y., & He, X. (2016). Production Trends of Food Crops: Opportunities, Challenges and Prospects to Improve Tanzanian Rural Livelihoods. *Natural Resources and Conservation*, *4*(4), 51–59. doi:10.13189/nrc.2016.040402

Montano, R. A. N. R., Sanquetta, C. R., Wojciechowski, J., Mattar, E., Corte, A. P. D., & Todt, E. (2017). Artificial Intelligence Models to Estimate Biomass of Tropical Forest Trees. *Polibits*, *56*(1), 29–37.

Morisset, J., Cunningham, V., & Haji, M. (2014). *Who Wants a Job?* The Magnetic Power of Cities.

Moussa, S., Kyereh, B., Tougiani, A. A., & Saadou, M. (2018). Carbon stocks of neem tree (Azadirachta indica A. Juss.) in different urban land use and land cover types in Niamey city. *South Asian Journal of Biological Research*, *1*(2), 153–165. http://aiipub.com/

Msuya, N., Masanja, E., & Temu, A. K. (2011). Environmental Burden of Charcoal Production and Use in Dar es Salaam, Tanzania. *Journal of Environmental Protection*, *02*(10), 1364–1369. https://doi.org/10.4236/jep.2011.210158

Msuya & Kidegesho. (2015). Mainstreaming Agroforestry Policy in Tanzania Legal Framework. *Agroforestry for Biodiversity and Ecosystem Services - Science and Practice, March*, 130–140. doi:10.5772/35859

Muthuraj, R., & Mekonnen, T. (2018). Recent progress in carbon dioxide (CO2) as feedstock for sustainable materials development: Co-polymers and polymer blends. *Polymer*, *145*, 348–373. https://doi.org/10.1016/j.polymer.2018.04.078

Mwaseba, D. L., Erkkilä, A., Friis-Hansen, E., Maro, A. H., & Maziku, J. D. (2020). Responsibilization in governance of non-industrial private forestry: Experiences from the Southern Highlands of Tanzania. *Forest Policy and Economics, 118*(April 2019), 102243. doi:10.1016/j.forpol.2020.102243

Nanang, D. M., Day, R. J., & Amaligo, J. N. (1997). Growth and yield of neem (Azadirachta indica A. Juss.) plantations in Northern Ghana. *Commonwealth Forestry Review*, *76*(245), 103–106.

Ngaga, Y. M. (2011). A Platform for Stakeholders in African Forestry. *Forest Pl Antations and Wo Odlots in Tanzania. 1*(16), 80.

Ngo, H. C., Hashim, U. R., Sek, Y. W., Kumar, Y. J., & Ke, W. S. (2019). Weeds detection in agricultural fields using convolutional neural network. *International Journal of Innovative Technology and Exploring Engineering*, *8*(11), 292–296. https://doi.org/10.35940/ijitee.K1327.0981119

Nogia, P., Sidhu, G. K., Mehrotra, R., & Mehrotra, S. (2016). Capturing atmospheric carbon: Biological and nonbiological methods. *The International Journal of Low Carbon Technologies*, *11*(2), 266–274. https://doi.org/10.1093/ijlct/ctt077

Noumi, V. N., Djongmo, V. A., Nyeck, B., Mbobda, R. B. T., & Zapfack, L. (2018). Vegetation structure, carbon sequestration potential and species conservation in four agroforestry systems in Cameroon (Tropical africa). *Acta Botanica Brasílica*, *32*(2), 212–221. https://doi.org/10.1590/0102-33062017abb0279

Odenberger, M., Kjärstad, J., & Johnsson, F. (2013). Prospects for CCS in the EU energy roadmap to 2050. *Energy Procedia*, *37*, 7573–7581. https://doi.org/10.1016/j.egypro.2013.06.701

Oladosu, Y., Rafii, M. Y., Abdullah, N., Magaji, U., Hussin, G., Ramli, A., & Miah, G. (2016). Fermentation Quality and Additives: A Case of Rice Straw Silage. *BioMed Research International, 2016*. doi:10.1155/2016/7985167

Olivier & Peters. (2020). Trends In Global Co 2 And Total Greenhouse Gas 2020 Report. *PBL Netherlands Environmental Assessment Agency, PBL public*(December), 85.

Partel, V., Charan Kakarla, S., & Ampatzidis, Y. (2019). Development and evaluation of a low-cost and smart technology for precision weed management utilizing artificial intelligence. *Computers and Electronics in Agriculture*, *157*(January), 339–350. https://doi.org/10.1016/j.compag.2018.12.048

Pedersen, R. H. (2017). The Political Economy of Private Forestry in Tanzania : a Review. In *Danish Institute for International Studies (DIIS)*. www.diis.dk

PFP. (2016). United Republic of Tanzania. In *Private Forestry Program*. doi:10.5089/9781513547442.002

Pye, S., Watkiss, P., Savage, M., & Goodwin, J. (2010). *Opportunities for Low Carbon Investment in Tanzania: An Assessment of Future Emissions Growth and Low Carbon Reduction Potential* (Issue December). SEI, Dew Point, Global Climate Adoption.http://www.economics-of-cc-in-tanzania.org/images/Tanzania_low_carbon_growth_assessment_v7.pdf

Pye-Smith, C. (2008). Farming Trees, Banishing Hunger How an agroforestry programme is helping smallholders in. *America*.

Rahman, M. F. F., Fan, S., Zhang, Y., & Chen, L. (2021). A comparative study on application of unmanned aerial vehicle systems in agriculture. *Agriculture (Switzerland), 11*(1), 1–26. doi:10.3390/agriculture11010022

REIT. (2016). Tanzania Climate Action Report for 2016. In Resilience and Economic Inclusion Team.

Rizos, B., & Drabik, R. T. (2018). The Role of Business in the Circular Economy: Markets, Processes and Enabling Policies. In *The Centre for European Policy Studies* (Issue March). https://www.scopus.com/inward/record.url?eid=2-s2.0-84949807776&partnerID=40&md5=736fc93fe5bc8635ff79c068db9c0e1b

Rojas-Downing, M. M., Nejadhashemi, A. P., Harrigan, T., & Woznicki, S. A. (2017). Climate change and livestock: Impacts, adaptation, and mitigation. *Climate Risk Management, 16*, 145–163. https://doi.org/10.1016/j.crm.2017.02.001

Rosle, R., Che'ya, N. N., Ang, Y., Rahmat, F., Wayayok, A., Berahim, Z., Ilahi, W. F. F., Ismail, M. R., & Omar, M. H. (2021). Weed detection in rice fields using remote sensing technique: A review. *Applied Sciences (Switzerland), 11*(22), 25. doi:10.3390/app112210701

Saka, M.G., Osho, J. S. A., & Iso, H. D. (2020). Assessment of biomass contents and carbon stock of plantation grown neem specie in Yola, Nigeria. *Journal of Research in Forestry, Wildlife & Environment, 12*(1), 284–289.

Seekings, J. (2020). International actors and social protection in Africa, 2000-2020. *University of Capetown, 2*(March), 1–17.

Shapa, M., Trojer, L., & Machuve, D. (2021). Mobile-based Decision Support System for Poultry Farmers: A Case of Tanzania. *International Journal of Advanced Computer Science and Applications, 12*(2), 584–590. https://doi.org/10.14569/IJACSA.2021.0120273

Singh, B., Strømman, A. H., & Hertwich, E. G. (2011). Comparative impact assessment of CCS portfolio: Life cycle perspective. *Energy Procedia, 4*, 2486–2493. https://doi.org/10.1016/j.egypro.2011.02.144

Soylu, S., & Çarman, K. (2021). Fuzzy logic based automatic slip control system for agricultural tractors. *Journal of Terramechanics, 95*, 25–32. https://doi.org/10.1016/j.jterra.2021.03.001

Suryawanshi, M. N., Patel, A. R., Kale, T. S., & Patil, P. R. (2014). Carbon sequestration potential of tree species in the environment of North Maharashtra University Campus, Jalgaon (MS) India. *Bioscience Discovery*, *5*(2), 175–179.

Swamila, M., Philip, D., Akyoo, A. M., Sieber, S., Bekunda, M., & Kimaro, A. A. (2020). Gliricidia agroforestry technology adoption potential in selected dryland areas of Dodoma Region, Tanzania. *Agriculture (Switzerland)*, *10*(7), 1–17. doi:10.3390/agriculture10070306

Tak, A. A., & Kakde, U. B. (2020). Analysis of carbon sequestration by dominant trees in urban areas of Thane city. *International Journal of Global Warming*, *20*(1), 1–11. https://doi.org/10.1504/IJGW.2020.104615

Tcvetkov, P., Cherepovitsyn, A., & Fedoseev, S. (2019). The changing role of CO2 in the transition to a circular economy: Review of carbon sequestration projects. *Sustainability (Switzerland)*, *11*(20), 1–19. doi:10.3390/su11205834

UNCTAD. (2019). Commodity Dependence, Climate Change and the Paris Agreement. In *Commodities and Development Report*. https://unctad.org/en/PublicationsLibrary/ditccom2019d3_en.pdf

United Republic of Tanzania (URT). (2021a). *2020 Tanzania in Figures*. Issue June.

United Republic of Tanzania (URT). (2021b). *National Environment Policy*. URT.

URT. (2011). *Tanzania Agriculture and Food Security Investment Plan (TAFSIP)* (Issue October 2011). URT.

URT. (2014). *Tanzania: Second national communication to the United Nations framework convention climate change*. URT .

URT. (2015a). National Forest Resources Monitoring and Assessment of Tanzania Mainland (NAFORMA). Main Results Report. pp.33.

URT. (2015b). *Tanzania Intended Nationally Determined Contributions (INDC)* (Issue 2007). URT. http://www4.unfccc.int/submissions/INDC/Published Documents/Turkey/1/The_INDC_of_TURKEY_v.15.19.30.pdf

URT. (2018). *Formal Sector Employment and Earnings Survey*. URT.

URT. (2019a). Health and Pollution Action Plan. In United Nations Industrial Development Organization, 2019 (Issue February).

URT. (2019b). *Tanzania National Nutrition Survey 2018*. Issue June.

URT. (2020). *National Climate Change Statistics Report 2019*. NBS.

van der Zwaan, B., & Gerlagh, R. (2016). Offshore CCS and ocean acidification: A global long-term probabilistic cost-benefit analysis of climate change mitigation. *Climatic Change*, *137*(1–2), 157–170. https://doi.org/10.1007/s10584-016-1674-5

van Minnen, J. G., Strengers, B. J., Eickhout, B., Swart, R. J., & Leemans, R. (2008). Quantifying the effectiveness of climate change mitigation through forest plantations and carbon sequestration with an integrated land-use model. *Carbon Balance and Management*, *3*, 1–20. https://doi.org/10.1186/1750-0680-3-3

Ventures, S. (2019). Artificial Intelligence in Tanzania, What's Happening . *The Medium*.

Viebahn, P., Vallentin, D., & Höller, S. (2015). Prospects of carbon capture and storage (CCS) in China's power sector - An integrated assessment. *Applied Energy*, *157*, 229–244. https://doi.org/10.1016/j.apenergy.2015.07.023

WB. (2018). Agricultural Pollution. Field Burning. *World Bank Group*, 1–7. https://documents.worldbank.org/curated/en/989351521207797690/pdf/124342-repl-WB-Knowledge-Burning.pdf%0Ahttps://openknowledge.worldbank.org/bitstream/handle/10986/29504/124342-BRI-p153343-PUBLIC-march-22-9-pm-WB-Knowledge-Burning.pdf?sequence=1&isAllowed=

WHO. (2016). *Ambient air pollution: A global assessment of exposure and burden of disease*. WHO.

WHO. (2018). Air pollution and child health. *Book, WHO/CED/PH*(NC-SA 3.0 IGO), 18.

WRI-CAIT. (2021). *HISTORICAL GHG EMISSION BY SECTOR IN TANZANIA (MTCO2e)*. WHO.

Zheng, H., Zhou, X., Cheng, T., Yao, X., Tian, Y., Cao, W., & Zhu, Y. (2016). Comparative analysis of vegetation indices, non-parametric and physical retrieval methods for monitoring nitrogen in wheat using UAV-based multispectral imagery, 7350–7353. IEEE.

KEY TERMS AND DEFINITIONS

Agroforestry: Is a farming system in which woody perennials are deliberately planted and managed on the same land unit with herbaceous crops and/or animals or aquatic life forms or insects arranged in a spatial or temporal sequence.

Artificial Intelligence: Is the application of programmed machines to perform tasks or processes which are normally performed by human beings who can normally reason, learn and solve problems.

Machine Learning: Is a process of programming a machine with application of computer science and data to imitate the way human beings learn through trial and error in order to improve the accuracy of machine performance.

Internet of Things: Refers to a network of computer associated devices connected to the internet to facilitate communication between the devices through the internet for performing various specific functions.

Carbon Recycling: Is the process of capturing carbon from the atmosphere through mechanical or biological processes and use it for socio-economic purposes without allowing it to go back to the atmosphere.

Bioeconomic Solution: Refers to a solution of an economic problem whose building blocks are derived from biological resources and processes of a renewable nature.

Carbon sequestration: Is the process of capturing carbon from the atmosphere and store it in the form of biomass or gas deposits underground.

Chapter 2
Machine Learning Tools for Ensuring Optimized Agribusiness Recovery Strategies

Athanasios Podaras
https://orcid.org/0000-0003-4573-8976
Technical University of Liberec, Czech Republic

Manomeet Gupta
https://orcid.org/0000-0001-8761-8628
Technical University of Liberec, Czech Republic

Dana Nejedlova
https://orcid.org/0000-0003-0992-4915
Technical University of Liberec, Czech Republic

ABSTRACT

The chapter presents machine learning approaches for business continuity and recovery optimization in agribusiness. Firstly, a mathematical method, entitled business continuity points (BCPTs) is tested with domain data for its potential to predict process recovery results, namely recovery time and criticality ranking of key operations. A 72.22% accuracy has been estimated. Then, decision tree prediction with 10-fold cross validation and random forest has been 92.31% accurate in classifying business functions as critical or not. Additionally, a new multi-approach and multi-class decision tree classifier with some of the BCPTs input variables is presented, with 55.36% accuracy, and 70.37% and 88.89% accuracy rates when boosted with

DOI: 10.4018/978-1-6684-4649-2.ch002

the 10 folds and the random forest. Finally, regression analysis techniques are used to improve the initial recovery time BCPTs formula. Exponential regression has been more precise compared to the quadratic model (R2exp=0.954, R2quad=0.85). Despite current data limitations, the inferred prediction patterns are robust and highly accurate in the given field.

INTRODUCTION

Modern agribusiness procedures are highly vulnerable to a multitude of threats including biological, human, environmental (Komarek et al, 2020) and technical (El Bilali & Allahyari, 2018). The agribusiness industry nowadays needs to face many crisis incidents which stem from the climate change such as unexpectedly extreme weather phenomena that emerge in various forms (e.g. significant fluctuations in drought, heat, etc.) (Lorencová et al, 2013; Davidson, 2016) and affect the global bioeconomy.

Furthermore, modern agribusiness has been highly influenced by the ongoing pandemic. This statement is highly emphasized by global agriculture authorities such as the Food and Agriculture Organization of the United Nations, stating that the COVID-19 outbreak has affected the "*global food systems, disrupting regional agricultural value chains, and posing risks to household food security*" (FAO, n.y). Academics have also underlined the fact that the pandemic had a negative impact on the conducting of "*basic and applied research in agricultural sciences*" (Hoogenboom, 2020). It is thus crucial for agribusiness domain experts to collaborate with academics and government institutions that serve the agricultural sectors and chart unified policies that can enhance resilience against the climatic, biological, human and even technical risks and ensure sustainability in the agrarian industry and its sub-sectors.

Multiple researchers have highlighted the importance of considering business continuity management policies to tackle crisis incidents in a sustainable manner (Culhane et al, 2018). Business continuity is aimed at ensuring the unstoppable operation to a minimum acceptable level of the critical business activities, services and supporting information and communication technology infrastructure regarding key organizational functions, via their rapid recovery after an unforeseen interruption.

Many studies focus on the importance of incorporating of big data processing and manipulation tools and techniques to chart efficient and preventive BCM strategies (Engemann & Henerson, 2014). Such strategies have been considered as a prerequisite towards the optimization of current agri-business continuity policies which are currently neglected or poorly defined (Hájek & Urbancova, 2013). Recently, a proactive recovery time effort estimation method has been proposed by academic researchers (Podaras et al, 2016, Podaras, 2018) named Business Continuity Points (BCPTs). The approach relies on process input elements such as human actors, technical actors and processes as well as recovery factors and scenarios that affect the restoration procedure in order to estimate criticality level and recovery time of each business function. According to the authors no similar predictive methods have been proposed. The method has been tested with success in the public higher education institutions (Podaras et al, 2021). Nevertheless it can be also tested for its effectiveness in the agrarian sector.

The goal of the present chapter is to suggest proactive optimized recovery policies of modern agribusiness critical operations with the assistance of the BCPTs, machine learning tools and methods and the incorporation of business continuity data from selected agrarian sectors. The effort focuses on the incorporation of agrarian BCM data to validate the BCPTs approach, machine learning techniques to investigate the predictive accuracy of the Business Continuity Points method in agribusiness and use specific BCPTs variables as input for further predictions.

The achievement of the above stated research objective relies on the following contributions:

- Business continuity data collection form selected agrarian sectors. Detailed business continuity data has been found only in regards to the dairy and the cattle feed yards sectors using the appropriate search strings in several prestigious databases and the google search engine
- Analysis of the BCPTs method with example and its current validation in terms of recovery time computation and recovery prioritization of the business functions included in the collected data
- Utilization of machine learning classification and regression techniques with the corresponding evaluation metrics to generate rule-based optimized recovery priority establishment for the agrarian sectors based on combined datasets datasets that stem from a) the above mentioned BCPTs validation process using the collected agrarian data and b) the empirical lab data used to develop the BCPTs method (APPENDIX).
- Explanation regarding current limitations of the present research which is mainly the small number of agribusiness BCM data
- Report the success rates of the inferred business continuity prediction models in optimizing the recovery procedure of disrupted agricultural functions in the given sectors.

The chapter is organized as follows: the section "background theory" refers to the utilized data searching and literature investigation databases, as well as the summarized analysis of the BCPTs method. The section "Tools and Methods" refers to the dataset formulation process based on the collected data and an explanatory example of how the BCPTs recovery complexity variables and recovery time effort (RTE) are computed for each of the involved agricultural operations. The section also includes a detailed delineation of all the utilized machine learning software tools, methods, algorithms, code, evaluation metrics approaches as they are delineated in the literature. The "Results and Discussion" section highlights the inferred machine learning prediction rules as well as their accuracy measurement. The BCPTs practical evaluation results are reported, its accuracy in predicting recovery priority levels (binary classification) is measured in comparison with the suggested by domain experts classification, and the multi-approach multiclass recovery prioritization using different BCM prioritization methods and specific recovery complexity input parameters are also investigated for their accuracy. Finally, the recovery time BCPTs formula is improved using regression analysis techniques and the collected data. Research limitations are reported as well in order to close the specific section. The "Conclusion" section briefly summarizes the research achievements and includes implications for further research in the future.

BACKGROUND THEORY

The current contribution proposes novel business continuity process optimization methods for the agribusiness industry, combining machine learning techniques and the BCPTs approach. Crucial part of the research has been considered a thorough investigation of the existing literature around the analysed topic to support novelty, robustness and practical value of the suggested prediction models. As a result the following target literature investigation objectives have been considered:

1. Obtain business continuity agribusiness data (BCM guides) with which the BCPTs method could be tested for its robustness
2. Detailed analysis of the BCPTs approach (equations, recovery complexity parameters, recovery factors) and its novelty comparing to similar (if any) recovery complexity approaches in BCM or business impact analysis in agriculture or other sectors
3. Related work regarding the use of business continuity, disaster and crisis management, processes' and business function's recovery policies in agribusiness in combination with machine learning and big data manipulation techniques in the agrarian field
4. Delineation and justification of the use of the selected machine learning tools and methods.

To fulfil the aforementioned literature overview objectives the following databases and search strings have been used and resulted in the below mentioned findings:

a) 'Business continuity template agriculture' search string has been used in google search engine to collect real BCM data in the agrarian sector and test practically the BCPTs method. Only two (PRPC 1(n.y), PRPC 2(n.y)). results have been considered as appropriate for the method's practical validation
b) 'Business Continuity Points' and 'Business Continuity Testing Points' in Researchgate platform to provide a link to the utilized method under evaluation for its predictive potential to optimize BCM procedures in agribusiness.
c) 'Machine learning repositories' search phrase in google has indicated datasets such as, Kaggle, UCI, Google's dataset, Microsoft Dataset, Awesome Public Dataset Collection, Government Datasets and Computer Vision Datasets. In these repositories no relevant data has been found, using the phrases 'business continuity agriculture', 'machine learning agriculture and business continuity'
d) Scientific publishing databases like Elsevier, IEEE, Springer, IGI Global, MDPI have been also explored using the search strings, 'machine learning agriculture', big data in agriculture', business continuity agriculture and machine learning', either directly in these databases or through google engine and researchgate platform. The relevant suggested articles are included in the present chapters, in the form of Backround theory, Related Work and Tools and Methods.

Related Work

Business continuity management is crucial for agribusiness. As stated in the introduction section, its importance has been addressed by multiple researchers and practitioners especially after the COVID-19 global pandemic outbreak (FAO, n,y). Additionally, another research states that that "emergency response that enables continuity of business for agriculture sectors effectively and rapidly ensures that disasters are handled in both an economically and environmentally sustainable way" (Culhane et al, 2018). However, even recent research outputs underline the ignorance of standard BCM policies in the domain. A study by (Urbanec and Urbancova, 2014) indicate that just 5% of the Czech agricultural companies rely on standard BCM norms. It is thus, important to chart BCM policies which rely on the incorporation of mathematical models for non-arbitrary business function recovery complexity parameter estimation tools (Podaras, 2015). The investigated literature in scientific databases revealed a limited number of mathematical models for business impact analysis and no mathematical model is based on machine learning approaches. Additionally, none all the mathematical models available address limited number of recovery complexity parameters and recovery factors (technical-eg. Complex technical infrastructure, environmental-eg. employee experience in BCM or motivation, unexpected-eg. severe conditions during a real recovery case). Finally, no such model has been applied in the agrarian industry.

For example (Torabi et al, 2014) suggested multiple criteria decision making for business impact analysis, an arbitrary recovery complexity parameters (actors, processes, recovery factors) definition based on domain experts' aspect. However, arbitrarily proposed BCM methods in a specific domain can be hardly applied to multiple domains. Moreover, (Sahebjamnia et al, 2015) use various recovery complexity parameter values but do not address the weight values or importance of each parameter. For example, managers are more important human actors than regular low-level employees. To tackle the aforementioned issues, the BCPTs tool has been proposed (Podaras, 2015; Podaras et al, 2016). The authors attempted to generate machine learning prediction rules using lab-generated own datasets (Podaras, 2018) but the results lack practical evidence, an issue which has been covered afterwards by (Podaras et al, 2021) using data from the field of higher education BCM.

The present work attempts to expand further the practical value of the BCPTs method by testing its effectivity in the agrarian sectors from which the current BCM data has been collected. Efficiency in manipulating agricultural big data volumes can result in optimized agribusiness decisions and "*smarter and more sustainable farming*" (Kamilaris et al., 2017). The use of machine learning techniques has been proved highly effective in optimized decision-making during critical conditions, for instance in security – related fields (Shaukat et. al, 2020 (1); Shaukat et. al, 2020 (2)), or in tackling unexpected power outages by severe weather conditions (Yang et al, 2020). In the agricultural field, representative machine learning application cases are mentioned in the 'Tools and Methods' section.

The Business Continuity Points Method

The approach focuses on the incorporation of multiple recovery complexity parameters, namely humans, systems and processes involved in the daily normal operation as well as the recovery process in case of failure of individual business functions, to estimate mathematically their criticality ranking (or recovery priority level) and the recovery time as proposed by BCM experts (Gibson 2011). The method is also based on various recovery strategy scenarios and technical, environmental and unexpected recovery factors that can influence the recovery time value and the priority level of its resumption (Podaras et al, 2016; Podaras, 2018). The method and its mathematical functions have been analytically presented and explained in the aforementioned published papers. A summarized mathematical expression is currently included:

Unadjusted Business Function Recovery Points (UBFRP): the parameter stems from the sum of human (HA), technical actors (TA) and involved processes (PR) and their importance (weight values W) in the recovery process (1):

$$UBFRP = \sum_{i=1}^{n}\left(\sum_{i=1}^{n}(HA*WHAi),\sum_{i=1}^{n}(TAi*WTAi),\sum_{i=1}^{n}(PRi*WPRi)\right) \quad (1)$$

Where i represents the ith element and n= number of elements (actors or processes)

Adjusted Business Function Recovery Points (ABFRP): the parameter is estimated as follows (2):

$$ABFRP = UBFRP * TRF * ERF * URF \quad (2)$$

Where TRF, ERF,URF are the Technical, Environmental and Unexpected Recovery Factors that affect the recover process and are used for formulating Simple, Average and Complex Recovery Scenarios (Podaras et al, 2016, Podaras, 2018).

Recovery Time Effort (RTE): The time required to resume an unexpectedly interrupted function is computed as follows (Podaras et al, 2016):

$$RTE = \frac{5000}{ABFRP^2} - 3 \quad (3)$$

TOOLS AND METHODS

Data Collection and Dataset Formulation

The data used in the present study has been extracted by the Panahandle Regional Planning Commission in Texas (PRPC 1(n.y), PRPC 2(n.y)). After applying the BCPTs method to the key business functions in both agrarian sectors the following computations have been inferred (Tab.1)

RTE Computation Example: In order to estimate the recovery time, since this information has not be directly provided by the BCM guide, the number of involved processes, the number of human and technical actors have been based on the criticality level of each function as indicated by domain experts.

The recovery scenario (TRF, ERF, URF) values have relied on the rules of BCPTs (Podaras, 2018)

If UBFRP<9.5 Points then Simple Scenario (TRF=ERF=URF=0.85)

If UBFRP between 9.5 and 21 points then Moderate Scenario (TRF, ERF, URF=1).
If UBFRP>21 points then Complex Scenario (TRF, ERF, URF= 1.15)

Table 1. The BCPTs Application in the Cattle/Dairy Sectors for 18 Key Business Functions (Source: Podaras, 2022)

Business Function	Sector (CF=Cattle Feedyards, DF-Dairy Farms)	UBFRP (BASED ON BCPTS) (Points)	ABFRP (BASED ON BCPTS) (Points)	RTE (BASED ON BCPTS) (Hours)
Cattle Movement and Holding	CF	9	5.53	160.67
Feed Delivery Operations	CF	16.5	16.50	15.37
Facility Maintenance	CF	15	15.00	19.22
Animal Health / Hospital Administration	CF	11	11.00	38.32
Business Administration / Personnel	CF	10	10.00	47.00
Capital and Investments	CF	4	2.46	825.58
Information technology	CF	3.5	2.15	1079.23
Environmental Management	CF	4	2.46	825.58
Parlor Operations	DF	11.5	11.50	34.81
Feeding/Watering	DF	10.5	10.50	42.35
Facility Maintenance	DF	13.5	13.50	24.43
Environmental Management	DF	13.5	13.50	24.43
Herd Management	DF	10.5	10.50	42.35
Calf Rearing	DF	8.5	5.22	180.49
Animal Health / Hospital Administration	DF	12	12.00	31.72
Information technology	DF	5.5	3.38	435.26
Business Administration / Personnel	DF	8.00	4.91	204.15
Capital and Investments	DF	7.00	4.30	267.56

An explanatory example is included regarding the *Cattle Movement and Holding* business function:

Number of Processes: 3, Process Weight (BCPTs)=1.5 (critical function)
Number of Human Actors: 3, Human Weight= 1.5 (critical function)
Number of Technical Actors (Applications): 0, weight=1.5 (critical function)
ABFRP Value (based on (2)): 5.53 points
RTE Value (based on (3): 160.67 hours

For *moderately critical functions, weight values=1* and for *least critical functions, weight values=0.5.*

The specific dataset has been used for investigating the validity of the BCPTs approach to the given sectors and to see how close the real estimations can be when compared to the empirical lab estimations (see APPENDIX). Moreover, the new data has been used to enrich the empirical data in order to implement more robust machine learning classification. Finally, regression – based RTE estimation using the ABFRP input has been also investigated.

R PACKAGE

R is a free language and one of the most popular languages used for the statistical analysis of data. It supports a large variety of techniques for statistical data processing in various fields and offers various graphical techniques to demonstrate the results. The advantage of R over other statistical languages is that it is open-source software that can be downloaded and used by anyone. As R is an open language many people have contributed to its wide and vivid library which allows its user to analyse any kind of data in any field needed. R can support a large variety of operations and is not only limited to a statistical system and is generally termed as the environment. R is also very flexible as it can be extended as required by the user by using packages that are also available to be downloaded by the user or even can be created by anyone. (R: What is R? 2019)

R packages are also used in the analysis of data in agribusiness. For example, R package is used to run an algorithm to analyse soil profile and compute stress induced on the soil due to agriculture traffic which can harm the soil and even on the crop. (De Lima et al., 2021) Another such example is the use of R package to calculate the potential threat to humans and the environment due to the use of pesticides according to the 'Pesticide Load' indicator in Denmark. (Möhring et al., 2021). Or another such research can be the use of R packages for analysis of climate adaptation in temperate fruit species. (Miranda et al., 2021)

CLASSIFICATION

Classification as defined by (Jiawei et al, 2011), is "*a form of data analysis that extracts models describing important data classes. Such models, called classifiers, predict categorical (discrete, unordered) class labels.*" Data Classification is a two-step process that includes the learning step and classification step. The learning step which is also called the training step consists of creating a classification algorithm based on the learning set. The learning set is created by random sampling the records from the database under study. Once the classifier is determined, it is tested upon the rest of the database and the accuracy of the classification is checked. If the accuracy is acceptable, the rules are then applied to new data and can be used to classify the data as required. A training set consists of the records and associated class labels.

Depending on the availability of class label the training step can also be divided into supervised learning or unsupervised learning. Supervised learning is when the class label is predetermined and is known whereas unsupervised learning is when the class label is not known, and the data must be clustered depending on the similarities between them. This form of unsupervised learning is known as clustering. In which the data is grouped into separate groups depending on the associations between them as there is no class label attached to those records. Once the classification algorithm is decided, the classification map or function can be known. Mapping can be depicted as a decision tree, classification rule, or formula. The mapping can then be further used to classify new data. Mapping helps the user to get further insight into the data and summarizes the data in a readable form.

The accuracy of the classifier is measured by the test data and not the learning data as, if the learning data was used to determine the accuracy of the classifier, then the classifier would over fit the data as classifier might try to fit the data based on some anomalies present in the testing set. As the testing set is independent of the training set, it can be used to measure the accuracy of the classifier.

DECISION TREES

Decision trees are a part of the classification and regression model, in which the data is divided into different classes based on similar features or any other similarities in the data associated with it. Decision tree consists of decision nodes, leaf nodes, and root nodes. Nodes are the points that differentiate the data based on a specific attribute. The leaf node is the result after segregation. Leaf node cannot be broken further. The time complexity of DT is $O(mn^2)$ with n = number of instances and m = number of attributes (Shaukat et al, 2020 (3)).

Figure 1. Example of data mapping (Source: own work)

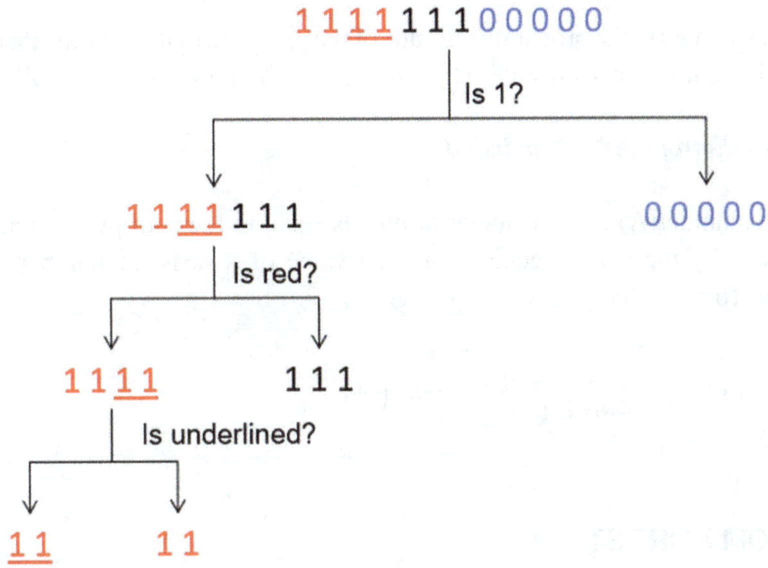

Figure 2. Example of decision tree (Source: own work)

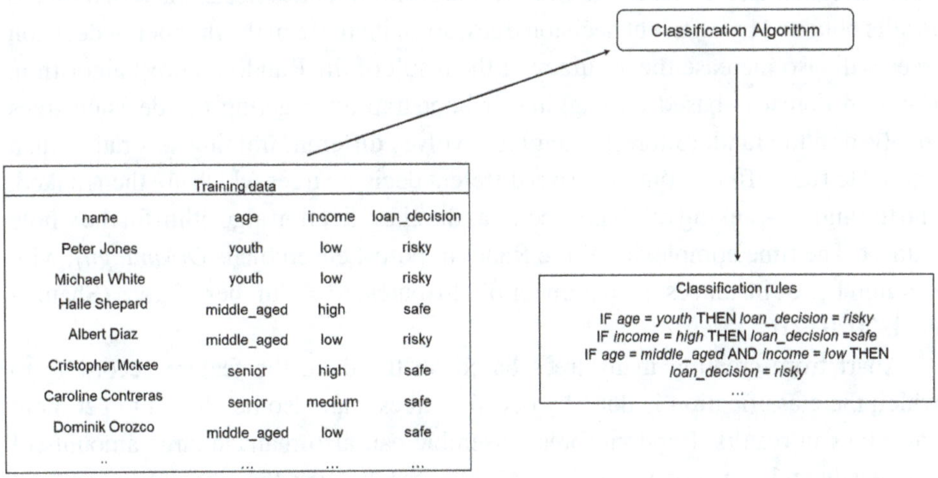

The decision tree classification approach uses the combination of *Entropy* impurity measure and the *Information Gain* which indicates the entropy reduction, to estimate the importance of each variable (Russell and Norvig, 2020). The *Entropy* impurity measure and the *Information gain* are estimated as follows (4):

$$Entropy(A) = \sum_{i=1}^{n} -P(V_i)*log_2 P(V_i) \quad (4)$$

where $P(V_i)$ refers to the probability a value of the given attribute *A* is i and there exist n possible values of this attribute. The *Information Gain* is estimated as follows (5):

$$Gain(A) = Entropy(A) - Remainder(A) \quad (5)$$

where Remainder(A) , or remainder of bits, is equal to the entropy of an attribute *A* weighted by the value frequency n_i for i value of *A* (*k* is the number of data instances) (6):

$$Remainder(A) = \sum_{i=1}^{n} \left(\frac{n_i}{k}\right) * Entropy(A)_i \quad (6)$$

RANDOM FOREST

Random Forest is a type of classification and regression model which is based around decision trees. Random Forest as the name suggests is the collection of many decision trees. The Random Forest algorithm makes decisions based on the results obtained by different decision trees, so an increase in the number of decision trees will also increase the accuracy of the result of the Random Forest algorithm. Random Forest is based on bagging or bootstrap aggregating the decision trees involved in the random forest. Bagging involves different training sets rather than one set to find different outputs from different decision trees which are then ranked, and the highest-ranking result is selected as the classification algorithm for the whole dataset. The time complexity of the Random Forest ensemble is *O(Mmlogn)*, with n = number of instances, m = number of attributes, M= Number of trees (Shaukat et al, 2020 (2)).

Apart from creating many trees based on the data, the feature according to which the classification is done by decision trees, can also be altered to get more variations in results. Random forest ensemble can also manage large amounts of data easily and can produce higher accuracy results over the single decision tree. Although, Random Forest takes a large computational volume of the machine and might require a longer time as compared to decision trees. CART and Random Forest are widely used algorithms for both classification and regression purposes. Specific examples that can be mentioned are the use of Random Forest for global and regional yield predictions (Jeong et al., 2016), evaluation of random forest for agricultural crop classification (Ok et al., 2012) and agricultural irrigation area predictions (Gao et al., 2021).

REGRESSION

Regression is another data – supported prediction tool like classification which is generally used for numeric data. Although other methods of numeric data analysis are developed, regression might be the most popular method used in the case of numeric data (Montgomery et al., 2013). There exist many types of regression models like Linear Regression, Logistic Regression, Ridge Regression, Lasso Regression, Polynomial Regression, etc. All these models are used for regression analysis depending on the application required. Regression is used for estimating the relationship between the dependent variable (outcome) and one or multiple independent variables (also known as predictor). Out of all the regression models, the linear regression model is the most used model out of it, where the most fitting linear line is to be found to the specific statistical model. Although regression analysis can predict the dependency of the dependent variable on the independent variable, there is still an error included. This means that the predictions made based on the algorithm are not the exact value but an estimate of the possible outcome. The training time complexity for the linear regression is $O(n*m^2 + m^3)$ while the testing time complexity is $O(m)$, with n = number of training examples, m = number of features (Kaggle, n.d.)

Regression finds its application in almost every field like economics, statistics, production, medical, agriculture, etc. The regression model has generally been used to make predictions based on the data acquired. Regression can also be used for the classification of data by using complex functions.

Regression model consists of following parameters:

- Unknown Parameter (β)
- Independent Variable (X_i)
- Dependent Variable (Y_i)
- Error term (e_i)

Here, i indicated the ith row.

Regression can be summarised by (7),

$$Y_i = f(X_i, \beta) + e_i \tag{7}$$

The objective of the model is to compute Y_i based on other independent variables. To carry out the regression analysis the f is to be determined to fit the data as close as possible. The function f is decided by the knowledge of the user or the most appropriate type of function is opted. This function can be linear, exponential, or polynomial and there are other more complex functions available as well which includes many applications in specific fields of study.

LOGISTIC REGRESSION

Logistic Regression is also known as logit regression is a type of regression used for binary responses. Unlike Linear or many other types of regression, logistical regression is based on discrete data rather than continuous and normally distributed data. This model aims to estimate the relationship between the binary variable (dependent variable) and the predictor. The dependent variable is generally in the form of 1 or 0 where 0 indicates the failure and 1 indicates success. (Mueller et al., 2005) The model can be defined by (8)

$$logit(p) = \ln\left(\frac{p}{1-p}\right) \tag{8}$$

Where p is the probability of the event occurring.

Logistic Regression is used widely in the space of data analysis as it is one of the most used method for analysing discrete and binomially distributed data. Some of its applications in agriculture fields are for creating erosion probability maps by using logistic regression (Mueller et al., 2005), analysis of accidents related to agricultural machinery farm-work accidents using logistic regression (Kim et al., 2014) or use of logistic regression for classification in agricultural ergonomics (Bhowmik et al., 2011). The training time complexity for the Logistic Regression is *O(n*m)* while the testing time complexity is *O(m)*, with n = number of training examples, m = number of features (Kaggle, n.d.).

K-FOLD CROSS VALIDATION

Cross Validation is a statistical method to estimate the accuracy of the prediction or classification model. It is used to compare the models based on the accuracy of the model. It can be easily applied to any model and is not too complex to use as well. Cross validation is generally used when there is limited data available to check the accuracy of the model. The time complexity for the k-folds CV is *O(kn)*, with n = sample size, k=number of folds (StackExchange Network, n.y.)..

Cross validation is often referred to as K – fold cross validation, where k stands for the number of iterations the model is required to be tested. If there are 10 sets to be created from the data available, k takes the value 10, and it becomes 10 – fold cross validation. Each observation is divided into sets as required and each observation in the dataset is used in one or the other group created so that the model designed can learn from all the data that is available. In this method, selected sets are used to formulate an algorithm (learning set) and the remaining sets are used to test the accuracy of the algorithm (predictor). The accuracy from the model can be then interpreted as required.

The sets are formulated by shuffling the available data and then the data is randomly picked to form k groups. The number of groups (k) is important for the results of cross validation to be useful. The wrong value of k might result in high variance values, unfit or over-fit model. So, the step to select the k value is important. Although there is no specific method to define the value of k in any given model, there are a few ways it can be decided. The most common way is to define k as 10, or the k can be decided so that there is enough data in each set so that it is statistically representative of the whole data or leave-one-out or leave-many-out CV can also be used. In leave-one-out, one dataset is left out to evaluate the predictor and others are used as training sets, whereas in leave-many-out higher proportion of test sets are left as compared to the training sets (e.g., 40%-60%, where 40% of data are used as training and 60% are used for testing). (Zafar & Salahuddin, 2009)

R SCRIPTS

In order to generate efficient and effective predictive patterns to optimize agribusiness procedures regarding business continuity in the given industry, the obtained data from two selected sectors has been processed and stored as .csv (comma separated value) spreadsheet form, and imported into the R-Package for rule induction, using the appropriate packages, functions and R-scripts. Depending on the predictive pattern, namely classification using CART, cross validation using 10-folds, tree boosting using random forest and regression analysis for predicting numeric values, specific R scripts need to be utilized.

Classification using CART:

```
library(rpart)
require(rpart)
library(rpart.plot)
require(rpart.plot)
r1=rpart(Y ~ X1+X2, data = training_set, method="class")
rpart.plot(r1)
#OR
prp(r1)
```

Confusion matrix

```
#caret for confusion matrix function
install.packages('caret')
library(caret)
require(caret)
#predicting with test data
pred=predict(r1,newdata= testing_set, type="class")
pred
#cf must have equal number of rows/columns
dat1 = factor(testing_set$Y, levels=c(0,1))
dat2 = factor(pred, levels=c(0,1))
cm = confusionMatrix(dat2,dat1)
cm
#evaluation metrics
accuracy_RP_CART <- sum(diag(cm$table))/sum(cm$table)
accuracy_RP_CART
```

Repeated 10-folds cv (Repeats=3)

```
install.packages('e1071')
library(e1071)
require(e1071)
library(caret)
require(caret)
require(rpart)
numFolds <- trainControl(method = "repeatedcv", number = 10,
repeats = 3)
cpGrid <- expand.grid(.cp = seq(0.01, 0.2, 0.01))
train(Y ~X1+X2, data = training_set, method = "rpart",
trControl = numFolds, tuneGrid = cpGrid)
```

```
r1cv<-rpart(Y ~X1+X2, data = training_set, method ="class",
cp=0.01)
```

Random Forest Ensemble (500 Trees, 3 trials)

```
install.packages("randomForest")
library(randomForest)
library(caret)
library(caTools)
require(caTools)
require(caret)
classifierRF<-randomForest(as.factor(Y) ~X1+X2,
data=training_set, ntree=500, mtry=3, importance=TRUE)
classifierRF
```

RESULTS AND DISCUSSION

Application of the Business Continuity Points Machine Learning Binary Criticality Classification to the Dairy and Cattle Sectors

According to the Business Continuity Points method, a binary criticality classification pattern can be applied for core enterprise business functions, having as input variable the Unadjusted Business Function Recovery Points (UBFRP) variable (Figure 3) (Podaras, 2018). The constructed decision tree regarding the criticality prioritization of core business functions relies on the Classification and Regression Trees (CART) algorithm. The algorithm uses the *Entropy* impurity criterion the *information gain*.

In order to confirm the applicability of the specific binary classification rule to the dairy and the cattle feed yards sectors, UBFRP and binary classification data collected from the given sectors have been used as validation set. The data concerns all the critical business activities at the given sectors. A set of 18 functions has been built. Given the fact that the empirical BCPTs data has been composed of 47 business functions created by the authors of the approach (APPENDIX 1), the currently collected data for validation is regarded as sufficient. The UBFRP has been estimated according to the BCPTs method for each individual function. The binary classification of each function as critical or non-critical has been defined by the domain experts from both agrarian sectors. The following dataset has been used as validation set for the BCPTs approach (Tab.2)

Figure 3. The decision tree for the BCPTs (Podaras, 2018) binary classification pattern

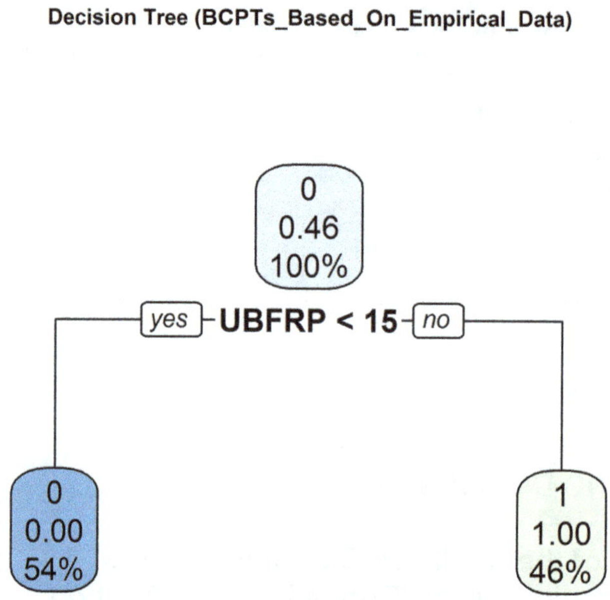

The following results have been inferred during the validation process in the R- Package:

Confusion Matrix and Statistics

Reference
Prediction **0 1**
0 11 5
1 0 2
Accuracy: 0.7222
Additionally the following evaluation metrics results have been estimated:
Precision: 0.6428571
Recall: 0.84375
F1-Score: 0.6296296

As a consequence the BCPTs method has been proved highly accurate for binary classification in the selected agrarian sectors. The next step regarding the present research, concerns the testing of multiple different recovery priority settlement strategies towards the more effective criticality ranking of key agrarian business activities.

Apart from evaluating the accuracy of BCPTs to predict a binary class (critical/

Table 2. The Validation set regarding the BCPTs binary criticality classification predictive potential (Source: own work Podaras, 2022)

UBFRP (Points)	RTE (Hours)	BINARY RANKING (Agrarian Domain Experts)	BINARY RANKING (BCPTs)
3.5	1079.23	0	0
4	825.58	0	0
4	825.58	0	0
5.5	435.26	0	0
7	267.56	0	0
8	204.15	0	0
8.5	180.49	0	0
9	160.67	1	0
10	47	0	0
10.5	42.35	1	0
10.5	42.35	0	0
11	38.32	0	0
11.5	34.81	1	0
12	31.72	0	0
13.5	24.43	1	0
13.5	24.43	1	0
15	19.22	1	1
16.5	15.37	1	1

non-critical) for individual agrarian business functions in terms of its recovery, an integrated dataset including the experimental as well as the real data has been investigated for its predictive accuracy. The set has been composed of 65 records (47 experimental own records (APPENDIX 1) and 18 based on the collected BCM data (Tab.2) from the dairy and the cattle feed yards sector. A similar decision tree has been inferred (Figure 4)

Cross validation using repeated 10 folds algorithm and decision tree boosting using random forest ensemble have been implemented using the R-Package. The following evaluation results have been inferred:

Figure 4. Decision tree classification of individual functions based on experimental and real data.

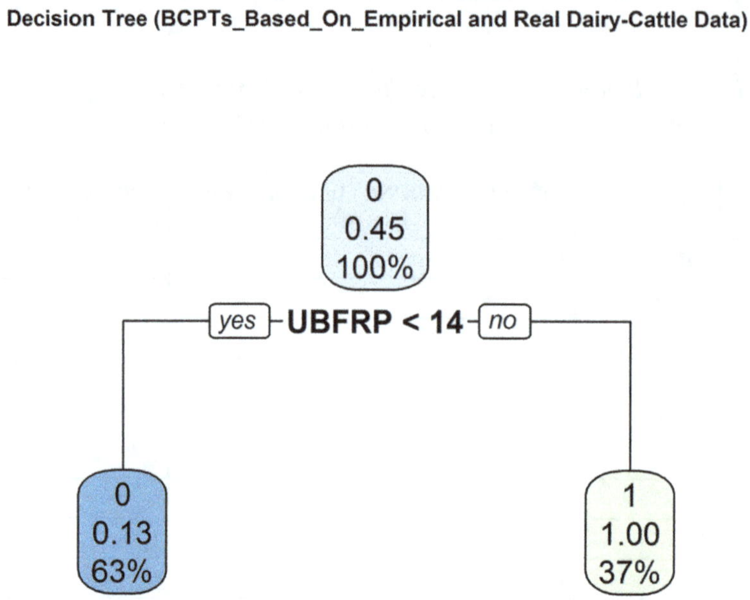

Confusion Matrix and Statistics

Reference
Prediction 0 1
0 36 5
1 0 24
Accuracy: 0.9231.

A comparative table including evaluation metrics for all the investigated classifiers is included (Tab.3). The illustrated results confirm the high predictive potential of the BCPTs algorithm in the given agricultural domain.

Multi-Class Machine Learning Recovery Priority Settlement based on Various Ranking Approaches

In order to develop more precise machine learning recovery priority prediction rules for crucial agribusiness operations in the dairy and cattle sectors, six different approaches have been utilized. (Gibson, 2011; Rittinghouse & Ransome, 2005;

Snedaker, 2007; Supriadi, & Sui Pheng, 2018; University of California San Francisco, 2016; Western Australian Government, 2009) for the settlement of recovery priorities and establish effective BCM policies. The idea of the current investigation relied on incorporating several different approaches as proposed by BCM experts and test their applicability to the selected agrarian industries.

Table 3. Comparison of the 3 classifiers regarding their predictive potential

		Evaluation Metrics			
		Accuracy	**Precision**	**Recall**	**F1-Score**
Classifier	**decision tree-CART**	0.9231	0.9137	0.9390	0.9203
	10 folds Cross Validation	0.9375	0.95	0.9285	0.9352
	Random Forest Ensemble	0.9375	0.9285	0.95	0.9352

A dataset of 108 records has been constructed (Podaras, 2021). Input variables (5 input columns), else called predictors or explanatory variables) that have been considered are the following:

- Sector
- Number of Processes
- Number of Users
- Number Applications
- Selected Approach

The output labelled class has been the *classification* (predicted variable or explained variable).

The following results have been inferred regarding the accuracy of the new classifier:

Confusion Matrix and Statistics

Reference
Prediction **1 2 3**
1 2 1 0
2 2 7 5
3 0 4 6
Accuracy: 0.5556
Precision: 0.5429293

Recall: 0.5888889
F1-Score: 0.5604396

Due to the fact that the predictive potential of the inferred decision tree classification pattern is relatively low, further boosting using K-FOLDS cross validation and Random Forest ensemble have been also investigated.

Figure 5. Decision Tree of the Recovery Priority Classification Pattern based on CART

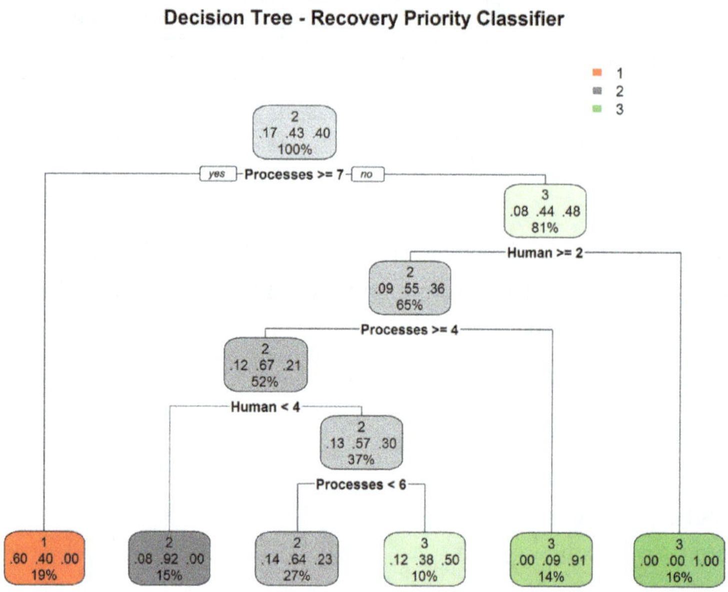

Figure 6. Variable Importance based on CART decision tree. Number of Processes and Human Actors are the most important input parameters that affect the Recovery Priority Classification of a given Function

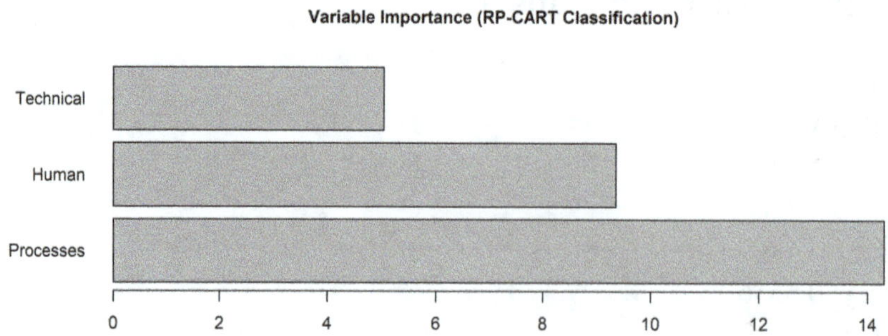

According to the inferred predictive pattern (decision tree) and the variable importance analysis regarding the input predictors, the most influential factor regarding the predicted output is the number of processes. The second is the number of human actors (users, managers and more) and third the number of technical actors (software applications). The rest of the variables appear to have no effect on the predictions. The decision tree reveals however that the factor *technical actors* is not considered as splitting node. We can thus support that the specific factor hardly affects the predictions of the classifier.

Nevertheless, it has been considered as crucial research step to test whether the predictive potential of the induced decision tree can be ameliorated. Initially the repeated 10-folds cross validation has been explored for its predictive potential, based on 3 repeated validation tests.

The following classifier's evaluation results have been conducted:

Confusion Matrix and Statistics (repeated 10-folds cross validation, repeats=3

Reference
Prediction **1 2 3**
1 2 2 0
2 1 11 0
3 0 5 6
Accuracy (rep. 10-folds CV): 0.7037
Precision(rep. 10-folds CV): 0.7592593
RecallCV (rep. 10-folds CV) = 0.6540404
F1-Score (rep. 10-folds CV): 0.6702148

Finally, the Random Forest ensemble has been investigated for its predictive potential and as an additional boosting to the initial decision tree classification pattern based on CART.

The following evaluation results regarding the predictions of the Random Forest ensemble have been inferred (Figure 7) and a summarized comparison of the three classifiers is illustrated (Figure 8):

Confusion Matrix and Statistics

Reference
Prediction **1 2 3**
1 3 0 0
2 1 10 0

3 0 2 11
Overall Statistics
Accuracy: 0.8889
Precision(random forests ensemble): 0.8611111
RecallCV (random forests ensemble): 0.9184149
F1-Score (random forests ensemble): 0.8811249

Figure 7. The random forest error plot of the recovery priority classifier (ntree=500, mtry=3). As the number of investigated trees grows the predictive error is reduced and stabilized

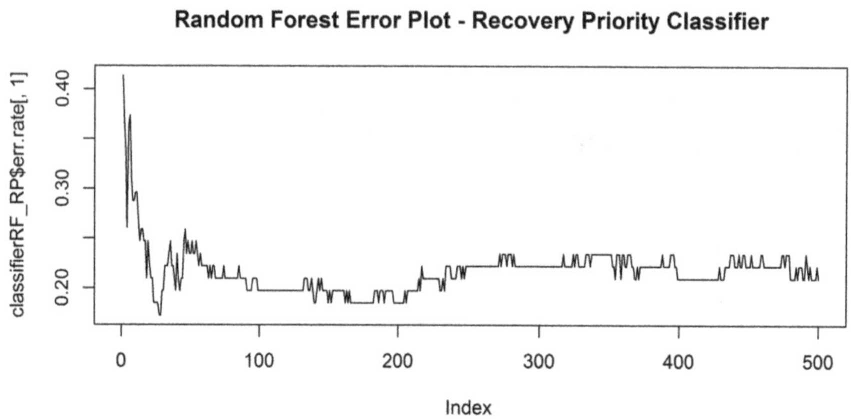

Table 4. Comparison of the 3 classifiers regarding their predictive potential

		Evaluation Metrics			
		Accuracy	**Precision**	**Recall**	**F1-Score**
Classifier	**decision tree-CART**	0.5556	0.5429	0.5889	0.5604
	10 folds Cross Validation	0.7037	0.7593	0.654	0.6702
	Random Forest Ensemble	0.8889	0.8611	0.9184	0.8811

From the above illustrated results, it is obvious that the random forest ensemble can significantly increase the predictive potential of the proposed classification pattern. Apart from the increased predictive potential of the random forest ensemble, the variable importance analysis differs from the one conducted by the initial decision tree classification (Figure 9). Comparing the two variable importance illustrations, it can be realized that the most important variable regarding both accuracy and predictions is the *number of processes*. However, the second most important variable for CART and the accuracy level of random forest is the human actors while in the predictions of the latter the second most important variable is the selected approach, and the human actors is third. This happens due to the multiple trees generated in random forest which is a process that can modify the variable importance. A secure remark is the fact that the accuracy of the model is not significantly affected when using the ensemble technique to boost the classifier and that human actors and number of processes play an important role in the formulation of the prediction pattern.

Figure 8. Comparison of the 3 investigated for their predictive potential classifiers

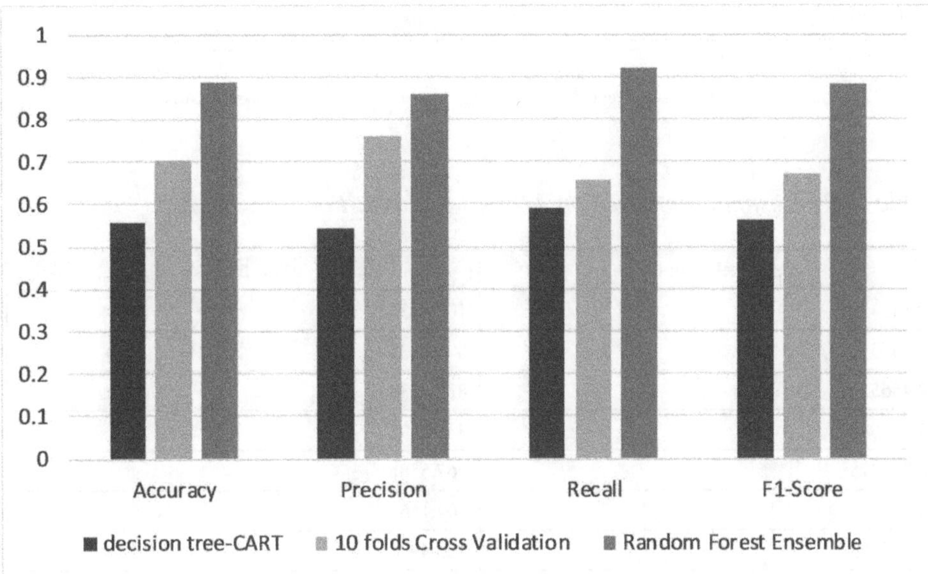

Regression Analysis to Predict Recovery Time based on BCPTs Input Parameters

An additional output regarding the current research is the possibility of the BCPTs parameters to be utilized as predictors regarding the Recovery Time of individual functions in the given agribusiness sectors. Using the BCPTs computations (Podaras et al, 2016) the following dataset has been constructed for the 18 business functions (Tab. 5):

Considering as input variable the ABFRP (Adjutsed Points) a specific recovery time predictive pattern has been created. The software package that has been used is the MS Excel (Figure 10)

Figure 9. Variable importance regarding accuracy (Mean Decrease Accuracy-How the accuracy is affected when removing a variable) and recovery priority predictions(Mean Decrease Gini- How the prediction is affected when removing a variable) based on random forest

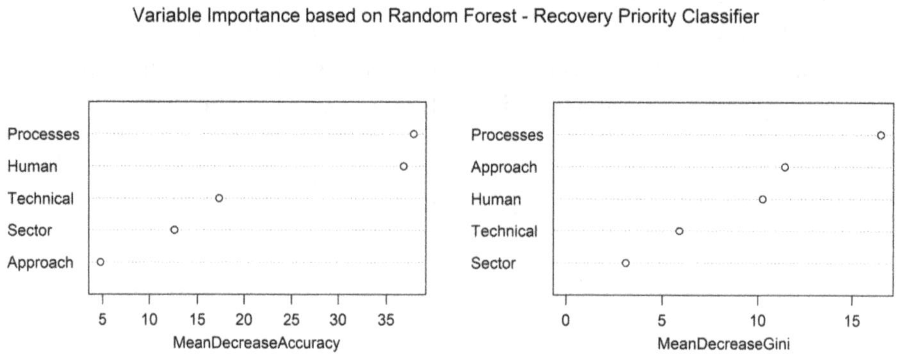

Table 5. RTE estimation based on ABFRP (Source: Podaras, 2022)

ABFRP(points)	RTE (hours)
2.149438	1079.232
2.4565	825.5839
2.4565	825.5839
3.377688	435.2592
4.298875	267.558
4.913	204.146
5.220063	180.4926
5.527125	160.6709
10	47
10.5	42.35147
10.5	42.35147
11	38.32231
11.5	34.80718
12	31.72222
13.5	24.43484
13.5	24.43484
15	19.22222
16.5	15.36547

The utilized evaluation metrics for performance reporting has been the R Squared (R^2) which is based on the residual's computation, that is the difference between the observed and the predicted by the new pattern RTE value. In both cases the reported R Squared values are very high which indicate satisfactory predictive potential (R^2_{ABFRP} (Quadratic Function) = 0.8428).

Nevertheless, due to the limited data records it has been attempted to further evaluate the model by estimating $R^2_{adjusted}$ and Root Mean Squared Error (RMSE) values. Based on the literature, it is possible to prevent overfitting in regression analysis when the number of records for each predictor is higher than 15 (Babyak, 2004) (18 is the current number) and when RMSE for training data is higher than RMSE for the testing data (Pošík, 2015).

The investigation has been implemented in R-Package. The estimated RMSE values regarding the training data(75% of the records) and the testing data (25% of the records) have been the following:

$RMSE_{TRAIN}$=117.87
$RMSE_{TEST}$= 36.79, which means that the second condition had been also fulfilled.

However, apart from the limited data, the present research could not be based on high ABFRP values (e.g. ABFRP>17 points) due to the fact that such values were not computed. It was thus not possible to investigate the applicability of the full empirical BCPTs data for ABFRP values higher than 17 points. As a result, the accuracy in recovery time predictions of the BCPTs method in the given sectors has been based on ABFRP input values within the interval 4.5-17 points based on the empirical data included in the study of (Podaras et al, 2018). The computed RMSE value when investigating the accuracy of BCPTs data over the real dataset has been estimated 109.23 which is very high and reveals potential overfitting of the quadratic model. The inferred interpolation of BCPTs validation over the real BCM data has been also used for illustrating the low accuracy of the quadratic model (Figure 11) in predicting similar RTE values compared to the empirical data within the same ABFRP interval.

Another issue which needed to be solved have been the negative RTE results obtained within specific ABFRP interval (e.g. Between 11 and 14 points) (Figure 11). To overcome this issue as well as the overfitting potential of the quadratic model the exponential regression model has been also investigated. The accuracy has been increased (Figure 12) which can be revealed from the increased R squared (0.954). Using again the R-Package further investigation has led to very low RMSE values for training and testing data based on 75% (trainining) and 25%(testing) splitting ratio ($RMSE_{train}$=0.316 > $RMSE_{test}$=0.15).

Figure 10. Regression Analysis for RTE estimation using ABFRP input variable (Quadratic Model)

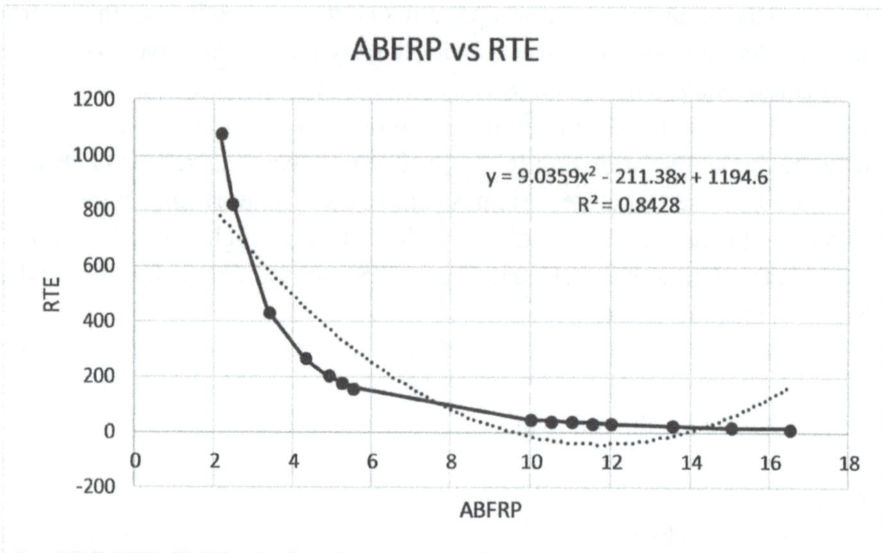

Figure 11. Interpolation of BCPTs (black curve and orange labels) over the real agrarian BCM labels (red labels) and the predictive regression plot (blue curve). Despite the high R squared value the predictive potential of BCPTs especially for high ABFRP values

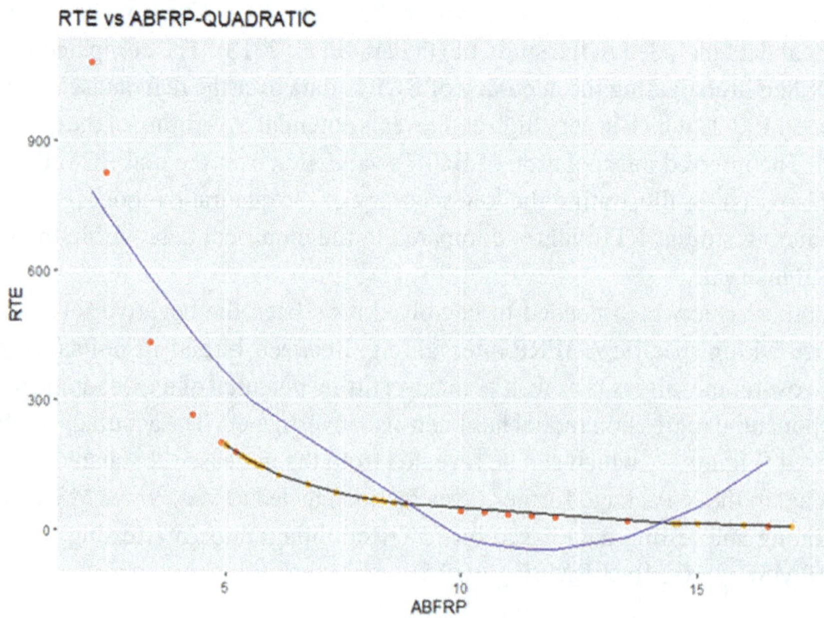

Figure 12. The investigation of the exponential regression potential to predict recovery time for agribusiness functions

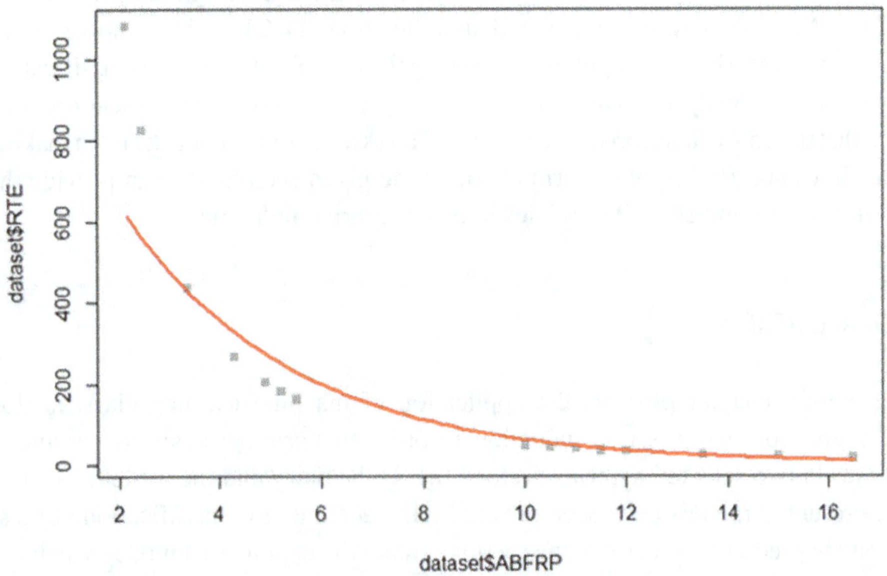

Finally, the obtained RMSE value regarding the validation of the BCPTs method when appllied in the agrarian sector using the exponential model has been also improved (RMSE=50.52), compared to the quadratic case (109.23). It is believed however that increasing the data records with more data from other agrarian sectors the exponential model can become more precise in its Recovery Time Estimation. More investigation could be performed using 2nd order exponential but since no real ABFRP values above 17 points have been computed this attempt has been avoided to prevent again the possibility to over-fit our model.

Limitations Regarding all Prediction Patterns

Despite the predictive potential of the inferred models, specific limitations can be reported. Firstly, the data that have been used for the construction of the pattern is limited. More observations are required to ensure predictive potential and at the same time high robustness regarding both classification and regression patterns and investigate further the accuracy of the suggested techniques considering possible cases of imbalanced data (Alam et al, 2020). Secondly the data utilized in the present chapter relies on the available BCM guide in the selected sectors. However, there is no confirmation based on discussions and interviews from the BCM or the domain experts (farmers and agronomists) whether the importance of the selected variables (weight values of the human actors, processes involved, technical actors/SW involved) is in agreement with the weighting values as assigned by the BCPTs method. Nevertheless, the data included in the utilized BCM guides imply that the specific weights should be applied when using the BCPTs approach for classification and regression analysis. In any case, it is strongly believed by the present research team that the constructed predictive patterns can assist in optimizing decision making procedures and BCM policy formulation at the given sectors and can provide the insights for optimized BCM policies in other agrarian industries.

CONCLUSION

The present chapter presents the application of machine learning classification and regression analysis rule induction in order to oprimize business continuity policies in two selected agrarian sectors, that is the dairy and the cattle feedyards. The presented models have been proved highly accurate for classification but less acurate in predicting precise recovery time except the exponential model which has inferred promissing estimations. The investigation relied on the possibility to apply the previously published BCPTs method to predict criticality ranking and recovery time for individual possibly interrupted business functions, in the selected sectors. The primary conclusions indicate that the BCPTs method can be applied with 72.22% accuracy for binary classification (critical business function/non-critical business function). Moreover, a comparison between quadratic polynomial regression and exponential regression to predict recovery time based on ABFRP BCPTs parameter using the collected data and the validation of the BCPTs when applied to the selected sectors indicated that the exponential model has been more accurate and less likely to overfit the data. However, due to the lack of ABFRP values in the real data a secure statement regarding the ideal model that can be used for this scope is currently avoided till further investigation using more BCM data is implemented in more agribusiness sectors. With the current data the exponential model has been proved accurate, less likely to over-fit but not so close to the BCPTs predictions (RMSE=50.52, R^2=0.954, $RMSE_{train}$=0.316 > $RMSE_{test}$=0.15).

Additionally, multiclass (3 classes for key business functions) classification as suggested by domain experts has been investigated. Multiclass classification has relied on 5 different recovery prioritization techniques and 3 different classifiers, namely the decision tree based on CART, the 10-folds cross validation and the Random Forest. The random forest has been the best classifier regarding performance based on accuracy, precision, recall and F1-score evaluation metrics (0.8889, 0.8611, 0.9184, 0.8811 respectively). The variable importance investigation indicated that the most important factors regarding the classification accuracy have been the number of processes, the number of human actors (users or managers) and the number of used applications (SW tools).

The presently conducted results imply that that the presented classification patterns based on a)BCPTs or b)simple number of recovery complexity elements can facilitate BCM policy optimization in the dairy and cattle sectors. Exponential regression pattern is also a promising recovery time predictive pattern for optimal BCM policies in the domain according to the validation results of BCPTs with the investigated function. In future, via the incoproration of more data and using feedback from domain experts via meetings and interviews, the current classification and regression patterns could be further boosted and applied to more agrarian industries.

REFERENCES

Alam, T. M., Shaukat, K., Luo, S., Hameed, I. A., Sarwar, M. U., Shabbir, S., Li, J., & Khushi, M. (2020). An Investigation of Credit Card Default Prediction in the Imbalanced Datasets. *IEEE Access: Practical Innovations, Open Solutions*, *8*, 201173–201198. doi:10.1109/ACCESS.2020.3033784

Babyak, M. A. (2004). What You See May Not Be What You Get: A Brief, Nontechnical Introduction to Overfitting in Regression-Type Models. *Psychosomatic Medicine*, *66*(3), 411–421. PMID:15184705

Bhowmik, A., & Ramasubramanian, V. & Chandrahas, & Kumar, A. (2011). Logistic Regression for Classification in Agricultural Ergonomics. *Advances in Applied Research.*, *3*, 163–170.

Culhane, M., Cardona, C., Goldsmith, T. J., St. Charles, K., Suskovic, G., Thompson, B., & Starkey, M. (2018). Building an all-hazards agricultural emergency response system to maintain business continuity and promote the sustainable supply of food and agricultural products. *Cogent Food & Agriculture*, *4*(1), 1550907. doi:10.1080/23311932.2018.1550907

Davidson, D. (2016). Gaps in agricultural climate adaptation research. *Nature Climate Change*, *6*(5), 433–435. doi:10.1038/nclimate3007

El Bilali, H., & Allahyari, M. S. (2018). Transition towards sustainability in agriculture and food systems: Role of information and communication technologies. *Information Processing in Agriculture*, *5*(4), 456–464. doi:10.1016/j.inpa.2018.06.006

Engemann, K. J., & Henderson, D. M. (2012). Business continuity and risk management: essentials of organizational resilience. Rothstein Associates Inc. Food and Agriculture Organization of the United Nations. (n.y). *Q&A: COVID-19 pandemic – impact on food and agriculture*. https://www.fao.org/2019-ncov/q-and-a/impact-on-food-and-agriculture/en/

Gao, G., Wang, M., & Huang, H. (2021). (PREPRINT). *Agricultural Irrigation Area Prediction Based on Improved Random Forest Model.* Advance online publication. doi:10.21203/rs.3.rs-156767/v1

Gibson, D. (2011). *Managing Risks in Information Systems*. Jones & Bartlett Learning.

Hájek, P. & Urbancova, Hana. (2013). Using of Business Continuity Standards in Agriculture, Industry and ICT. *AGRIS On-Line Papers in Economics and Informatics*, *5*, 55–67.

Hoogenboom, G. (2020). Agronomic data and the coronavirus epidemic. *The Journal of Agricultural Science*, *158*(3), 161–162. doi:10.1017/S0021859620000623

Jeong, J. H., Resop, J. P., Mueller, N. D., Fleisher, D. H., Yun, K., Butler, E. E., Timlin, D. J., Shim, K. M., Gerber, J. S., Reddy, V. R., & Kim, S. H. (2016). Random Forests for Global and Regional Crop Yield Predictions. *PLoS One*, *11*(6), e0156571. doi:10.1371/journal.pone.0156571 PMID:27257967

Jiawei, H., Micheline, K., & Jian, P. (2011). *Data Mining Concepts and Techniques, Third Edition*. Elsevier.

Kaggle (n.d). Computational Complexity of Machine Learning Models –II. *Kaggle*. https://www.kaggle.com/general/263127

Kamilaris, A., Kartakoullis, A., & Prenafeta-Boldú, F. X. (2017). A review on the practice of big data analysis in agriculture. *Computers and Electronics in Agriculture*, *143*, 23–37. doi:10.1016/j.compag.2017.09.037

Kim, B., Yum, S., Kim, Y.-Y., Yun, N., Shin, S.-Y., & You, S. (2014). An Analysis of Factors Relating to Agricultural Machinery Farm-Work Accidents Using Logistic Regression. *Journal of Biosystems Engineering*, *39*(3), 151–157. doi:10.5307/JBE.2014.39.3.151

Komarek, A. M., De Pinto, A., & Smith, V. H. (2020). A review of types of risks in agriculture: What we know and what we need to know. *Agricultural Systems*, *178*, 102738. doi:10.1016/j.agsy.2019.102738

Lorencová, E., Frélichová, J., Nelson, E., & Vačkář, D. (2013). Past and future impacts of land use and climate change on agricultural ecosystem services in the Czech Republic. *Land Use Policy*, *33*, 183–194. doi:10.1016/j.landusepol.2012.12.012

Miranda, C., Urrestarazu, J., & Santesteban, L. G. (2021). An R package for climate adaptation assessment of temperate fruit species. *Computers and Electronics in Agriculture*, *180*, 105879. doi:10.1016/j.compag.2020.105879

Montgomery, D. C., Peck, E. A., & Vining, G. G. (2013). Introduction to Linear Regression Analysis. In *Wiley Series in Probability and Statistics* (5th ed.). John Wiley & Sons.

Möhring, N., Kudsk, P., Jørgensen, L. N., Ørum, J. E., & Finger, R. (2021). An R package to calculate potential environmental and human health risks from pesticide applications using the 'Pesticide Load' indicator applied in Denmark. *Computers and Electronics in Agriculture*, *191*, 106498. doi:10.1016/j.compag.2021.106498

Mueller T.G., Cetin H., Fleming R.A., Dillon C.R., Karathanasis A.D, Shearer S.A.. Erosion probability maps: Calibrating precision agriculture data with soil surveys using logistic regression. (2005), *Journal of Soil and Water Conservation November, 60* (6), 462-468.

Ok, A. O., Özlem, A., & Oguz, G. (2011). Evaluation of random forest method for agricultural crop classification. *European Journal of Remote Sensing*, *45*(1), 421–432. doi:10.5721/EuJRS20124535

Pahandle Regional Services (n.d.). *The PRPC 1*. PRS. http://webcache.googleusercontent.com/search?q=cache:583KGJS00EwJ:theprpc.org/Programs/RegionalServices/Cattle-Feedyard-Industry/PRPC_CattleFeed/bcdoc/Texas%20Cattle%20Feedyard%20BCG%20Supplement.docx+&cd=3&hl=en&ct=clnk&gl=cz

Pahandle Regional Services (n.d.). *The PRPC 2*. PRS. http://webcache.googleusercontent.com/search?q=cache:OayhnqhEWRcJ:www.theprpc.org/Programs/RegionalServices/Dairy-Industry/prpc_dairy/bcdoc/Texas%20Dairy%20Farm%20BCG%20Template.docx+&cd=2&hl=en&ct=clnk&gl=cz

Podaras, A. (2018). Measuring the Accuracy Levels Regarding the Dual Business Function Criticality Classifier. *IEEE Access: Practical Innovations, Open Solutions*, *6*, 41598–41606. doi:10.1109/ACCESS.2018.2860976

Podaras, A., Antlová, K., & Motejlek, J. (2016). Information management tools for implementing an effective enterprise business continuity strategy. *E+M. Ekonomie a Management*, *19*(1), 165–182. doi:10.15240/tul/001/2016-1-012

Podaras, A., Moirogiorgou, K., & Zervakis, M. (2021). Regression-Based Recovery Time Predictions in Business Continuity Management: A Public College Case Study. In *Handbook of Research on Global Challenges for Improving Public Services and Government Operations*. IGI Global. doi:10.4018/978-1-7998-4978-0.ch020

Podaras, A. (2021): Multi-Approach Classification for Cattle and Dairy Business Functions for Business Continuity Policy Formulation. figshare. [Dataset]. doi:10.6084/m9.figshare.17429579.v1

Podaras, A. (2022). BCM_AGRO_DATA. figshare. [Dataset]. doi:10.6084/m9.figshare.19500164

Pošík, P., & Pošík, P. (n.y.). Czech Technical University in Prague, Faculty of Electrical Engineering Department of Cybernetics. *Bias-variance trade-off. Crossvalidation. Regularization.* https://cw.fel.cvut.cz/old/_media/courses/ae3m33ui/lectures/biasvariance-slides.pdf

de Lima, R. P., da Silva, A. R., & da Silva, Á. P. (2021). An R package for simulation of soil compaction induced by agricultural field traffic. *Soil & Tillage Research, 206*, 104824. doi:10.1016/j.still.2020.104824

Rittinghouse, J. W., & Ransome, J. F. (2005). Business Impact Analysis (BIA) - Prioritization and Classification of Critical Business Functions. In: Business Continuity and Disaster Recovery for Infosec Managers (pp. 83–86). Elsevier Press.

Russell, S., & Norvig, P. P. (2020). Artificial Intelligence: A Modern Approach (4th ed.). Pearson.

Sahebjamnia, N., Torabi, S. A., & Mansouri, A. (2014). Integrated Business Continuity and Disaster Recovery Planning: Towards Organizational Resilience. *European Journal of Operational Research, 242*(1), 261–273. doi:10.1016/j.ejor.2014.09.055

Shaukat, K., Luo, S., Varadharajan, V., Hameed, I. A., Chen, S., Liu, D., & Li, J. (2020). Performance Comparison and Current Challenges of Using Machine Learning Techniques in Cybersecurity. *Energies, 13*(10), 2509. doi:10.3390/en13102509

Shaukat, K., Luo, S., Varadharajan, V., Hameed, I. A., & Xu, M. (2020). A Survey on Machine Learning Techniques for Cyber Security in the Last Decade. *IEEE Access: Practical Innovations, Open Solutions, 8*, 222310–222354. doi:10.1109/ACCESS.2020.3041951

Shaukat, K., Luo, S., Chen, S., & Liu, D. (2020). Cyber Threat Detection Using Machine Learning Techniques: A Performance Evaluation Perspective," In: *Proc. 2020 International Conference on Cyber Warfare and Security (ICCWS)*, 1-6. 10.1109/ICCWS48432.2020.9292388

Snedaker, S. (2007).Business Impact Analysis. In: *Business Continuity and Disaster Recovery for IT Professionals*. Elsevier.

Stackexchange Network. (n.d). Cross Validated. *Stackexchange*. https://stats.stackexchange.com/questions/450322/what-is-the-time-complexity-of-k-fold-cross-validation

Supriadi, L. S. R., & Sui Pheng, L. (2018). Business Continuity Management (BCM). In *Business Continuity Management in Construction. Management in the Built Environment* (pp. 41–73). Springer.

Torabi, S. A., Rezaei, H., & Sahebjamnia, N. (2014). A New Framework for Business Impact Analysis in Business Continuity Management (with a case study). *Safety Science*, *68*, 309–323. doi:10.1016/j.ssci.2014.04.017

University of California San Francisco IT Service Management. (2016). *IT Service Continuity Management, Business Impact Analysis Process Activity*. UCSF. https://itsm.ucsf.edu/sites/itsm.ucsf.edu/files/ITSM%20BIA%20Process%20v1.2-1.pdf

Urbanec, J., & Urbancova, H. (2014). Adoption of business continuity management standards in Czech organizations. *Scientia Agriculturae Bohemica*, *45*(1), 66–74. doi:10.7160ab.2014.450109

Western Australian Government. (2009). *Business Continuity Management Guidelines (2nd Edition)*. WAG. https://www.complianceonline.com/articlefiles/Western%20Australia_BCM_Guidelines.pdf

Yang, F., Watson, P., Koukoula, M., & Anagnostou, E. N. (2020). Enhancing Weather-Related Power Outage Prediction by Event Severity Classification. *IEEE Access: Practical Innovations, Open Solutions*, *8*, 60029–60042. doi:10.1109/ACCESS.2020.2983159

Zafar, M., & Salahuddin, K. (2009). On the Use of K-Fold Cross-Validation to Choose Cutoff Values and Assess the Performance of Predictive Models in Stepwise Regression. *The International Journal of Biostatistics*, *5*(1).

APPENDIX

Table 6. Empirical Lab Dataset (Own Computations) for constructing the BCPTs Model (IVL=Impact Value Levels, NO=Non-Critical Function, YES=Critical Function (Source, own work)

BF_ID	UBFRP	IVL-Based on UBFRP	SCENARIO	RTE	IVL-Based on RTE
1	8.10	L4(NO)	SIMPLE	199.06	L4(NO)
2	8.10	L4(NO)	AVG	73.20	L4(NO)
3	8.12	L4(NO)	SIMPLE	198.06	L4(NO)
4	8.12	L4(NO)	AVG	72.83	L4(NO)
5	8.15	L4(NO)	SIMPLE	196.59	L4(NO)
6	8.15	L4(NO)	AVG	72.27	L4(NO)
7	8.16	L4(NO)	SIMPLE	196.10	L4(NO)
8	8.16	L4(NO)	AVG	72.09	L4(NO)
9	8.17	L4(NO)	SIMPLE	195.61	L4(NO)
10	8.17	L4(NO)	AVG	71.96	L3(NO)
11	8.70	L4(NO)	SIMPLE	172.15	L4(NO)
12	8.70	L4(NO)	AVG	63.05	L3(NO)
13	8.80	L4(NO)	SIMPLE	168.19	L4(NO)
14	8.80	L4(NO)	AVG	61.56	L3(NO)
15	8.90	L4(NO)	SIMPLE	164.36	L4(NO)
16	8.90	L4(NO)	AVG	60.12	L3(NO)
17	9.00	L4(NO)	SIMPLE	160.67	L4(NO)
18	9.00	L4(NO)	AVG	58.72	L3(NO)
19	9.10	L4(NO)	SIMPLE	157.09	L4(NO)
20	9.10	L4(NO)	AVG	57.37	L3(NO)
21	9.20	L4(NO)	SIMPLE	153.63	L4(NO)
22	9.20	L4(NO)	AVG	56.07	L3(NO)
23	9.30	L4(NO)	SIMPLE	150.28	L4(NO)
24	9.30	L4(NO)	AVG	54.81	L3(NO)
25	9.40	L4(NO)	SIMPLE	147.03	L4(NO)
26	9.40	L4(NO)	AVG	53.58	L3(NO)
27	10.00	L3(NO)	SIMPLE	129.57	L4(NO)
28	10.00	L3(NO)	AVG	47.00	L3(NO)
29	11.00	L3(NO)	SIMPLE	106.56	L4(NO)

continues on following page

Table 6. Continued

BF_ID	UBFRP	IVL-Based on UBFRP	SCENARIO	RTE	IVL-Based on RTE
30	11.00	L3(NO)	AVG	38.32	L3(NO)
31	12.00	L3(NO)	SIMPLE	89.06	L4(NO)
32	12.00	L3(NO)	AVG	31.72	L3(NO)
33	13.00	L3(NO)	SIMPLE	75.44	L4(NO)
34	13.00	L3(NO)	AVG	26.58	L3(NO)
35	13.29	L3(NO)	SIMPLE	72.05	L4(NO)
36	13.29	L3(NO)	AVG	25.30	L3(NO)
37	13.30	L3(NO)	SIMPLE	71.94	L3(NO)
38	13.30	L3(NO)	AVG	25.26	L3(NO)
39	13.35	L3(NO)	SIMPLE	71.38	L3(NO)
40	13.35	L3(NO)	AVG	25.05	L3(NO)
41	13.36	L3(NO)	SIMPLE	71.27	L3(NO)
42	13.36	L3(NO)	AVG	25.01	L3(NO)
43	13.60	L3(NO)	SIMPLE	68.67	L3(NO)
44	13.60	L3(NO)	AVG	24.03	L3(NO)
45	13.70	L3(NO)	SIMPLE	67.63	L3(NO)
46	13.70	L3(NO)	AVG	23.63	L2(YES)
47	14.00	L3(NO)	SIMPLE	64.63	L3(NO)
48	14.00	L3(NO)	AVG	22.56	L2(YES)
49	14.40	L3(NO)	SIMPLE	60.93	L3(NO)
50	14.40	L3(NO)	AVG	21.11	L2(YES)
51	14.50	L2(YES)	AVG	20.78	L2(YES)
52	14.50	L2(YES)	COMPLEX	7.28	L2(YES)
53	14.60	L2(YES)	AVG	20.45	L2(YES)
54	14.60	L2(YES)	COMPLEX	7.14	L2(YES)
55	14.99	L2(YES)	AVG	19.25	L2(YES)
56	14.99	L2(YES)	COMPLEX	6.62	L2(YES)
57	15.00	L2(YES)	AVG	19.22	L2(YES)
58	15.00	L2(YES)	COMPLEX	6.60	L2(YES)
59	16.00	L2(YES)	AVG	16.50	L2(YES)
60	16.00	L2(YES)	COMPLEX	5.44	L2(YES)
61	17.00	L2(YES)	AVG	14.30	L2(YES)
62	17.00	L2(YES)	COMPLEX	4.47	L2(YES)
63	18.00	L2(YES)	AVG	12.43	L2(YES)

continues on following page

Table 6. Continued

BF_ID	UBFRP	IVL-Based on UBFRP	SCENARIO	RTE	IVL-Based on RTE
64	18.00	L2(YES)	COMPLEX	3.67	L2(YES)
65	19.00	L2(YES)	AVG	10.85	L2(YES)
66	19.00	L2(YES)	COMPLEX	2.98	L2(YES)
67	20.00	L2(YES)	AVG	9.50	L2(YES)
68	20.00	L2(YES)	COMPLEX	2.40	L2(YES)
69	20.74	L2(YES)	AVG	8.62	L2(YES)
70	20.74	L2(YES)	COMPLEX	2.02	L2(YES)
71	20.76	L2(YES)	AVG	8.60	L2(YES)
72	20.76	L2(YES)	COMPLEX	2.01	L2(YES)
73	20.78	L2(YES)	AVG	8.57	L2(YES)
74	20.78	L2(YES)	COMPLEX	2.00	L1(YES)
75	21.00	L1(YES)	AVG	8.33	L2(YES)
76	21.00	L1(YES)	COMPLEX	1.90	L1(YES)
77	22.00	L1(YES)	AVG	7.33	L2(YES)
78	22.00	L1(YES)	COMPLEX	1.46	L1(YES)
79	23.00	L1(YES)	AVG	6.45	L2(YES)
80	23.00	L1(YES)	COMPLEX	1.08	L1(YES)
81	24.00	L1(YES)	AVG	5.68	L2(YES)
82	24.00	L1(YES)	COMPLEX	0.75	L1(YES)
84	25.00	L1(YES)	COMPLEX	0.45	L1(YES)
83	25.00	L1(YES)	AVG	5.00	L2(YES)
86	26.00	L1(YES)	COMPLEX	0.19	L1(YES)
85	26.00	L1(YES)	AVG	4.39	L2(YES)
88	26.80	L1(YES)	COMPLEX	0.00	L1(YES)
87	26.80	L1(YES)	AVG	3.96	L2(YES)
90	31.50	L1(YES)	COMPLEX	0.00	L1(YES)
89	31.50	L1(YES)	AVG	2.03	L2(YES)
92	31.60	L1(YES)	COMPLEX	0.00	L1(YES)
91	31.60	L1(YES)	AVG	2.00	L1(YES)
94	31.65	L1(YES)	COMPLEX	0.00	L1(YES)
93	31.65	L1(YES)	AVG	1.99	L1(YES)

Chapter 3
The Future of Artificial Intelligence in Agricultural Field:
A Bibliometric Analysis

Ahmed Karmaoui
https://orcid.org/0000-0003-3881-4029
Université Moulay Ismail, Meknès, Morocco

ABSTRACT

Artificial intelligence (AI) is well known for diverse applications in many fields including agriculture. The use of AI in the agricultural field has allowed for optimization in real-time and with less cost and effort. This chapter presents a bibliometric analysis of the future of AI in agriculture. This analysis was carried out using the online Scopus tool and VOSviewer software. The outputs show an increasing trend of documents from 1979 to 2021 and 2020 is the most productive year in terms of quantity. Computer science, engineering, Agricultural and Biological sciences, and environmental science are the main subjects while the United States, China, and India are the most influential in the field. United States, China, and India are the most influential in the field, while article and conference papers are the top two doctypes. For the trend, research in this field is turning toward machine learning, the Internet of things, deep learning, agricultural robots, learning algorithms, big data, data analysis, image processing, support vector machines, and automation.

DOI: 10.4018/978-1-6684-4649-2.ch003

INTRODUCTION

The agricultural field is constantly developing thanks to new technologies such as artificial intelligence (AI) which is of growing concern to academics and planners. Worldwide, automation in agriculture is an emerging subject (Talaviya et al., 2020). Studies dealing with the applications of AI in the field are numerous (Meshram et al., 2021; Syifa et al., 2020). In fact, AI is changing agriculture in its different life cycle stages (Karmaoui, 2022a). Raparelli and Bajocco, (2019) stated the art of Machine learning (ML) in the agriculture domain. Shaikh et al., 2022 describe the roles of artificial intelligence, ML, and sensors used in agriculture and review the challenges of ML, robotics, and Internet of things (IoT) devices and the use of drones in crop surveillance and managing crop yield. IoT has made different categories of sensor-based equipment in the agriculture sector (Wongchai et al., 2022a). This technology offers smart solutions for management systems and resources to accomplish many tasks. Particularly, in yield optimization, harvesting, irrigation, plant nutrition and protection based on the collection, and analysis of data. AI has hugely helped this economic sector to overcome the associated issues such as climate change (CC), extreme events (droughts, fires, and floods), water quality and quantity, salinity, and desertification. Otherwise, Leal Filho et al., 2022 reported that AI is helpful in CC teaching, SDGs implementation, and research.

AI provides many tools and techniques such as unmanned aeronautical vehicles (UAVs), agriculture robots, ML, soil moisture and raindrop sensors, and weed control automated system. Meshram et al., (2021) explored the importance of support vector machine (SVM), Random Forest, ML, decision trees, learning systems, the Convolution Neural Network (CNN), and support vector machine (s). Hayat, (2016) highlighted the role, the advantages (cost-efficient, small size, and growing demand), and the prospects of UAVs. UAVs are used to detect infected trees (Syifa et al., 2020), in soil erosion estimation and control (Chidi et al., 2022), map agricultural subsurface drainage systems (Allred et al., 2021), and improve the separation accuracy of different agricultural crops and land areas (Malamiri et al., 2021). Other techniques are frequently applied such as weeding, spraying, and irrigation that allow for saving water, pesticides and herbicide, and conserve soil quality (Talaviya et al., 2020).

This current paper explored international collaboration, the future prospects, with a focus on the importance of artificial intelligence for agriculture. A bibliometric review and a networking analysis were used as approaches through the VOSviewer software and the online Scopus database tool. This method was used in several domains including geology (Karmaoui, 2022b), mathematics (Verma et al., 2021), geography (Wang & Liu, 2014), and socio-ecological systems (Karmaoui et al., 2022). Data about international collaboration, publication quality, research structure, and trending topics and technologies in the field were analyzed. In this study, the specific methods, tools, and models employed were highlighted.

DATA AND METHODOLOGY

This chapter presents a bibliometric analysis of the future of AI in agriculture. This analysis was carried out using the online Scopus tool and VOSviewer software. The approach used in this study concerns the diffusion of knowledge (Liu et al., 2017). It was conducted in a large number of studies to explore international collaboration, the co-authorship, the evolution of publications in time (Karmaoui, 2023), the trending analysis (Sereno et al., 2022),

The keywords used in the search are (TITLE-ABS-KEY (future AND of AND artificial AND intelligence)) AND (in AND agriculture) (Figure 1).

Figure 1. The structure of the search using the key terms (future AND of AND artificial AND intelligence)) AND (in AND agriculture). a, Scopus database; b, VOSViewer tool. Source: The authors

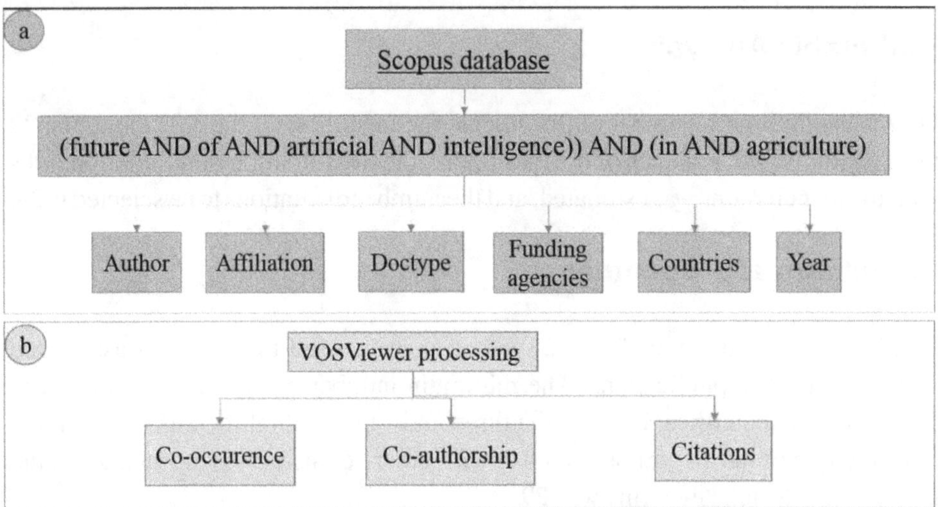

The Scopus online tool yielded 855 document results. This database is the World's Largest (Schotten et al., 2017). The 855 retrieved documents from 1979 to 2021 were processed using the VOSviewer tool. The acquired data was exported in the form of CSV excel including titles, abstracts, full references, keywords, funding details, affiliations, publishers, subjects, years, authors, number of citations, and countries. Haut du formulaire the VOSviewer software was developed by Van Eck & Waltman, (2010), which uses a clustering algorithm that visualizes keywords based on occurrence approach (Leydesdorff & Rafols, 2012). Several analysis methods were used including co-occurrence, citation, and co-authors.

Co-occurrence Analysis

Co-occurrence is the type of analysis used for all keywords of the retrieved publications. The minimum number of occurrences was set at 10. Of the 8,290 keywords, 198 meet the thresholds (Appendix 1). For each of the 198 keywords, the total strength of the co-occurrence links with other key terms was estimated and the number of selected key terms was 198.

Citation Analysis

To represent the most relevant publications in the field, the minimum number of citations considered in this analysis is 50. Of the 855 documents, 52 meet the threshold. For each of the 52 key terms, the total strength of the citation links with to other citations was estimated, and the number of citations to be selected is 52.

Authorship Analysis

Four is the minimum number of documents for an author. Of the 3,037 authors, 29 meet the threshold. For each of the 29 authors, the total strength of the citation links with to other citations was estimated, and the number of citations to be selected is 29.

Co-authors and Countries

In this section, "co-authors" is the type of analysis used for countries associated with the retrieved publications. The minimum number of occurrences was set at 10. Of the 103 countries, 29 meet the thresholds. For each of the 29 countries, the total strength of the co-authorship links with other countries was estimated and the number of selected key terms was 29.

RESULTS

The outputs of the Scopus tool show an increasing trend of documents from 1979 to 2021 and 2020 is the most productive year in terms of quantity. Computer science, engineering, Agricultural and Biological sciences, and environmental science are the main subjects (Figure 2).

Fountas, S., Hoogenboom, G., and Ilapakurti, A. are the top three most productive authors, United States, China, and India are the most influential in the field, while article and conference papers are the top two doctypes (Figure 3). However, the National Natural Science Foundation of China, the National Science Foundation, and the European Commission are the principal funding agencies in the field.

Figure 2. The outputs of the Scopus tool. a, trend of documents from 1979 to 2021; b, the principal subjects using the key terms (future AND of AND artificial AND intelligence)) AND (in AND agriculture). Source: The authors

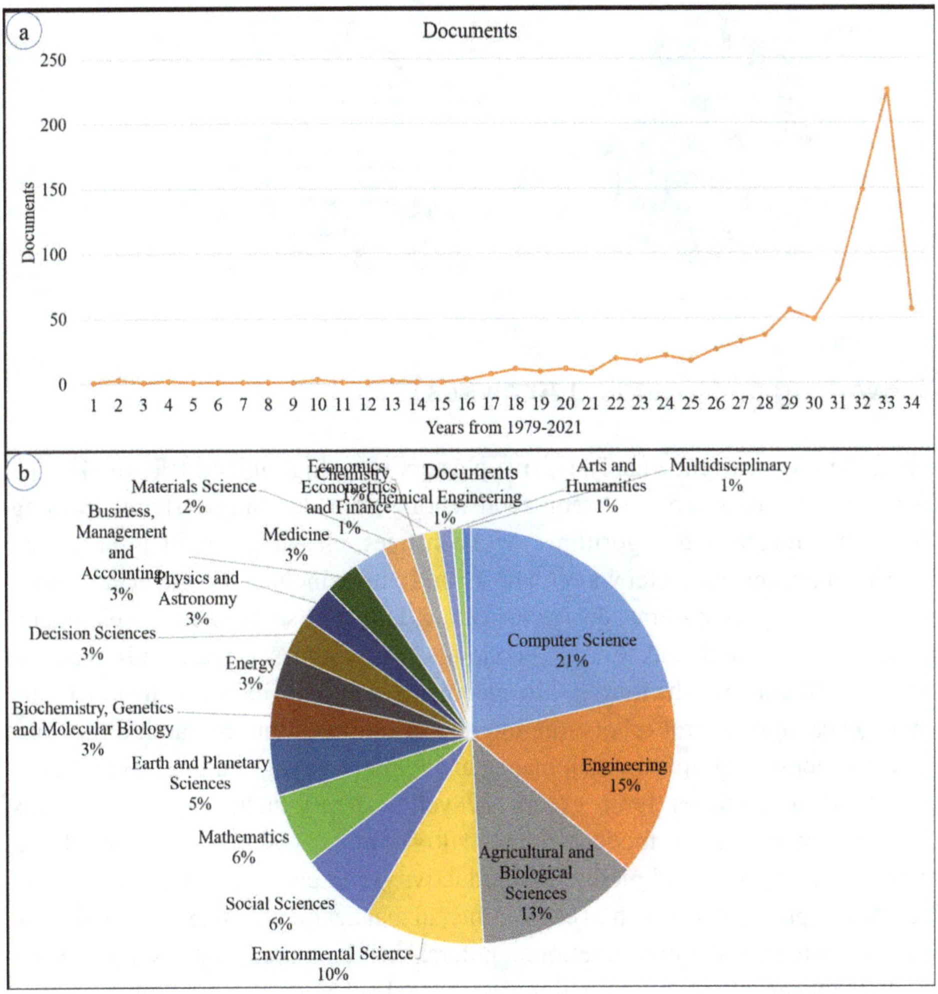

Figure 3. The outputs of the Scopus tool. a, the authors; b, the countries; c, the principal doctypes; and d, the funding sponsors using the key terms (future AND of AND artificial AND intelligence)) AND (in AND agriculture). Source: The authors

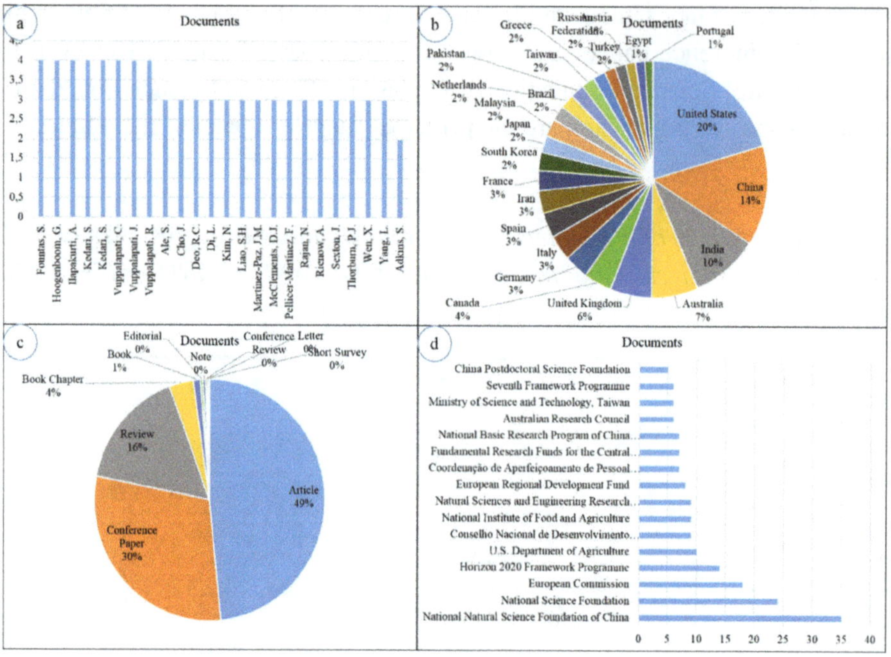

Networking and Trending Analyses

Using the co-occurrence analysis, five clusters were highlighted (Figure 4). The first cluster visualized 52 red items representing terms learning systems, learning algorithms, forecasting, algorithms, decision trees, classification of information, deep learning, and computer vision. The second cluster includes 46 green items such as decision support systems, decision-making, land use, sustainable development, planning, agricultural land, forestry, economics, ecosystem services, land cover, and models… (Figure 4). The third group encompasses 39 blue items, such as artificial intelligence, agricultural robots, the internet of things, future research directions, big data, monitoring, information management, precision agriculture, and robotics (Figure 4). The cluster fours groups 37 yellow items including the keywords climate change, climate models, crops, cultivation, irrigation, soils, agricultural management, agricultural modeling, productivity, and data set… The fifth cluster includes 24 purple items such as environmental management, computer simulation, water resources, water conservation, aquifers, surface waters, water supply, water resource management, water quality, rivers, and hydrological modeling.

Figure 4. Network visualization with an occurrence of 10 keywords, the search key terms (future AND of AND artificial AND intelligence)) AND (in AND agriculture) using VOS viewer software. Source: Processed by the authors

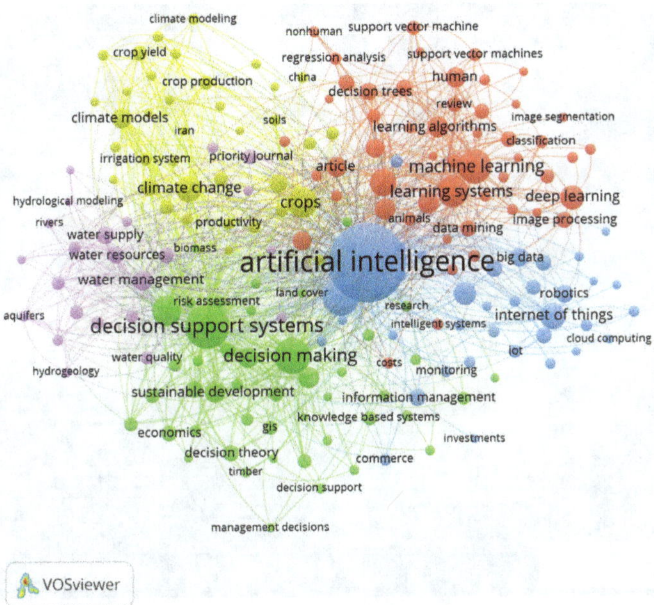

The used software provides the density analysis that visualizes the hottest keywords in the field (Figure 5). In addition to the principal key terms artificial intelligence, this representation shows a high occurrence of the concepts such as machine learning, learning systems, and decision support systems, the internet of things, decision-making, CC, and deep learning.

For trending analysis, research in this field is turning toward machine learning, the internet of things, deep learning, agricultural robots, learning algorithms, big data, data analysis, image processing, support vector machines, and automation (Figure 6).

For the authors, the overlay visualization showed that Wang J., Chen S., Yang J., and Zhou Y. are the trending authors from Around the Globe in the field (Figure 7). However, the network visualization of the countries of co-authors highlighted the USA, China, and India the top countries the most influencing (Figure 8) while India is the emerging country in the field (Figure 9).

In terms of quantity, Zhang X. (with 7 documents) and Zhang L., (7 documents) are the most productive authors while in terms of quality, Liu X. (151 citations) and Wang L. (131 citations) are the most cited authors in the field (Table 1).

Figure 5. Density analysis (heat map) with an occurrence of 10 using key words (future AND of AND artificial AND intelligence)) AND (in AND agriculture) using VOS viewer software and VOSviewer software. Source: Processed by the authors

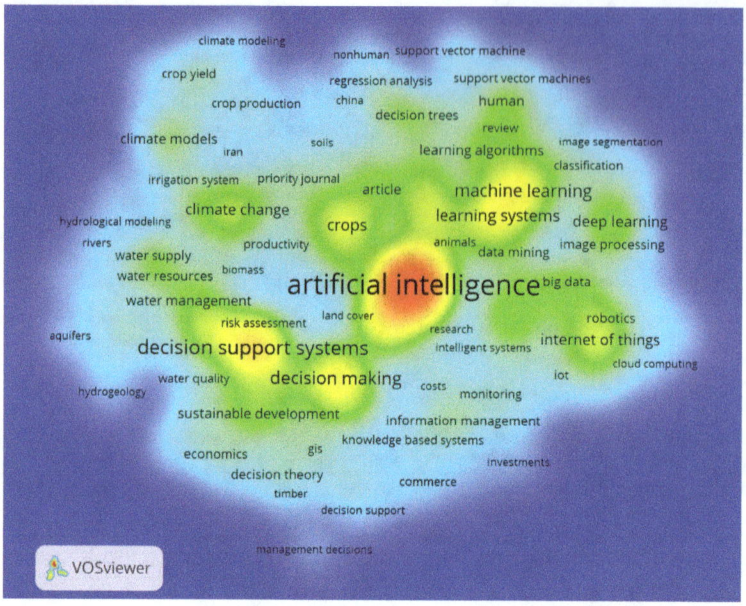

Figure 6. Trending analysis (Overlay visualization) with an occurrence of 10 using key words (future AND of AND artificial AND intelligence)) AND (in AND agriculture) using VOS viewer software and VOSviewer software. Source: Processed by the authors

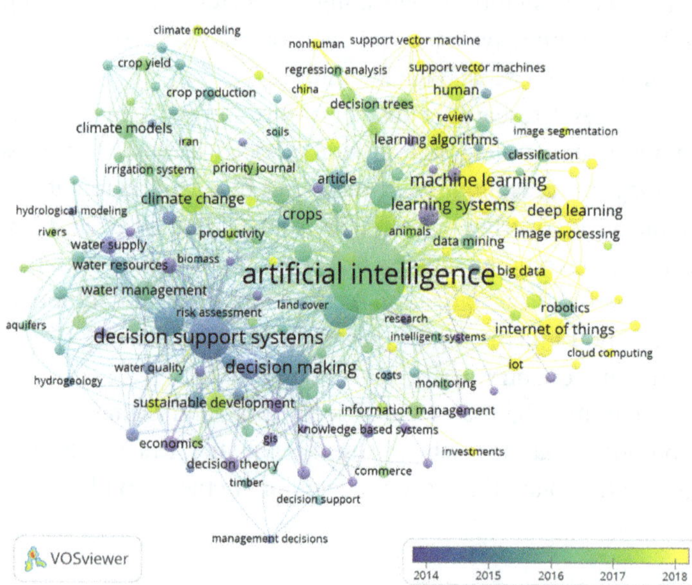

Figure 7. Trending analysis (Overlay visualization) of the authors with an occurrence of 4 using key words (future AND of AND artificial AND intelligence)) AND (in AND agriculture) using VOS viewer software and VOSviewer software. Source: Processed by the authors

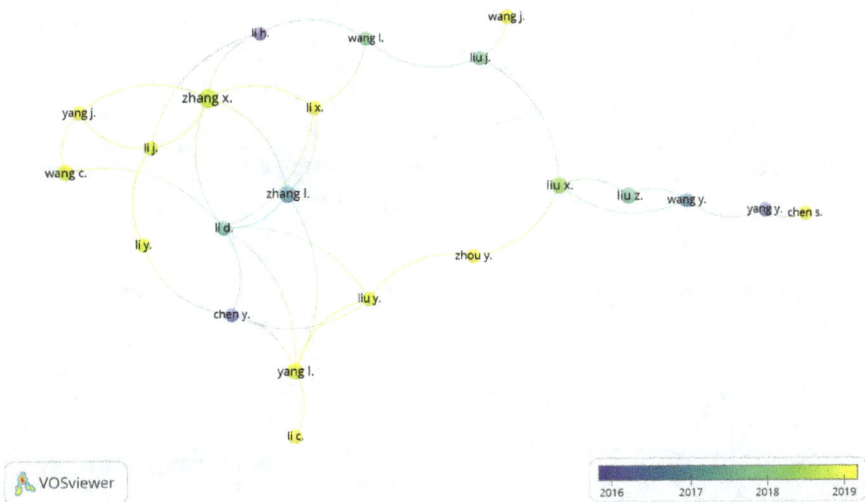

Figure 8. Network visualization, of the countries of co-authors, the search key terms (future AND of AND artificial AND intelligence)) AND (in AND agriculture) using VOS viewer software. Source: Processed by the authors

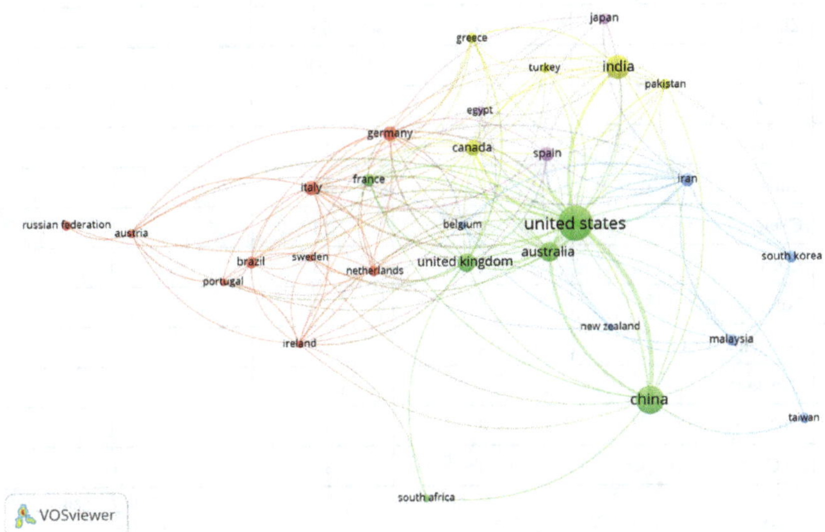

Figure 9. Trending analysis (Overlay visualization) of co-authors countries using key words (future AND of AND artificial AND intelligence)) AND (in AND agriculture) using VOS viewer software and VOSviewer software. Source: Processed by the authors

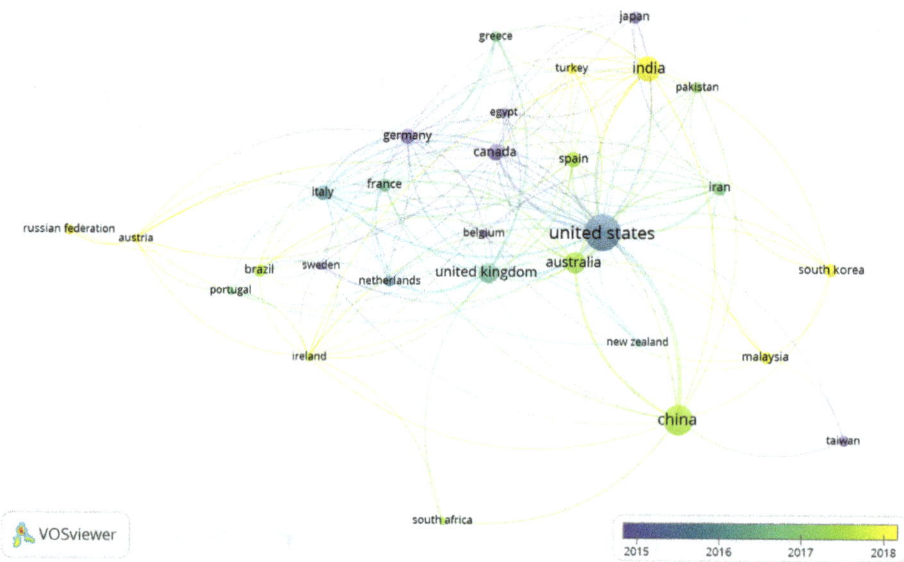

Table 1. The authors, their number of publications, citations, and the associated total link strength processed by VOSviewer. Source: the authors

Id	Author	Documents	Citations	Total Link Strength
2981	Zhang X.	7	79	7
2972	Zhang L.	6	85	6
1517	Liu X.	5	151	4
1519	Liu Z.	5	66	2
2777	Wang C.	5	17	2
2891	Yang L.	5	30	5
487	Chen S.	4	23	1
489	Chen Y.	4	23	4
812	Fountas S.	4	124	0
1061	Hoogenboom G.	4	42	0
1129	Ilapakurti A.	4	0	16
1276	Kedari S.	4	0	16
1458	Li C.	4	31	1

continues on following page

Table 1. Continued

Id	Author	Documents	Citations	Total Link Strength
1459	Li D.	4	25	8
1463	Li H.	4	12	3
1464	Li J.	4	30	5
1472	Li X.	4	21	4
1474	Li Y.	4	34	3
1505	Liu J.	4	69	3
1518	Liu Y.	4	24	4
2764	Vuppalapati C.	4	0	16
2765	Vuppalapati J.	4	0	16
2766	Vuppalapati R.	4	0	16
2782	Wang J.	4	14	1
2784	Wang L.	4	131	3
2791	Wang Y.	4	29	3
2889	Yang J.	4	92	3
2895	Yang Y.	4	30	2
3009	Zhou Y.	4	33	2

Concerning the quality of documents using citation analysis of the VOSviewer software, the findings show that the documents Huang (2015), Liao, (2005b), and Wolfert (2017) are the most cited works conducted in the field (Table 2).

Table 2. The documents the most cited and the associated links processed by VOSviewer. To represent the most relevant publications in field, a 50 is the minimum number of citations considered in this study. Source: the authors

Id	Document	Citations	Links
652	Huang G. (2015)	988	0
823	Liao S.-H. (2005b)	620	0
538	Wolfert S. (2017)	581	1
702	Chaouachi A. (2013)	417	1
839	Lebaron B. (1999)	411	0
498	Amasyali K. (2018)	368	0
833	Liao S.-H. (2003)	350	1
812	Scott S.M. (2006)	317	0
450	Mohammadi M. (2018)	314	0

continues on following page

Table 2. Continued

Id	Document	Citations	Links
591	Khatami R. (2016)	218	0
619	Yaseen Z.M. (2015)	183	0
460	Chlingaryan A. (2018)	179	0
691	Ines A.V.M. (2013)	160	0
761	Becker-Reshef I. (2010)	157	0
665	Tsai C.-W. (2014)	151	0
629	Fountas S. (2015)	119	0
645	Van Der Linden S. (2015)	116	0
590	Scharr H. (2016)	114	0
688	Lee M. (2013)	111	0
771	Mauser W. (2009)	110	0
834	Gotts N.M. (2003)	105	0
711	Papadakis P. (2013)	103	0
804	Huang Y. (2007)	93	0
687	Fallah-Mehdipour E. (2013)	90	0
435	Goap A. (2018)	87	0
569	Singh A. (2016)	84	0
562	Yu N. (2016)	81	0
844	Kondo N. (1996)	78	0
474	Fallah S.N. (2018)	77	1
622	Pastick N.J. (2015)	76	0
634	Dronova I. (2015)	75	0
820	Kohler T.A. (2005)	75	0
391	Rehman T.U. (2019)	72	0
750	Mercurio M.D. (2010)	70	0
837	Pavlou A.K. (2000)	70	0
516	Nunes I. (2017)	68	0
369	Azareh A. (2019)	67	0
443	Wäldchen J. (2018)	67	0
459	Nabavi-Pelesaraei A. (2018)	66	0
476	Ghritlahre H.K. (2018)	66	0
821	Liao S.-H. (2005a)	66	1
819	Pydipati R. (2005)	62	0
578	Hellsmark H. (2016)	59	0
289	Klerkx L. (2019)	57	1

continues on following page

Table 2. Continued

Id	Document	Citations	Links
576	Chen Z. (2016)	56	0
383	Lu Y. (2019)	55	0
835	Filser J. (2002)	54	0
847	Aishima T. (1991)	53	0
542	Amirkhani A. (2017)	52	0
635	Zacepins A. (2015)	52	0
779	Ramachandra T.V. (2009)	52	0
514	Chen M. (2017)	50	0

DISCUSSION AND CONCLUSION

Traditional agricultural approaches are no longer satisfactory to meet the increasing food demands of the global population because of population growth, climate change, and traditional practices (Wongchai et al., 2022a). From here comes the importance of intelligent solutions to meet the growing needs of the global population and tackle the most serious challenges such as climate change. Artificial Intelligence is a set of smart technologies that help achieve various human tasks in real-time and with low effort and cost. It is well known for diverse applications in many fields including agriculture. According to Javaid et al., (2022), Artificial Intelligence (AI) has been extensively employed in several tasks:

- Analyze data for farmers and offer data on weather forecasts.
- Monitor growing conditions and soil,
- Use healthier crops and manage pests,
- Improve management of the food supply chain,
- Choose the good seed and the ideal time to cultivate seeds.
- Help farmers produce more with less resources,
- Enhance crop quality, and decrease time to reach the market,
- Suggest the wright nutrients to improve the quality of the soil,
- Offer information on the health of the crops and the nutrients
- Offer access to data and analytics methods to improve efficiencies, reduce waste in biofuel, and reducing the negative environmental effects.

The use of AI in the agricultural field has allowed us to optimize productivity, support global food security, and tackle CC. Worldwide, a large number of experts emphasized that climate change is among the most studied and controverted topics.

Coeckelbergh, (2021) reported that AI can be used to mitigate CC. Jung et al. (2021) reported the potential of artificial intelligence to enhance the resilience of agriculture production systems. AI contributes to increasing climate resilience and adaptation, turning towards innovative solutions (Leal Filho et al., 2022). ML for example is an AI technology used to predict weather and climate-related events (Gaitán, 2020). However, to improve productivity and quality of modern agriculture, precision agriculture (Management strategy) uses advanced sensors marriage with information technology (Singh & Singh, 2020).

This chapter presents a bibliometric analysis of the future of AI in agriculture. This analysis was carried out using the online Scopus tool and VOSviewer software. The outputs of the Scopus tool show an increasing trend of documents from 1979 to 2021 and 2020 is the most productive year in terms of quantity. This positive trend in terms of publications in the field was highlighted by several studies (Javaid et al., 2022; Couliably et al., 2022). Computer science, engineering, Agricultural and Biological sciences, and environmental science are the main subjects while the United States, China, and India are the most influential in the field. Fountas, S., Hoogenboom, G., and Ilapakurti, A. are the top three most productive authors. One of the most influencial works of Fountas, S. was published in 2015 and deals with farm management information systems presenting current situations and future perspectives (Fountas et al., 2015). This study presents the advancements in the functionality of commercial and academic farm management information systems.

AI technologies are being developed to predict precisely how long it will take for fruit maturity, forecast productivity and weather, and sales. The findings showed that research in this field is turning toward the use of machine learning, the internet of things, deep learning, agricultural robots, learning algorithms, big data, data analysis, image processing, support vector machines, and automation. The principal outputs of the trending analysis highlighted ML, a branch of artificial intelligence that refers to computer programming that allows computers to learn. A large number of studies dealing with ML were identified. Meshram et al., (2021) considered it ideal to increase productivity and reduce losses while Gaitán, (2020) emphasized its role to predict the weather. The Internet of things (IoT) is another technology highlighted as an emerging tool employed to optimize agriculture. In fact, IoT is gaining more attention due to its ability to exchange data between systems using primarily the internet. It refers to physical objects or technologies that assure connections. In the agricultural sector, this technology has made different categories of sensor-based equipment (Wongchai et al., 2022a). For data analysis, deep learning is a useful tool in terms of acceleration and improvement of data processing (Couliably et al., 2022). Wongchai et al., (2022b) proposed a new method to monitor agricultural farms and predict crop disease using deep learning architectures. Otherwise, the agricultural robots concept was raised in the trend analysis. It was introduced in parallel with the emergence of smart agriculture as a response to meet the needs of a growing population. In fact, agricultural robots pick fruits, harvest vegetables, and collect aerial images (to assess and monitor crop health). Support vector machines were found as another trending technology used in agriculture. It is an ML algorithm employed to acquire solutions to support better crop management (Kok et al., 2021). Many other interesting concepts were highlighted in this trending analysis including learning algorithms, big data, data analysis, image processing, and automation.

Considering all these applications, AI technologies use abiotic data, images, and ML algorithms to monitor crop health, fruit ripeness, irrigation, plant nutrition, and growing conditions, and improve management. The agricultural sector benefits also from the advancement of satellite imagery, drones, and connectivity to enhance yield and income supporting environmental sustainability, which in turn help to achieve sustainable development goals.

REFERENCES

Allred, B., Martinez, L., Fessehazion, M. K., Rouse, G., Koganti, T., Freeland, R., Eash, N., Wishart, D. B., & Featheringill, R. (2021). Time of day impact on mapping agricultural subsurface drainage systems with UAV thermal infrared imagery. *Agricultural Water Management, 256*, 107071. doi:10.1016/j.agwat.2021.107071

Chidi, C. L., Zhao, W., Thapa, P., Paudel, B., Chaudhary, S., & Khanal, N. R. (2022). Evaluation of traditional rain-fed agricultural terraces for soil erosion control through UAV observation in the middle mountain of Nepal. *Applied Geography (Sevenoaks, England), 148*, 102793. doi:10.1016/j.apgeog.2022.102793

Coeckelbergh, M. (2021). AI for climate: Freedom, justice, and other ethical and political challenges. *AI and Ethics, 1*(1), 67–72. doi:10.100743681-020-00007-2

Couliably, S., Kamsu-Foguem, B., Kamissoko, D., & Traore, D. (2022). Deep learning for precision agriculture: A bibliometric analysis. *Intelligent Systems with Applications, 200102*, 200102. Advance online publication. doi:10.1016/j.iswa.2022.200102

Fountas, S., Carli, G., Sørensen, C. G., Tsiropoulos, Z., Cavalaris, C., Vatsanidou, A., Liakos, B., Canavari, M., Wiebensohn, J., & Tisserye, B. (2015). Farm management information systems: Current situation and future perspectives. *Computers and Electronics in Agriculture, 115*, 40–50. doi:10.1016/j.compag.2015.05.011

Javaid, M., Haleem, A., Khan, I. H., & Suman, R. (2022). *Understanding the potential applications of Artificial Intelligence in Agriculture Sector*. Advanced Agrochem. doi:10.1016/j.aac.2022.10.001

Jung, J., Maeda, M., Chang, A., Bhandari, M., Ashapure, A., & Landivar-Bowles, J. (2021). The potential of remote sensing and artificial intelligence as tools to improve the resilience of agriculture production systems. *Current Opinion in Biotechnology*, *70*, 15–22. doi:10.1016/j.copbio.2020.09.003 PMID:33038780

Karmaoui, A. (2022a). Network Analysis on Artificial Intelligence in Agriculture, a Bibliometric Review. In J. Marx Gómez & M. R. Lorini (Eds.), *Digital Transformation for Sustainability. Progress in IS*. Springer. doi:10.1007/978-3-031-15420-1_20

Karmaoui, A. (2022b). Ordovician-Cambrian Palaeontological Heritage of Zagora Province: A Bibliometric Analysis from 1984 to 2020 (Anti-Atlas, Morocco). *Geoheritage*, *14*(2), 55. doi:10.100712371-022-00695-8

Karmaoui, A. (2023). Flood vulnerability in developing countries, international collaboration, and network visualization: A Bibliometric analysis. In V. O. Oladokun, D. Proverbs, O. A. Adebimpe, & T. Adedeji (Eds.), *Handbook of Flood Risk Management in Developing Countries* (p. 384). Routledge. https://www.routledge.com/Handbook-of-Flood-Risk-Management-in-Developing-Countries/Oladokun-Proverbs-Adebimpe-Adedeji/p/book/9780367749743

Karmaoui, A., El Jaafari, S., Chaachouay, H., & Hajj, L. (2022b). The socio-ecological system of the pre-Sahara zone of Morocco: A conceptual framework to analyse the impact of drought and desertification. *GeoJournal*, *87*(6), 4961–4974. doi:10.100710708-021-10546-8

Kok, Z. H., Shariff, A. R. M., Alfatni, M. S. M., & Khairunniza-Bejo, S. (2021). Support vector machine in precision agriculture: A review. *Computers and Electronics in Agriculture*, *191*, 106546. doi:10.1016/j.compag.2021.106546

Leal Filho, W., Wall, T., Mucova, S. A. R., Nagy, G. J., Balogun, A. L., Luetz, J. M., & Gandhi, O. (2022). Deploying artificial intelligence for climate change adaptation. *Technological Forecasting and Social Change*, *180*, 121662. doi:10.1016/j.techfore.2022.121662

Leydesdorff, L., & Rafols, I. (2012). Interactive overlays: A new method for generating global journal maps from Web-of-Science data. *Journal of Informetrics*, *6*(2), 318–332. doi:10.1016/j.joi.2011.11.003

Liu, J. G., Zhou, Q., Guo, Q., Yang, Z. H., Xie, F., & Han, J. T. (2017). Knowledge diffusion of dynamical network in terms of interaction frequency. *Scientific Reports*, *7*(1), 1–7. doi:10.103841598-017-11057-8 PMID:28883456

Malamiri, H. R. G., Aliabad, F. A., Shojaei, S., Morad, M., & Band, S. S. (2021). A study on the use of UAV images to improve the separation accuracy of agricultural land areas. *Computers and Electronics in Agriculture*, *184*, 106079. doi:10.1016/j.compag.2021.106079

Meshram, V., Patil, K., Meshram, V., Hanchate, D., & Ramkteke, S. D. (2021). Machine learning in agriculture domain: A state-of-art survey. *Artificial Intelligence in the Life Sciences*, *1*, 100010. doi:10.1016/j.ailsci.2021.100010

Raparelli, E., & Bajocco, S. (2019). A bibliometric analysis on the use of unmanned aerial vehicles in agricultural and forestry studies. *International Journal of Remote Sensing*, *40*(24), 9070–9083. doi:10.1080/01431161.2019.1569793

Schotten, M., El Aisati, M., Meester, W. J., Steiginga, S., & Ross, C. A. (2017). A brief history of Scopus: The world's largest abstract and citation database of scientific literature. Research analytics: Boosting university productivity and competitiveness through. *Scientometrics*, *31*. https://www.taylorfrancis.com/chapters/brief-history-scopus-world-largest-abstract-citation-database-scientific-literature-michiel-schotten-hamed-el-aisati-wim-meester-susanne-steiginga-cameron-ross/e/10.1201/9781315155890-3

Sereno, D., Oury, B., Geiger, A., Vela, A., Karmaoui, A., & Desquesnes, M. (2022). Isothermal Nucleic Acid Amplification to Detect Infection Caused by Parasites of the Trypanosomatidae Family: A Literature Review and Opinion on the Laboratory to Field Applicability. *International Journal of Molecular Sciences*, *23*(14), 7543. doi:10.3390/ijms23147543 PMID:35886895

Shaikh, T. A., Rasool, T., & Lone, F. R. (2022). Towards leveraging the role of machine learning and artificial intelligence in precision agriculture and smart farming. *Computers and Electronics in Agriculture*, *198*, 107119. doi:10.1016/j.compag.2022.107119

Singh, N., & Singh, A. N. (2020). Odysseys of agriculture sensors: Current challenges and forthcoming prospects. *Computers and Electronics in Agriculture*, *171*, 105328. doi:10.1016/j.compag.2020.105328

Talaviya, T., Shah, D., Patel, N., Yagnik, H., & Shah, M. (2020). Implementation of artificial intelligence in agriculture for optimisation of irrigation and application of pesticides and herbicides. *Artificial Intelligence in Agriculture.*, *4*, 58–73. doi:10.1016/j.aiia.2020.04.002

Verma, R., Lobos-Ossandón, V., Merigó, J. M., Cancino, C., & Sienz, J. (2021). Forty years of applied mathematical modelling: A bibliometric study. *Applied Mathematical Modelling*, *89*, 1177–1197. doi:10.1016/j.apm.2020.07.004

Wang, J., & Liu, Z. (2014). A bibliometric analysis on rural studies in human geography and related disciplines. *Scientometrics*, *101*(1), 39–59. doi:10.100711192-014-1388-2

Wongchai, A., Jenjeti, D., Priyadarsini, A. I., Deb, N., Bhardwaj, A., & Tomar, P. (2022b). Farm monitoring and disease prediction by classification based on deep learning architectures in sustainable agriculture. *Ecological Modelling*, *474*, 110167. doi:10.1016/j.ecolmodel.2022.110167

Wongchai, A., Shukla, S. K., Ahmed, M. A., Sakthi, U., Jagdish, M., & kumar, R. (2022a). Artificial intelligence-enabled soft sensor and internet of things for sustainable agriculture using ensemble deep learning architecture. *Computers & Electrical Engineering*, *102*, 108128. doi:10.1016/j.compeleceng.2022.108128

KEY TERMS AND DEFINITIONS

Artificial Intelligence (AI): are advanced computing technologies, introduced by John McCarthy since 1955 and are used in many economic fields including agriculture. They helped to achieve human tasks in real time and with low effort and cost.

Artificial neural networks (ANNs): named also neural networks refers to an algorithm or a hardware that simulate or inspired by the human brain functioning.

Machine Learning: a scientific branch of artificial intelligence is a computer programming that allows to computers to learn to learn.

The Internet of things (IoT): refers to physical objects or technologies that assures the connections and exchange of data between systems using primarily internet.

Deep Learning: a set of methods for modeling and processing data in a field

Antenna: is a device allowing the exchange of electromagnetic waves such as GPS antenna.

APPENDIX

Table 3. Occurrence of keywords on the future of AI in agriculture field using 855 documents from 1979 to 2021 and VOSviewer tool. The key terms classified by occurrences and total link strength. Source: the authors

Id	Keyword	Occurrences	Total link strength
501	Artificial intelligence	659	4127
1802	Decision support systems	204	1971
223	Agriculture	144	1105
1784	Decision making	143	1272
4452	Machine learning	122	891
4301	Learning systems	105	832
1604	Crops	100	962
1799	Decision support system	98	1121
2843	Forecasting	79	619
3982	Internet of things	76	465
1095	Climate change	74	738
1831	Deep learning	63	340
198	Agricultural robots	60	399
2885	Forestry	50	422
4289	Learning algorithms	48	381
6285	Remote sensing	48	441
7392	Sustainable development	48	418
5133	Neural networks	47	301
487	Article	46	479
3497	Human	44	320
4195	Land use	44	480
8079	Water management	43	537
1113	Climate models	42	478
5379	Optimization	41	359
711	Big data	39	236
520	Artificial neural network	38	262
8120	Water supply	36	469
4030	Irrigation	35	454
1622	Cultivation	34	397

continues on following page

Table 3. Continued

Id	Keyword	Occurrences	Total link strength
6473	Robotics	34	161
5832	Precision agriculture	33	256
2209	Ecosystems	32	321
8111	Water resources	32	419
1816	Decision trees	31	328
3772	Information management	31	256
291	Algorithm	30	337
1344	Computer simulation	30	302
3644	Image processing	30	188
5842	Prediction	30	296
606	Automation	28	153
1072	Classification (of information)	28	219
2190	Economics	28	274
3983	Internet of things (iot)	28	221
1728	Data mining	27	191
1810	Decision theory	27	231
5900	Priority journal	26	287
6407	Review	26	174
3093	Geographic information systems	25	295
3527	Humans	25	196
7380	Surveys	25	165
1592	Crop yield	24	285
2167	Ecology	24	232
1808	Decision supports	23	215
5247	Numerical model	23	265
7385	Sustainability	23	242
297	Algorithms	22	173
1354	Computer vision	22	136
1579	Crop production	22	274
1800	Decision support system (dss)	22	234
8052	Water conservation	22	283
3133	Gis	21	220
4127	Knowledge based systems	21	124
7350	Support vector machine	21	206
7826	United states	21	265

continues on following page

Table 3. Continued

Id	Keyword	Occurrences	Total link strength
8107	Water resource	21	284
365	Animals	20	147
1071	Classification	20	164
2577	Expert systems	20	109
2965	Future prospect	20	205
6984	Soil moisture	20	231
2170	Economic and social effects	19	206
2871	Forest management	19	217
3249	Groundwater	19	260
4356	Literature review	19	144
5712	Plants (botany)	19	154
8138	Watersheds	19	289
1723	Data handling	18	124
2829	Food supply	18	155
4012	Iot	18	139
4047	Irrigation system	18	214
7358	Support vector machines	18	149
8087	Water pollution	18	212
8092	Water quality	18	202
388	Antennas	17	105
1210	Commerce	17	125
5659	Planning	17	165
7528	Technology	17	100
2993	Fuzzy logic	16	100
3922	Intelligent robots	16	66
4890	Monitoring	16	118
5967	Productivity	16	142
6249	Regression analysis	16	166
6425	Risk assessment	16	138
6730	Sensors	16	146
6944	Software	16	162
169	Agricultural machinery	15	132
744	Biodiversity	15	126
1521	Costs	15	106
3262	Groundwater resources	15	220

continues on following page

Table 3. Continued

Id	Keyword	Occurrences	Total link strength
4482	Maize	15	206
4842	Modeling	15	130
5213	Nonhuman	15	137
6120	Rain	15	158
7816	Uncertainty analysis	15	144
8236	Wireless sensor networks	15	101
457	Aquifers	14	216
1047	China	14	97
1106	Climate effect	14	193
1474	Convolutional neural network	14	66
1512	Cost effectiveness	14	119
1921	Diagnosis	14	95
2318	Embedded systems	14	83
2440	Environmental impact	14	164
2644	Farms	14	123
6058	Quality control	14	101
7020	Soils	14	137
8282	Zea mays	14	194
1112	Climate modeling	13	144
1691	Data acquisition	13	124
1767	Database systems	13	98
2450	Environmental management	13	160
2922	Fruits	13	71
3058	Genetic algorithms	13	100
3629	Image analysis	13	112
3632	Image classification	13	93
3929	Intelligent systems	13	68
4200	Land use change	13	139
4462	Machine learning techniques	13	132
4558	Mapping	13	141
5515	Pattern recognition	13	52
5764	Population statistics	13	100
5854	Predictive analytics	13	114
6332	Research	13	78
7928	Vegetation	13	131

continues on following page

Table 3. Continued

Id	Keyword	Occurrences	Total link strength
8155	Weather forecasting	13	123
166	Agricultural land	12	135
175	Agricultural modeling	12	141
189	Agricultural productions	12	113
1796	Decision support	12	124
1974	Digital storage	12	128
1990	Disasters	12	94
2544	Evapotranspiration	12	117
4513	Management information systems	12	104
6241	Regional planning	12	116
6447	Rivers	12	196
6713	Sensor	12	134
6799	Simulation	12	156
7204	State of the art	12	64
7247	Stochastic systems	12	88
7373	Surface waters	12	168
170	Agricultural management	11	157
1696	Data analytics	11	91
1738	Data processing	11	111
3318	Harvesting	11	105
3575	Hydrogeology	11	173
3782	Information systems	11	132
4838	Model validation	11	113
4857	Models	11	94
5923	Problem solving	11	80
7641	Timber	11	115
514	Artificial intelligence technologies	10	50
687	Behavioral research	10	66
780	Biomass	10	78
903	Calibration	10	148
1139	Cloud computing	10	69
1263	Complex networks	10	96
1476	Convolutional neural networks	10	55
1529	Covid-19	10	59
1746	Data set	10	78

continues on following page

Table 3. Continued

Id	Keyword	Occurrences	Total link strength
1786	Decision making process	10	102
1807	Decision support tools	10	95
1900	Developing countries	10	71
2080	Drought	10	104
2203	Ecosystem service	10	136
2455	Environmental monitoring	10	112
2669	Feature extraction	10	65
2969	Future research directions	10	54
3056	Genetic algorithm	10	99
3197	Grain (agricultural product)	10	113
3583	Hydrological modeling	10	129
3653	Image segmentation	10	65
3749	Industry 4.0	10	49
3783	Information technology	10	48
3799	Innovation	10	45
3981	Internet of thing (iot)	10	71
4009	Investments	10	40
4024	Iran	10	133
4177	Land cover	10	127
4510	Management decisions	10	93
4728	Methodology	10	98
5107	Network architecture	10	69
5167	Nitrogen	10	100
6367	Reservoirs (water)	10	123
7076	Spatial analysis	10	103
7330	Supervised learning	10	100
7466	Systems analysis	10	64
7545	Technology transfer	10	90
8113	Water resources management	10	140

Section 2

Ecological and Socio-Economic Techniques to Optimize Agriculture in Context of Climate Change

Chapter 4
Assessment of Carbon Balance in Olive Cultivation Systems in Morocco:
Toward an Intelligent Solution in Climate Change Mitigation

Oulbi Sara
National Institue of Agronomic Research, Morocco

Kassout Jalal
https://orcid.org/0000-0002-3702-2751
National Institue of Agronomic Research, Morocco

Abdelaziz Bouizgaren
National Institute of Agronomic Research, Morocco

Nabil El Khatri
National Institute of Agronomic Research, Morocco

Lhassane Sikaoui
International Olive Council (IOC), Spain

Rachid Hadria
National Institute of Agronomic Research, Morocco

ABSTRACT

Climate change is expected to greatly alter and modify all life forms on Earth. Therefore, maintaining natural and agroecosystems sustainability is one of the major scientific concerns in the current era. Reducing greenhouse gas (GHG) emissions represents an urgent need to mitigate the effects and consequences of climate change.

DOI: 10.4018/978-1-6684-4649-2.ch004

The present chapter analyzes carbon footprint and sequestration in different olive plantations, from extensively managed to super-intensively managed plantations, in three geographic districts of Marrakech-Safi region in Morocco. Olive tree plantations showed an important capacity to store CO2, especially in high-density plantations, however, the carbon footprint is still high compared with extensively managed plantations. The chapter contributes to an innovative solution to reduce GHG and, consequently, to maintain the sustainability of the agricultural sector in Morocco under the current context of global changes.

INTRODUCTION

Climate change has undergone significant acceleration in recent decades mainly as a result of human activities. Therefore, the ongoing global changes, due to global warming and anthropogenic pressures, are likely to affect all life forms on Earth (Pörtner et al., 2022). This is the case for the Mediterranean region, which is likely to experience drastic effects from climate change impacts (Schilling et al., 2012). Current predictions suggest a significant decrease in precipitations with increases in temperatures and extreme climatic events (Driouech et al., 2010; Lionello and Scarascia, 2018). Therefore, these changes will impose challenging conditions for natural ecosystems (Matesanz and Valladares 2014; Lloret et al. 2016). In addition, the increasing drought is likely to affect Mediterranean agricultural activities and productivity which greatly depend on precipitation as the main water source (Schilling et al., 2012). However, to ensure the sustainability of natural and agricultural ecosystems in light of the growing human global population, several international conventions and protocols, such as Kyoto Protocol, were proposed to limit climate warming by reducing greenhouse gas (GHG) emissions resulting from human activities. Accordingly, reducing CO_2 emissions and increasing carbon sequestration are highly recommended to reduce greenhouse gas emissions at the global scale (Lal, 2004; Paustian et al., 2016). Indeed, the natural processes that maintain the global carbon cycle (Figure 1) are now insufficient to keep up with the accelerating pace of anthropogenic emissions.

Agriculture and livestock, besides fossil fuels, are considered to be the major sources of GHG emissions and contribute to deforestation and land degradation for at least 15% of GHG emissions (Van der Werf et al., 2009; World Bank, 2012). However, adequate agricultural management practices can balance CO_2 emissions and store it as carbon (C) in the biomass and soil. In fact, land use and agricultural management can either increase or decrease soil C stocks (Lal, 2008). Consequently, increasing C storage in soil and biomass is proposed as a sustainable solution to mitigate the consequences of increased atmospheric CO_2 concentration (Lal,

2004). However, it is worth mentioning that C sequestration is used to describe the processes by which CO_2 is removed from the atmosphere and stored in the ocean, terrestrial environments, and geological formations. Accordingly, at the terrestrial level and through their life cycle, plants absorb CO_2 from the atmosphere and release the oxygen gas (O_2), in which a proportion of CO_2 is returned to the atmosphere while another part is stored in various organic compounds. This latter is referred to as carbon sequestration. However, the C footprint is a measurement indicator for anthropogenic CO_2 emissions (Peters, 2010).

Carbon sequestration is becoming of great interest in research dealing with climate change and aiming to reduce GHG emissions and their impact on different terrestrial ecosystems. Therefore, the enhancement of C sequestration is urgently needed to mitigate global warming (Smith et al., 2004). In this context, tree plantation constitutes an effective solution looking to their higher capacity, compared with herbaceous crops, in net C fixation within tree parts and soil over long periods (Proietti et al., 2016). Furthermore, fruit trees have a central role in several economic, ecological, and cultural aspects, which is the case of the olive tree (Kaniewski et al., 2012; Proietti et al., 2016). In fact, the olive represents a remarkable tree species mainly distributed around the Mediterranean region in its spontaneous form (e.g. oleaster tree), and the cultivated varieties are widely found in many areas worldwide (Besnard et al., 2018; Kassout et al., 2021). Actually, olive remain the most important oil tree crop in temperate areas, with 10.2 million ha under cultivation (FAO, 2019). However, its cultivation has been extended to several areas such as Argentina, Chile, China, and the United States (FAO, 2019). In the last decades, olive oil and olive table production has increased significantly, mainly because of the development of modern orchards and management practices. This situation has led to distinguishing between traditional (extensive) and intensive olive systems, depending on the tree density (Fernández-Escobar et al., 2013). The traditional olive system is characterized by low plant density that varies between 50 to 160 trees ha^{-1}, while intensive and super-intensive systems are characterized by high to very high plant densities which vary around 350 and 700 trees ha^{-1} in the intensive system and between 1200 and 2500 trees ha^{-1} in the orchard under the super-intensive system. Following this variability in terms of plant density, agricultural management varies considerably from one system to another one.

As a matter of fact, olive orchards under the traditional olive growing system are generally observed in rainfed areas, in this type of system farmers use few inputs in terms of fertilization, irrigation, phytosanitary products, ploughing and energy. In contrast, the agricultural management of olive orchards under both intensive and super-intensive systems is based on higher input quantities. Therefore, the carbon footprint of a functional unit of production can vary significantly between the culture systems.

Several studies have quantified the C sequestration of olive plantations. For instance, Sofo et al. (2005) studied the amount of CO_2 captured by young and mature olive plantations and showed that they captured 2.74 and 9.54 Mg C ha^{-1} yr^{-1}, respectively. Thus, Lopez-Bellido et al. (2016) showed that the net C balance of olive trees under intensive and super-intensive plantations was important with values of 2.05 and 4.10 Mg C ha^{-1} yr^{-1}, respectively. The obtained results highlight the positive and important role that olive trees can play in C sequestration, and therefore, in limiting GHG emissions. In this context, the main objective of this chapter is to evaluate the C balance in olive tree production systems in Morocco and suggest adequate solutions and recommendations to improve this balance in the context of the current global changes.

Figure 1. The global carbon cycle and its overall perturbation caused by anthropogenic activities averaged between 2011 and 2020 (Canadell et al., 2022).

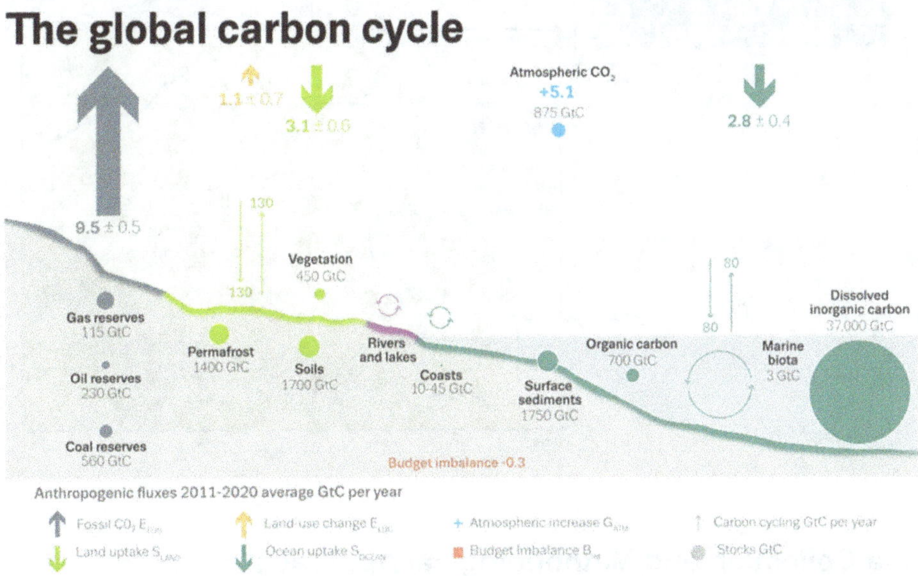

MATERIALS AND METHODS

Study Area

This study was conducted in the Marrakech-Safi region, one of the most important agricultural regions of Morocco (Figure 2). This region is under a semi-arid climate with oceanic influence characterized by important climatic variability. Most of the annual rainfall occurs between late November and March and ranging from 190 mm in the plains to 800 mm in mountainous areas, with almost no precipitation during the hot summer period. Agriculture in the region represents 22% of the total agricultural area in Morocco and is mainly represented by cereals (78%), tree plantations (9.5%), fodder plants (1.8%), and vegetables (1.2%). Olive orchards cover about 87 598ha of the total agricultural area in the Marrakech-Safi region (ORMVA, 2018).

Figure 2. Geolocalisation of the study area of Marrakech-Safi region (black line). The three concerned districts are Al Haouz, Essaouira, and Marrakech.

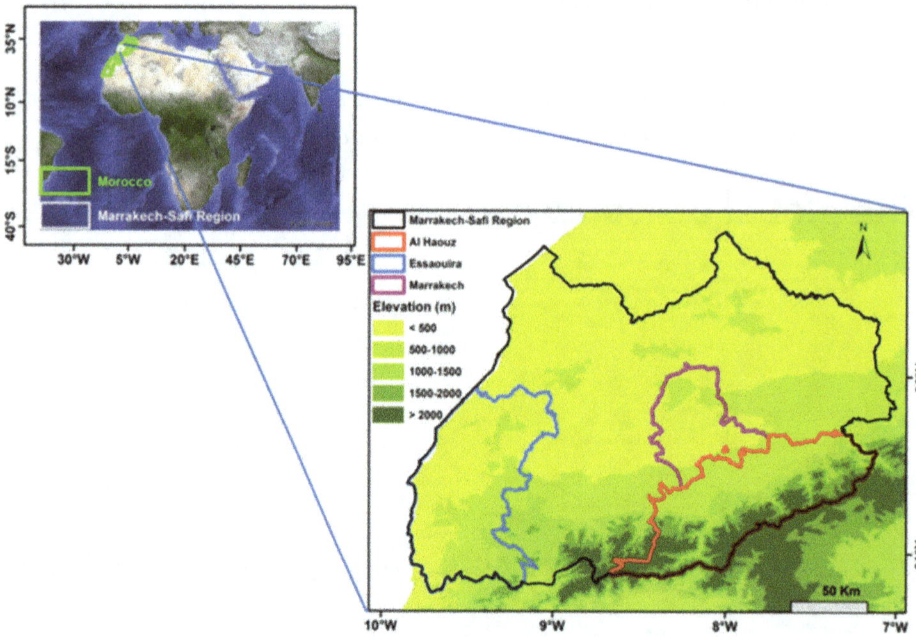

Data Collection and Methodological Approach

To achieve the study goals, field surveys with farmers and olive oil producers were conducted in three different locations of the Marrakech-safi region between February and June 2021: (i) Marrakech district, (ii) Haouz district, and (iii) Essaouira district (Figure 2). A total of 100 olive orchards were inspected. The investigated olive farms were classified into four types: extensive farms under rainfed conditions, irrigated extensive farms, irrigated intensive and super-intensive farms (Figure 3). Accordingly, the conducted surveys were mainly used in order to evaluate energy and environmental loads related to the production of olive oil. Then, the obtained information from field surveys was used as inputs for the 'Carbon balance' application (http://carbonbalance.internationaloliveoil.org/fr), which is widely used to evaluate carbon balance in olive orchards. Thus, carbon stock in soil was measured by analyzing soil samples from the investigated farms according to their type and from the three-studied districts. Accordingly, C stock in soil was analyzed using the following formula (Lopez-Bellido et al., 2016):

$$SCOS = TCOS * S * P * DA * 10^{-3}$$

where *SCOS* is the stock of soil organic carbon; *TCOS* is the organic carbon content (%); *S* is the soil surface situated in olive tree shadow (m^2); *P* is the depth of sampled soil (m), and *DA* is the bulk density of soil (t/m^3) which was determined using the core method. Consequently, the soil C stock obtained from samples reflects the C quantities disseminated by olive trees in the soils through their root system or dead leaves, branches, and fruits. The obtained SCOS values were converted to a functional unit (1L of olive oil) by dividing SCOS on the oil produced by each single olive tree. Then, obtained C stock was converted to CO_2 ($C = 0.272\ CO_2$) and used to estimate the carbon balance between emitted and stocked CO_2.

RESULTS AND DISCUSSIONS

Farm Structure and Diversity in Marrakesh-Safi Region

The investigated farms showed different olive plantation types depending on their geographic location. For instance, in Haouz district (53% of total farms) 49.06% were under the intensive regime, 37.74% and 13.21% were extensive (irrigated) and super intensive, respectively. In Marrakech (35% of total farms), 48.57%, 40%, and 11.43% were, respectively, extensive (irrigated), intensive and super-intensive farms. Hence, in Essaouira district (12% of total farms) extensive farms under irrigation represent 33.33%, while under rainfed conditions they represent 16.67%. Thus, intensive and super-intensive represents 41.67% and 8.33%, respectively (Figure 4). The farmer's choice of olive plantation types differs from one district to another, which could be explained by the differences found in terms of climate, soil type, and irrigation water availability. Generally, the obtained results show that extensive farms under irrigation were the dominant olive farms in Marrakech-Safi region (Figure 4). This is not surprising, looking that extensive farms were considered the traditional olive farming system in the Mediterranean region, well adapted to dry conditions of the region with densities of up to 160 trees/ha and poorly mechanized (Fernández-Escobar et al., 2013; Lorite et al., 2018). However, in the study region, extensive farms under

Figure 3. Different olive growing systems found within the study area. a: extensive, b: intensive, c: super intensive olive plantations.

rainfed conditions were found only in Essaouira region, which could be explained by the fact that this district is under the oceanic influence, which could decrease the effect of aridity found in Haouz and Marrakech districts. From another hand, intensive farms under irrigation represent an important fraction of olive farms in the region, ranging from 40% to 49%, which attests to a clear progression toward high-density orchards in the study area. In fact, the development of agricultural practices (e.g. irrigation, harvesting) besides the dissemination of new olive cultivars has changed the olive production systems (Fernández-Escobar et al., 2013).

Looking at olive plantations, our results show an important diversity in cultivars planted. Figure 5 highlight the used cultivars and their diversity. Not surprisingly, Picholine marocaine seems to be the dominant cultivar in olive orchards of Marrakesh-safi region (43%), followed by Picholine languedoc (21%). Thus, local Moroccan varieties (18%), such as Bakhboukh Beldi and Bouchouk, are highly appreciated in the region. Hence, Haouzia (4%) and Menara (3%) two Moroccan local new varieties are found in the studied area. The olive variety Picholine marocaine, also identified by the farmers under the denomination of 'Zitoune', is widely spread in Moroccan agroecosystems and largely appreciated by local farmers (Hmimsa and Ater, 2008; Ater et al., 2016; Boselli et al., 2020). These findings are similar to those of Ater et

Figure 4. Structure and diversity of studied olive plantations in the Marrakech-Safi region.

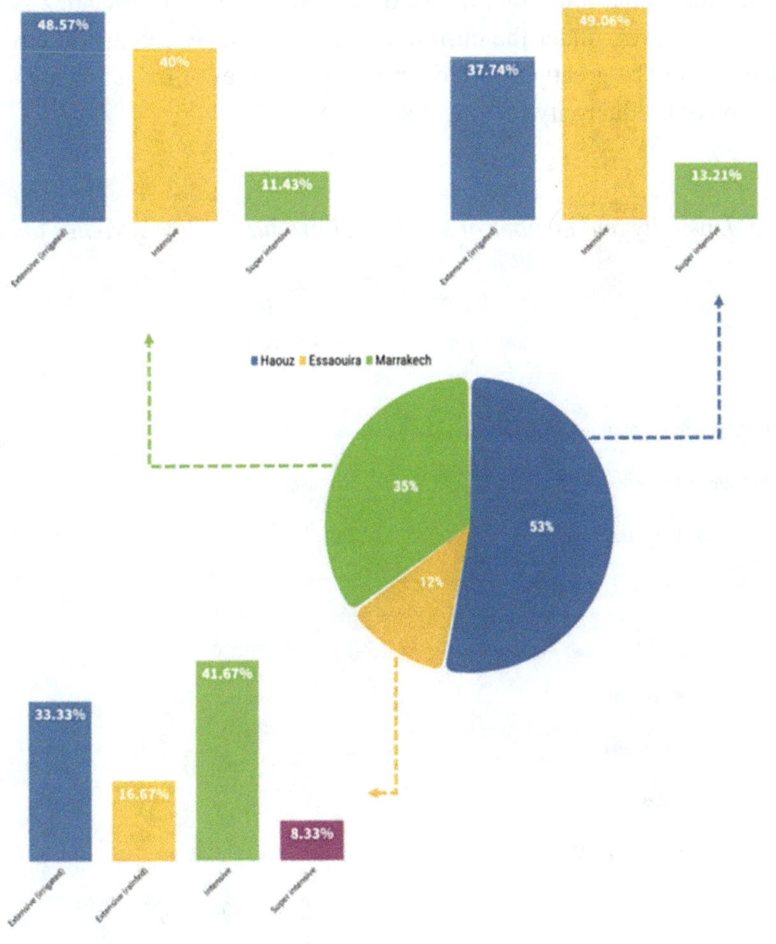

al. (2016) showing that Picholine marocaine (PM) is the dominant olive variety in Moroccan traditional agroecosystems. The widespread of PM could be explained by its high plasticity under various environmental conditions found within Moroccan territory (Khadari et al., 2008). Moreover, the important phenotypic variability found in PM could largely contribute to its tolerance of limited water availability in traditional and extensive agroecosystems (Kassout et al., 2016; Razouk et al., 2022). Hence, our results show that local varieties represent an important fraction of olive plantations in the study area (Figure 5), which are locally selected by farmers and well adapted to local conditions (Khadari et al., 2008; Ater et al., 2016). On the other hand, the introduction of new olive varieties used for intensive and super-intensive plantations, such as Arbequina and Kroneiki, could be viewed as a response to the development of agricultural practices and the strong tendency to mechanize the harvest, thus it's greatly related to the important demand of olive products (Fernández-Escobar et al., 2013). However, under the current context of climate change, intensive olive plantations could be greatly affected, and new measures of irrigation are needed to maintain olive productivity (Lorite et al., 2018).

Figure 5. Diversity and number of local and introduced olive varieties found in the study area.

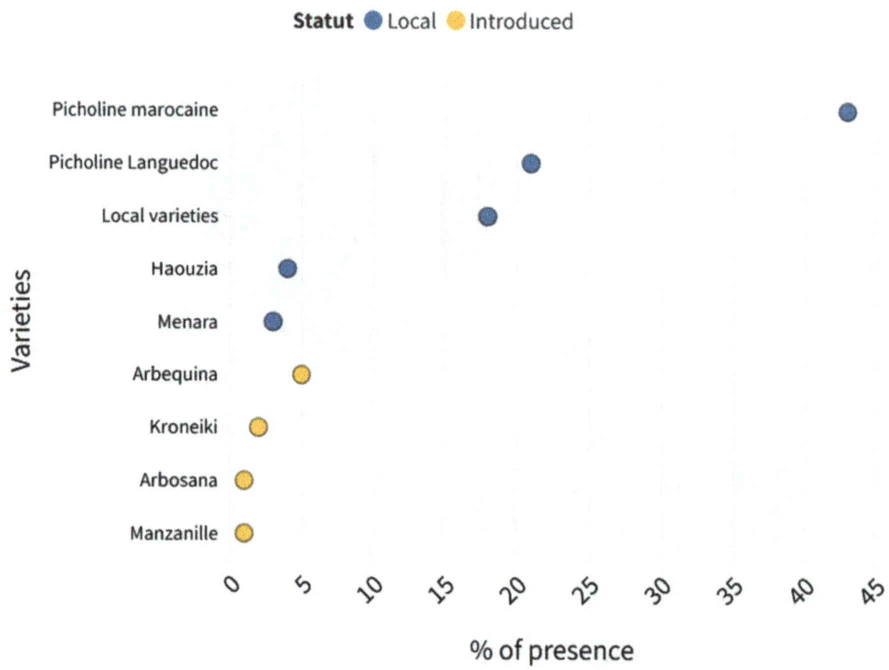

Olive Systems and Agricultural Inputs in Marrakech-Safi Region

Table 1 resumes the main agricultural inputs within olive systems in the three studied districts. In general, the olive plantations in Essaouira district have the highest yield production with minimum water consumption. In fact, Essaouira district is under the oceanic influence and characterized by plain land type (Altitude <500, see Figure 2), which could create less stressful climatic conditions and facilitate the mineralization and decomposition of fertilizer, and consequently explain the important yield production. However, Essaouira plantations show a higher consumption of mineral and organic fertilizer compared with other districts of Marrakech and Haouz, which could be linked to the dominance of extensive olive farms (54.41%) and the use of traditional agricultural practices, compared to more controlled management for intensive farms. Looking at olive systems, intensive plantations clearly show high yield production compared with extensive and super-intensive systems, but with important water consumption as well. The use of new olive varieties, such as Arbequina, has probably increased yield production per farm for the intensive system. However, when looking at oil productivity (Table 2), it seems that super-intensive system score the best results with a mean of 19.8% compared to intensive (15.5%), extensive under irrigation (17.6%), and extensive under rainfed conditions (15.6%). Thus, water consumption, fuel for machines, and the amount of fertilizer used to produce 1L of oil were lower in the super-intensive system than in other systems of the study area.

Table 1. Summary of agricultural inputs in different olive systems of the studied area.

District	System	Fruit yield (Kg/ha)	Water consumption (m³/ha)	Fuel for machine consumption (L/ha)	Fertilizer applied (Kg/ha)	Manure applied (Kg/ha)
Marrakech	E.I	8648.2	2854.7	2.3	55.3	380
	IN	10799.6	2918.6	2.3	39.7	355.7
	S.I	5999.8	2262.5	12.2	47.1	87.5
Haouz	E.I	6376.8	3007.8	0.9	28.2	585.7
	IN	6196.9	3743.1	1.6	41	542.7
	S.I	4945.8	4265.9	14.1	75.5	0.0
Essaouira	E.I	7080	53	5.1	42.5	900
	E.R	6750	0	0.3	7.5	1110
	IN	34747.5	538	3.0	62	1968
	S.I	25500	1958	4.6	50	120

E.I: extensive (irrigated), E.R: extensive rainfed, IN: intensive; S.I: super intensive.

Table 2. Oil production and inputs needed for 1L oil production.

	Yield (%)	Density (Tree/ha)	Cultivated area (m^2)	Water consumption (m^3)	NPK (g)	Waste (Kg)	Fuel (ml)	Mill waste (Kg)	Waste-water (Kg)
E.I	15.6	35	13.20	0	29.7	1.09	0.97	1.95	2.17
E.R	17.6	97	6.07	1.61	74.3	0.48	0.6	2.60	1.92
IN	15.5	242	2.92	0.91	37.6	0.26	0.7	4.91	0.26
S.I	19.8	1396	0.53	0.18	11.4	0.016	0.4	3.84	0.20

E.I: extensive (irrigated), E.R: extensive rainfed, IN: intensive; S.I: super intensive.

Carbon Footprint and Sequestration in Olive Farms of the Marrakesh-Safi Region

The data collected through the surveys allowed us to calculate indicators related to the carbon footprint, CO_2 sequestration, and carbon balance of each farm using the application "Carbon Balance" of the International Olive Oil Council. Figure 6 clearly shows significant differences in C footprint either for the studied district or for the olive systems in Marrakech-Safi region. Our results show that Haouz region scored the highest value in C footprint to produce 1L of olive oil with 3.7 Kg of CO_2-eq, followed by Essaouira (3.09 Kg of CO_2-eq) and Marrakech (2.02 Kg of CO_2-eq) (Figure 6a). Regarding plantation density, super intensive olive orchards were characterized by the highest values in C footprint with 4.66 Kg of CO_2-eq to produce 1L of oil, followed by intensive (3.02 Kg of CO_2-eq), extensive under irrigation (2.64 Kg of CO_2-eq) and extensive under rainfed conditions (2.52 Kg of CO_2-eq) (Figure 6b). The observed differences in C footprints between districts indicate a remarkable diversity in the agricultural practices adopted by olive farmers and for olive oil production specific to each area. These results are in line with previous findings showing the different patterns in C footprint depending on olive systems and geographic regions (Proietti et al., 2014; Lopez-Bellido et al., 2016). Moreover, the observed differences in C footprint between the studied olive plantations might be explained by the management practices such as the tillage system, use of organic or mineral fertilizers, and mechanization (Parras-Alcantara et al., 2013; Fernandez-Romero et al., 2016). However, is worth mentioning that the heterogeneity of the studied olive plantations makes it difficult to compare them with the integration of all factors and parameters explaining those differences.

Figure 6. Carbon footprint in to the studied districts (a) and olive systems plantations (b).

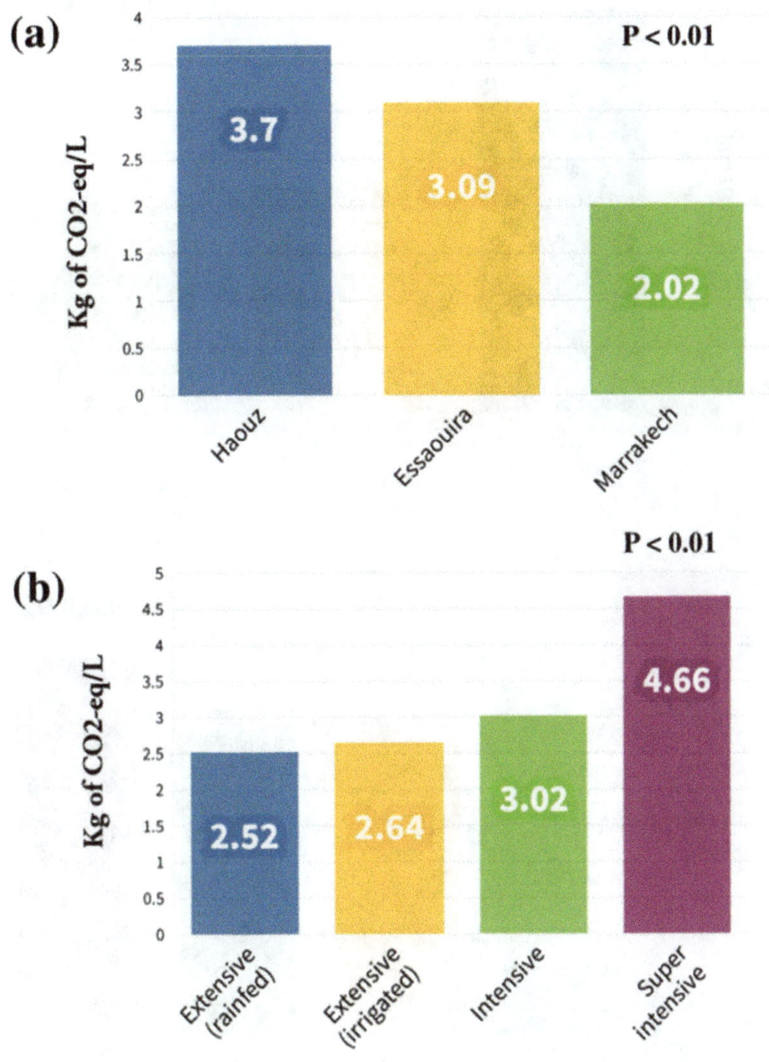

In terms of carbon sequestration, the obtained results show that the studied regions are significantly different ($P < 0.01$) in terms of retained C to produce 1L of olive oil (Figure 7). Haouz district is the highest region in C sequestration with 9.42 Kg of CO_2-eq per liter of olive oil produced, followed by Marrakech (5.27 Kg of CO_2-eq) and Essaouira (4.92 Kg of CO_2-eq) (Figure 7a). In fact, the observed trends in Haouz district could be explained by the favorable conditions for olive

Figure 7. Carbon sequestration in the studied districts (a) and olive plantations (b).

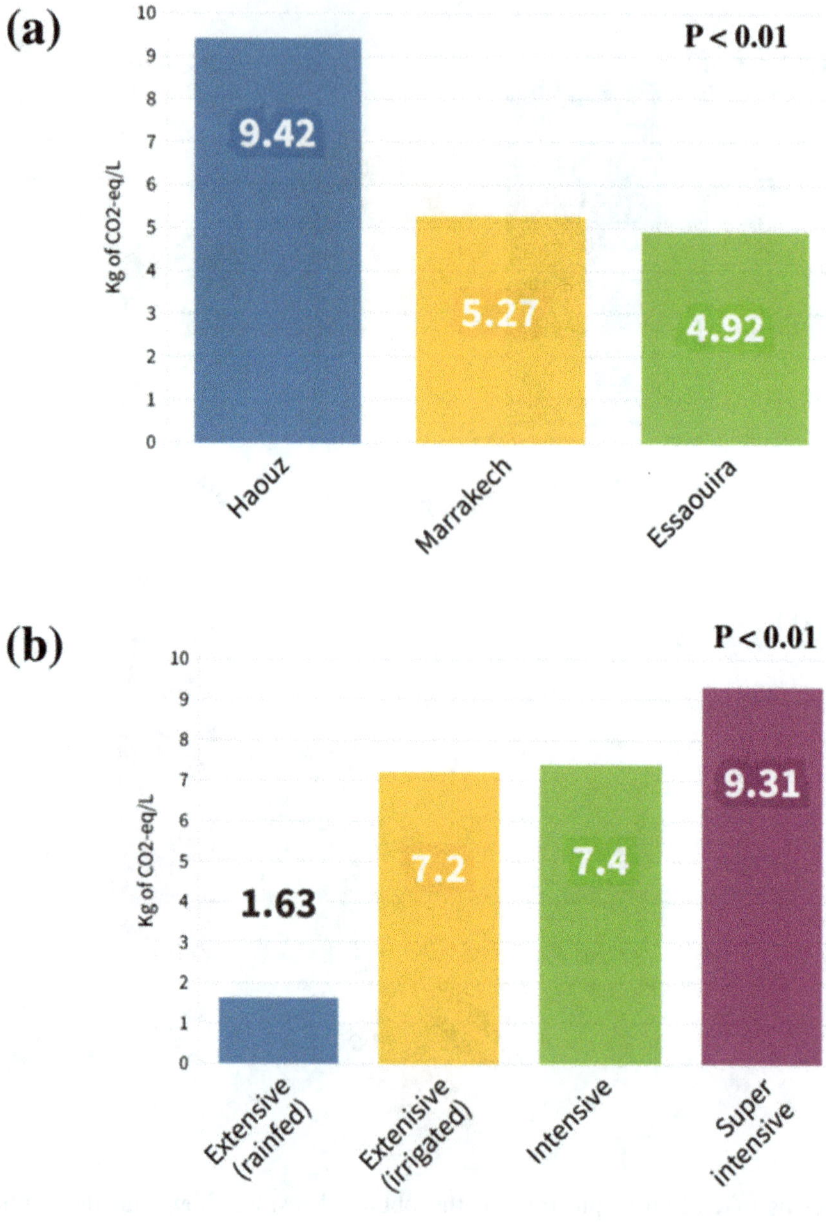

growth and development in terms of climate, soil, and water resource availability compared with the other districts of Marrakech and Essaouira. On other hand, super-intensive plantations seem to be the most efficient system in C sequestration with a mean value of up to 9.31 Kg of CO_2-eq (Figure 7B). Thus, intensive and irrigated extensive plantations are showing a similar trend in C sequestration with 7.4 and 7.2 Kg of CO_2-eq to produce 1L of oil. Extensive plantations under rainfed conditions are likely the lowest system in terms of C sequestration (1.63 Kg of CO_2-eq) (Figure 7b). Lopez-Bellido et al. (2016) showed that C sequestration in super-intensive olive plantations was significantly higher than in extensive and intensive plantations. Thus, Sofo et al. (2005) demonstrated that the density of olive plantations affect significantly the C sequestration. The similar C sequestration between extensive (irrigated) and intensive olive plantations could be explained by the age of the trees in extensive systems which are older than those observed in intensive plantations, thus with the water availability, that enhances shoots and roots development and therefore higher biomass.

Regarding the carbon balance in the studied plantations and districts of Marrakech-Safi, the obtained results showed a general tendency toward negative balance to produce 1L of olive oil, which attests that C sequestration is higher than C footprint. Figure 8 shows that Haouz district contributed the highest C balance with a reduction of 5.7 kg of carbon that is sequestered within olive plantations. Thus, Marrakech (-3.21 Kg of CO_2-eq) and Essaouira (-1.82 Kg of CO_2-eq) contributed to the C balance (Figure 8a). Except for extensive plantations under rainfed conditions, all olive systems contributed with a negative C balance, with the highest values reported in the super intensive plantations (-4.66 Kg of CO_2-eq) followed by extensive under irrigation (-4.56 Kg of CO_2-eq) and intensive (-4.36 Kg of CO_2-eq) plantations (Figure 8b). These findings highlight that olive tree plantations have an important potential in C sequestration, which was previously reported in olive orchards in Spain (Lopez-Bellido et al., 2016) and Italy (Sofo et al., 2005). Thus, our results highlight the potential of the olive tree, especially under high-density plantations, in storing C and their role in CO_2 sinks.

Figure 8. Carbon balance in the studied districts (a) and olive plantations (b).

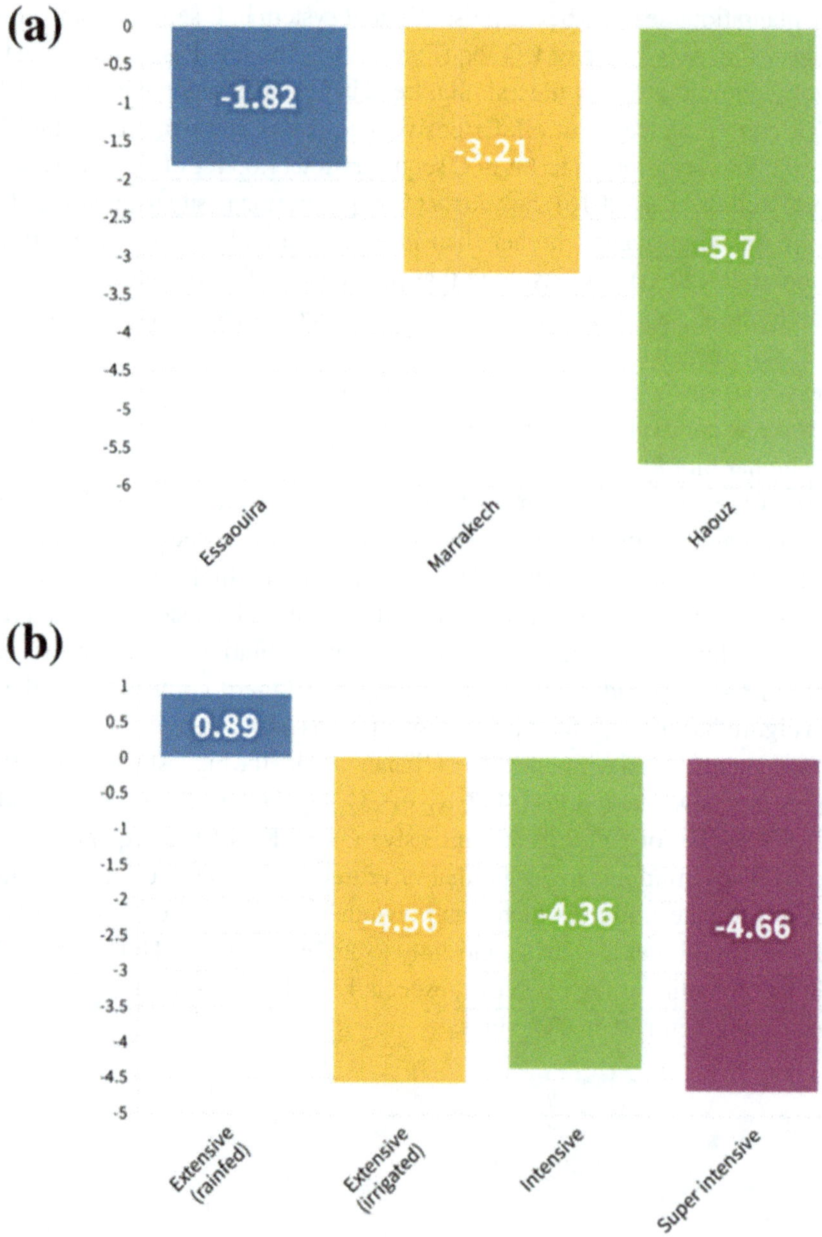

CONCLUSION

In the present chapter, the potential of olive tree plantations for storing carbon has been confirmed within different plantation types. Therefore, our findings highlight the importance of olive plantations, not only from an economic point of view but also from an environmental view. Hence, this study clearly showed that olive trees can be reliably used for reducing the impact of greenhouse gas and, consequently, to mitigate climate change effects. The C balance in the olive tree can present a positive impact on olive oil production in the Marrakech-Safi region, and therefore, contribute to increasing C sinks within olive farms.

This chapter has highlighted the ability of the olive tree to reduce current greenhouse gas emissions and, consequently, to significantly balance carbon sinks under future climate change scenarios. In addition, improved agricultural management practices in olive orchards can significantly increase the carbon balance and the capacity of the olive tree to store carbon. The adoption of good agricultural management of olive plantations and the increase of cultivation areas could have a significant effect on CO2 sinks and therefore an effective way to counteract the negative effects of global warming.

REFERENCES

Ater, M., Barbara, H., & Kassout, J. (2016). Importance des variétés locales, de l'oléastre et des pratiques traditionnelles de l'oléiculture dans la région de Chefchaouen (Nord du Maroc). L 'oléiculture au Maroc de la préhistoire à nos jours: pratiques, diversité, adaptation, usages, commerce et politiques. CIHEAM.

Besnard, G., Terral, J. F., & Cornille, A. (2018). On the origins and domestication of the olive: A review and perspectives. *Annals of Botany*, *121*(3), 385–403. doi:10.1093/aob/mcx145 PMID:29293871

Boselli, V., Ouallali, A., Briak, H., Houssni, M., Kassout, J., El Ouahrani, A., & Michailidi, E. M. (2020). System dynamics applied to terraced agroecosystems: The case study of assaragh (Anti-Atlas Mountains, Morocco). *Water (Basel)*, *12*(6), 1693. doi:10.3390/w12061693

Canadell, J. G., Monteiro, P. M., Costa, M. H., Da Cunha, L. C., Cox, P. M., Alexey, V., et al. (2021). *Global carbon and other biogeochemical cycles and feedbacks*. Cambridge University Press.

Driouech, F., Déqué, M., & Sánchez-Gómez, E. (2010). Weather regimes—Moroccan precipitation link in a regional climate change simulation. *Global and Planetary Change*, *72*(1-2), 1–10. doi:10.1016/j.gloplacha.2010.03.004

FAO. (2019). *The statistical database*. FAOSTAT.

Fernández-Escobar, R., De la Rosa, R., Leon, L., Gomez, J. A., Testi, L., Orgaz, F., & Masallem, M. (2013). Evolution and sustainability of the olive production systems. *Options Méditerranéennes*, *106*, 11–42.

Fernández-Romero, M. L., Parras-Alcántara, L., Lozano-García, B., Clark, J. M., & Collins, C. D. (2016). Soil quality assessment based on carbon stratification index in different olive grove management practices in Mediterranean areas. *Catena*, *137*, 449–458. doi:10.1016/j.catena.2015.10.019

Hmimsa, Y., & Ater, M. (2008). Agrodiversity in the traditional agrosystems of the Rif mountains (North of Morocco). *Biodiversity (Nepean)*, *9*(1-2), 78–81. doi:10.1080/14888386.2008.9712890

Kaniewski, D., Van Campo, E., Boiy, T., Terral, J. F., Khadari, B., & Besnard, G. (2012). Primary domestication and early uses of the emblematic olive tree: Palaeobotanical, historical and molecular evidence from the Middle East. *Biological Reviews of the Cambridge Philosophical Society*, *87*(4), 885–899. doi:10.1111/j.1469-185X.2012.00229.x PMID:22512893

Kassout, J., Ater, M., Ivorra, S., Barbara, H., Limier, B., Ros, J., Girard, V., Paradis, L., & Terral, J.-F. (2021). Resisting aridification: Adaptation of sap conduction performance in Moroccan wild olive subspecies distributed over an aridity gradient. *Frontiers in Plant Science*, *12*, 1288. doi:10.3389/fpls.2021.663721 PMID:34276726

Kassout, J., Barbara, H., Ivorra, S., Terral, J. F., & Ater, M. (2016). Etude préliminaire de la variation de caractères anatomiques du bois d'une forme spontanée et de sept variétés traditionnelles d'olivier (Olea europaeaL.) de la région Nord du Maroc (Chefchaouen et Ouazzane). L'oléiculture au Maroc de la préhistoire à nos jours: pratiques, diversité, adaptation, usages, commerce et politiques, eds M. Ater, L. Essalouh, H. Ilbert, A. Moukhli, and B. Khadari (Montpellier: CIHEAM, Options Méditerranéennes: Série A. Séminaires Méditerranéens, (118), 181-189.

Khadari, B., Charafi, J., Moukhli, A., & Ater, M. (2008). Substantial genetic diversity in cultivated Moroccan olive despite a single major cultivar: A paradoxical situation evidenced by the use of SSR loci. *Tree Genetics & Genomes*, *4*(2), 213–221. doi:10.100711295-007-0102-4

Lal, R. (2004). Soil carbon sequestration impacts on global climate change and food security. *science, 304*(5677), 1623-1627.

Lal, R. (2008). Carbon sequestration. *Philosophical Transactions of the Royal Society of London. Series B, Biological Sciences, 363*(1492), 815–830. doi:10.1098/rstb.2007.2185 PMID:17761468

Lionello, P., & Scarascia, L. (2018). The relation between climate change in the Mediterranean region and global warming. *Regional Environmental Change, 18*(5), 1481–1493. doi:10.100710113-018-1290-1

Lloret, F., de la Riva, E. G., Pérez-Ramos, I. M., Marañón, T., Saura-Mas, S., Díaz-Delgado, R., & Villar, R. (2016). Climatic events inducing die-off in Mediterranean shrublands: Are species' responses related to their functional traits? *Oecologia, 180*(4), 961–973. doi:10.100700442-016-3550-4 PMID:26801493

Lopez-Bellido, P. J., Lopez-Bellido, L., Fernandez-Garcia, P., Muñoz-Romero, V., & Lopez-Bellido, F. J. (2016). Assessment of carbon sequestration and the carbon footprint in olive groves in Southern Spain. *Carbon Management, 7*(3-4), 161–170. doi:10.1080/17583004.2016.1213126

Lorite, I. J., Gabaldón-Leal, C., Ruiz-Ramos, M., Belaj, A., De La Rosa, R., León, L., & Santos, C. (2018). Evaluation of olive response and adaptation strategies to climate change under semi-arid conditions. *Agricultural Water Management, 204*, 247–261. doi:10.1016/j.agwat.2018.04.008

Matesanz, S., & Valladares, F. (2014). Ecological and evolutionary responses of Mediterranean plants to global change. *Environmental and Experimental Botany, 103*, 53–67. doi:10.1016/j.envexpbot.2013.09.004

ORMVA (2018). *Office National de Mise en Valeur Agricole.* Ministère de l'Agriculture, de la Pèche Maritime, du Développement Rural et Eaux et Forêts du Maroc.

Parras-Alcántara, L., Martín-Carrillo, M., & Lozano-García, B. (2013). Impacts of land use change in soil carbon and nitrogen in a Mediterranean agricultural area (Southern Spain). *Solid Earth, 4*(1), 167–177. doi:10.5194e-4-167-2013

Paustian, K., Lehmann, J., Ogle, S., Reay, D., Robertson, G. P., & Smith, P. (2016). Climate-smart soils. *Nature, 532*(7597), 49–57. doi:10.1038/nature17174 PMID:27078564

Peters, G. P. (2010). Carbon footprints and embodied carbon at multiple scales. *Current Opinion in Environmental Sustainability*, *2*(4), 245–250. doi:10.1016/j.cosust.2010.05.004

Pörtner, H. O., Roberts, D. C., Adams, H., Adler, C., Aldunce, P., Ali, E., & Birkmann, J. (2022). [Impacts, adaptation and vulnerability. *IPCC Sixth Assessment Report.*]. *Climatic Change*.

Proietti, P., Sdringola, P., Brunori, A., Ilarioni, L., Nasini, L., Regni, L., Pelleri, F., Desideri, U., & Proietti, S. (2016). Assessment of carbon balance in intensive and extensive tree cultivation systems for oak, olive, poplar and walnut plantation. *Journal of Cleaner Production*, *112*, 2613–2624. doi:10.1016/j.jclepro.2015.10.009

Proietti, S., Sdringola, P., Desideri, U., Zepparelli, F., Brunori, A., Ilarioni, L., Nasini, L., Regni, L., & Proietti, P. (2014). Carbon footprint of an olive tree grove. *Applied Energy*, *127*, 115–124. doi:10.1016/j.apenergy.2014.04.019

Razouk, R., Hssaini, L., Alghoum, M., Adiba, A., & Hamdani, A. (2022). Phenotyping Olive Cultivars for Drought Tolerance Using Leaf Macro-Characteristics. *Horticulturae*, *8*(10), 939. doi:10.3390/horticulturae8100939

Schilling, J., Freier, K. P., Hertig, E., & Scheffran, J. (2012). Climate change, vulnerability and adaptation in North Africa with focus on Morocco. *Agriculture, Ecosystems & Environment*, *156*, 12–26. doi:10.1016/j.agee.2012.04.021

Smith, P. (2004). Carbon sequestration in croplands: The potential in Europe and the global context. *European Journal of Agronomy*, *20*(3), 229–236. doi:10.1016/j.eja.2003.08.002

Sofo, A., Nuzzo, V., Palese, A. M., Xiloyannis, C., Celano, G., Zukowskyj, P., & Dichio, B. (2005). Net CO2 storage in Mediterranean olive and peach orchards. *Scientia Horticulturae*, *107*(1), 17–24. doi:10.1016/j.scienta.2005.06.001

Van der Werf, G. R., Morton, D. C., DeFries, R. S., Olivier, J. G., Kasibhatla, P. S., Jackson, R. B., Collatz, G. J., & Randerson, J. T. (2009). CO2 emissions from forest loss. *Nature Geoscience*, *2*(11), 737–738. doi:10.1038/ngeo671

World Bank. (2012). *Carbon sequestration in agricultural soils*. World Bank.

Chapter 5
Arbuscular Mycorrhizal Fungi for Enhanced Production and Crop Yield Under Changing Climate

Abderrahim Boutasknit
https://orcid.org/0000-0002-0311-2441
Cadi Ayyad University, Morocco

Mohamed Anli
https://orcid.org/0000-0002-7028-709X
Cadi Ayyad University, Morocco

Raja Ben-Laouane
https://orcid.org/0000-0002-1656-0462
Cadi Ayyad University, Morocco

Abdessamad Fakech
Cadi Ayyad University, Morocco

Youssef Ait Rahou
Cadi Ayyad University, Morocco

Said Wahbi
Cadi Ayyad University, Morocco

Mohamed Ait-El-Mokhtar
https://orcid.org/0000-0001-8096-9927
Hassan II University, Morocco & Cadi Ayyad University, Morocco

ABSTRACT

Climate change has negatively affected crop productivity and induced intensive degradation of various ecosystems components. Eco-friendly management practices, such as the use of arbuscular mycorrhizal fungi (AMF), are becoming a crucial tool to improve crop yield and productivity under changing climate. AMF are generally known as biofertilizers, which can improve plant growth and development and

DOI: 10.4018/978-1-6684-4649-2.ch005

provide protection to plants against environmental constraints such as soil poverty, drought, salinity, as well as metal toxicity. AMF are natural root symbionts that provide essential mineral nutrients to plants in order to improve their growth under normal and stressed conditions. AMF's role as bio-fertilizers can potentially fortify plants development against a changing environment. The present chapter will provide current comprehensive knowledge on AMF and their influence on plants growth and yield through improving nutrient uptake as well as soil characteristics, which can contribute to the maintenance of ecosystem health and sustainability under harsh environmental conditions.

INTRODUCTION

Improvements in the agricultural production capacity are expected to produce more to feed a growing population and to reap the benefits from the agricultural sector (Agnolucci et al., 2020). The challenges of the agroecosystem are increasingly getting serious. Farmers are faced with the problems of unpredictable climate change (Ait-El-Mokhtar et al., 2019a, 2020c), declining farmland, shrinking agricultural areas, depleted soil resources (Buras et al., 2020), lack of balanced soil nutrition, and declining crop responses to phytosanitary products (Olivares et al., 2019). As a result, concerns about the sustainable improvement of crop productivity and soil health are increasing (Garcia et al., 2019). Soil is the main source of nutrients and the true reservoir of bioresources for crop production and development (Kour et al., 2020). This ecosystem is rich in a wide variety of beneficial microorganisms that serve as active agents for carbon, nitrogen and phosphorus sequestration, recycling, and supplying for plants (Meglouli et al., 2019; Quiroga et al., 2020). The rhizosphere has many advantages for plants growth performance through microorganisms that improve phosphorus and nitrogen uptake (Zhang et al., 2018; Ben-Laouane et al., 2020b). Since soil productivity is declining and environmental stresses are significantly affecting agricultural production, plants productivity promotion and soil quality enhancement are key challenges to ensure sustainable agriculture for the world growing population (Agnolucci et al., 2020). However, to ensure better crop productivity in agricultural soils, chemical inputs are used regularly (Mahfudz et al., 2019) even if these fertilizers are not the optimal solution for maintaining sustainable crop productivity and soil fertility (Rivera-Becerril et al., 2017). To improve soil and crop productivity while minimizing chemical inputs, several potential biological practices were investigated. Among these, the use of arbuscular mycorrhizal fungi (AMF) as biofertilizers is getting considerable attention (Ben-Laouane et al., 2019; Ait-El-Mokhtar et al., 2020a; Anli et al., 2020b) and has to be better exploited to ensure sustainable agriculture (Boutasknit et al., 2020a; Meddich et al., 2020), reduced chemical inputs and protected crop and soil productivity (Anli et al., 2020a; Ben-Laouane et al., 2020b).

Soil fertility constitutes an important component in the ecosystem, and it is linked to higher enzymatic activity and soil microorganism's abundance, which are indicators of soil biological quality (Gałązka et al., 2020). AMF play a crucial role in maintaining biological activities and soil stability (Ben-Laouane et al., 2020b; Gałązka et al., 2020). These microorganisms are able to colonize plant roots, including the most economically valuable crops such as fruit trees (Anli et al., 2020c), cereals (Bernardo et al., 2019), legumes (Ben-Laouane et al., 2019) and vegetables (Renaut et al., 2020). AMF have the capacity to enhance plant growth and production (Symanczik et al., 2018), absorb and translocate soil nutrients to the host plant roots (Boutasknit et al., 2020b), in exchange for carbon (Pan et al., 2020), and positively affect crop tolerance to environmental stresses (Anli et al., 2020a; Boutasknit et al., 2020b; Toubali et al., 2020). Furthermore, AMF symbiosis enhance plant physiology especially water content, leaf water potential and photosynthetic pigment as well as stomatal conductance (Symanczik et al., 2018; Boutasknit et al., 2020b).

AMF widely occur in arid and semi-arid regions (Meddich et al., 2017; Ait-El-Mokhtar et al., 2020b), where, salinity and drought negatively affect the formation and function of mycorrhizal symbiosis (Abdel Latef et al., 2019a; Ait-El-Mokhtar et al., 2020a). This symbiotic relationship was found to be more effective when plants are subjected to drastic environment (Begum et al., 2020; Ben-Laouane et al., 2020a). To protect themselves from the damaging effects of changing environmental conditions, plants develop physiological and biochemical strategies. Among the tools used within plants, there is antioxidant enzyme activities enhancement (Abdel Latef et al., 2019a; Al Kharusi et al., 2019; Ait-El-Mokhtar et al., 2020a). These antioxidant enzymes like superoxide dismutase (SOD), catalase (CAT), peroxidase (POD) and ascorbate peroxidase (APX) as well as polyphenoloxidase (PPO) play a key role on plant adaptation and protection under drastic conditions (Abdel Latef et al., 2019a; Ait-El-Mokhtar et al., 2019b; Ben-Laouane et al., 2019). However, AMF inoculation as biofertilizers significantly enhances these enzymes activity in plants subjected to abiotic and biotic stresses (Ait-Rahou et al., 2020; Boutasknit et al., 2020b). The improvement of antioxidant enzymes activity by AMF, represents an adaptation mechanism to limit the harmful impact of the reactive oxygen species (ROS) accumulation in plants (Ben-Laouane et al., 2020b; Golubkina et al., 2020), and improves photosynthetic and growth performances under environmental stresses (Balestrini et al., 2020; Boutasknit et al., 2020b).

In this context, this chapter will provide a comprehensive knowledge on AMF inoculation and its impacts on soil characteristics as well as plants growth and productivity under climate change through improving nutrients and water uptake and general plant physiology and biochemistry, which could be a positive influence of these microorganisms to maintain ecosystem health and sustainability.

AMF SYMBIOSIS ESTABLISHMENT

Mycorrhizal fungi belong to the phylum and order *Glomeromycota* and *Glomerales*, respectively. They can form endomycorrhizal symbioses with their host plants (Sangwan and Prasanna, 2021). Endomycorrhizal symbiosis is a more common type of mycorrhizal association between AMF and their host plants. In this type of association, fungal hyphae develops in the apoplastic space of the root cortex forming highly branched arbuscules and vesicles (Diagne et al., 2020). The former are the nutrient exchange interface with the host plant and the latter are storage structures (Field and Pressel, 2018). In this relationship, the host plants provide the fungi with carbon in exchange for water and nutrients. AMF form symbiotic associations with 80% of vascular plants and are involved in various functions ranging from accelerating nutrient uptake to improving plant development and soil properties (Basu et al., 2018). A wide range of plants, including angiosperms, gymnosperms, pteridophytes, mosses, etc., can exhibit this type of symbiotic partnership between fungi and plants (Rimington et al., 2018).

Although fungal spores can germinate in the absence of the host plant, the presence of the host plant is necessary for the symbiosis process. The fungus and the host plant must communicate to develop the symbiosis process. As a result, both symbionts produce signaling molecules, which leads to the activation of their symbiotic genes and thus the development of new tissues for the symbiotic process (Mateus et al., 2019). It has been reported that there is a kind of plant hormones called strigolactones produced by plant roots, which are capable of activating fungal genes and lead to the production of fungal hyphae (Lanfranco et al., 2018). Spore growth results in the production of an extensive network of hyphae and the fungal structures of AMF, namely arbuscules and vesicles (Giovannini et al., 2020) . The extensive hyphae network can extend into the soil even beyond the rhizospheric zones, absorbing water and nutrients.

AMF IMPACTS ON SOIL PHYSICO-CHEMICAL AND BIOLOGICAL CHARACTERISTICS

AMF Impacts on Soil Chemical Properties

Soil pH and Electrical Conductivity

Soil pH affects different inorganic elements solubility, leading to their accumulation in more or less available forms for plants. Soil pH strongly affects the accessibility of mineral nutrients, mainly because it affects the microbiological performance and, ultimately, the adaptability of different plant species. Most bacteria -on which nitrogen fixation, nitrification and some organic matter decomposition processes depend- prefer a sub-acid or mildly alkaline environment (pH 6.8-7.2). Fungi prefer an acidic environment and under such conditions they ensure organic compounds decomposition. Variations in soil pH can alter the concentration of many nutrients and toxic ions, hence the importance of AMF, which can directly affect soil chemical characteristics by regulating pH and electrical conductivity (EC) (Gałązka et al., 2017). They reduce rhizospheric soil pH and EC through the release of organic acids into the soil (Chen et al., 2017), that play important role in increasing the availability of N (nitrogen), P (phosphorus), and K (potassium) in soil (Symanczik et al., 2018; Anli et al., 2020a).

AMF and Soil Nutritional Quality

It is a well-known fact that plants in organic agriculture are often in nutrient deficient conditions due to the limited amount of nutrients in the soil or their low solubility. One of the roles of AMF is to improve the availability of nutrients from soil to plants through increasing soil cation exchange (Pylak et al., 2019). It was demonstrated that AMF could improve soil nutritional quality (Ben-Laouane et al., 2020a) (Table 1). In fact, AMF can supply up to 80, 25, 10, 25, and 60% of plant P, N, K, Zn (zinc), and Cu (copper), respectively (Symanczik et al., 2018; Ben-Laouane et al., 2020a). Improving P acquisition by AMF can be carried out through different mechanisms such as solubilization (Etesami et al., 2021). A large quantity of P is fixed in organic form or as metal complexes in soils (Andrino et al., 2021). Colonization plant by AMF could induce the secretion of organic acids such as oxalate or citric acid from the roots, which hydrolyze the insoluble organic or inorganic phosphorus to its available form, accessible to plants (Zhang et al., 2018). AMF may also be able to induce the release of phosphatase from P-solubilizing bacteria (Etesami et al., 2021).

On the other hand, AMF inoculation can play an important role in carbon storage in the soil (Gałązka et al., 2020). In fact, they assure their needs through a demand for photosynthates transferred by the host plant, which could result in a drainage of 20% of the host plant carbon to the soil (Agnihotri et al., 2022). This indirectly contributes to the process of C sequestration by storing it in soils (Gałązka et al., 2020; Quiroga et al., 2020). Furthermore, AMF hyphae excretes glomalin, a long-chain carbon glycoprotein, which has been shown to be difficult to digest in soils and therefore glomalin-C would be sequestered in soils for a long period of time (Ben-Laouane et al., 2020b; Gałązka et al., 2017, 2020).

Table 1. Physicochemical analysis of soil before and after alfalfa cultivation and AMF application under an open field condition (Ben-Laouane et al., 2020a)

	Soil properties		
Before the experiment		**After the experiment**	
	-	Ct	AMF
Texture	Sandy-silty- clay		
pH	8.2 ± 0.2a	8.0 ± 0.0b	8.0 ± 0.0b
Electrical Conductivity (mS cm^{-1})	0.3 ± 0.0a	0.1 ± 0.0bc	0.1 ± 0.0bc
Nitrogen content (mg/kg)	90 ± 1.6b	200 ± 18.6a	252 ± 1.6a
Carbon content (%)	0.4 ± 0.0c	0.7 ± 0.0b	0.9 ± 0.0a
Organic matter (%)	0.7 ± 0.0c	1.2 ± 0.0b	1.5 ± 0.0a
Limestone content (%)	16.3 ± 0.5a	15.0 ± 0.5a	15.8 ± 0.9a
Phosphorus content (ppm)	20.6 ± 0.0c	23.9 ± 0.3b	27.0 ± 0.9a

Ct: non inoculated alfalfa control plants; AMF: alfalfa treated with arbuscular mycorrhizal fungi. Means (±SE) with different letters are significantly different according to Tukey's HSD test, after performing one-way ANOVA ($P<0.05$)

AMF and Soil Toxic Ions

AMF could potentially play a crucial role in minimizing soils contamination by organic compounds or heavy metals (Riaz et al., 2021). Thanks to their mineral scavenging abilities, they have the capacity to support plant growth in polluted soils, a potential capability widely known as bioremediation (Umar, 2017). Among the ways they do this: either they can build up and store toxic metal ions, thus protecting their host against the contaminant, or they can supply them to the host together with the essential mineral elements like Cu and Zn, leading to a build-up of heavy metals in the host (Chen et al., 2018). In the first scenario, plant growth can be promoted in a contaminated soil, with minimal crop contamination. In the second situation, plants can be collected and destroyed to minimize the heavy metal load of the location (phytoextraction). Both of these approaches, of course, need heavy metal-tolerant AMF, and phytoextraction also needs host plants that are very tolerant to heavy metal toxicity, and at the same time can generate a high shoot biomass in order to accumulate significant quantities of heavy metals (Chen et al., 2018).

AMF Impacts on Soil Physical Properties

AMF and Soil Structure

Soil structure is one of the most essential soil ecological functions (Gałązka et al., 2020) since it is strongly related to soil water status and gas and nutrient exchange (Mariotte et al., 2018; Quiroga et al., 2020). For a good plant growth, better soil structure is required. Soil structure is determined by size and arrangement of soil pores and particles (Gałązka et al., 2017). In general, soil particles can be aggregated into microaggregates (<0.25 mm) using soil bonding agents, which then aggregate into macroaggregates (>0.25 mm).

AMF could improve soil physical and chemical characteristic by improving soil structure and soil aggregation (Gałązka et al., 2020). Due to the extent and thickness of their mycelial network, AMF fit into pores between microaggregates and form stable macroporous aggregates that allow air and water penetration, while preventing soil erosion (Sharma et al., 2021). Additionally, glomalin secreted into the soil by AMF (Figure 1) (Ben-Laouane et al., 2020a) could play a key role in this process. Glomalin is a hydrophobic and thermotolerant glycoprotein, which makes it slowly biodegradable by soil microorganisms (Gałązka et al., 2020). When it binds to clay particles to which it gives strong cohesion, it is even more recalcitrant to biological decomposition, thus ensuring strong structural soils stability (Gałązka et al., 2020). A positive correlation has been shown between glomalin and the stability of soil aggregates (Gałązka et al., 2020). At the same time, higher hyphal length and spore density can contribute to the enmeshing of the microaggregates into macroaggregates (Gałązka et al., 2020). AMF facilitate the formation of stable aggregates in the soil,

Figure 1. Average values of easily extractable glomalin- related soil protein (EE-GRSP) (A) and total glomalin-related soil protein (T-GRSP) (B) of Medicago sativa, submitted to different treatments: control (Ct) and arbuscular mycorrhizal fungi. Bars with different letters are significantly different as determined by Tukey's HSD test, after performing one-way ANOVA (P < 0.05) (Ben-Laouane et al., 2020a)

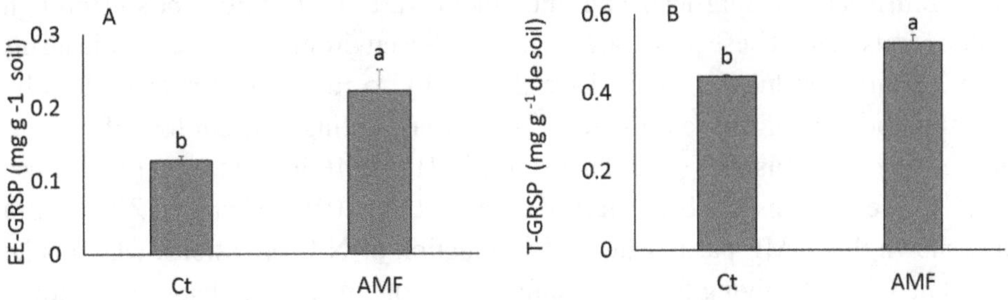

thus increasing soil moisture retention properties (Zhang et al., 2019). Mycorrhizal-mediated soil aggregation might also provide physical protection for organic carbon from microbial decomposition (Gałązka et al., 2020).

Consequently, AMF and their host plant (roots) are consistently found to be a crucial force in driving soil aggregation. However, understanding the characteristics of roots and AMF could also have great importance. According to Soh et al. (2019), this knowledge could lead to innovative initiatives such as seed mixtures, or fungal inoculums, suitable for regeneration of degraded lands, which would optimize the coverage of soil aggregation characteristics within the available group of plant species. It could also set priorities for conservation efforts by predicting which ecosystems are most likely to be degraded due to invasive species and impending global changes.

AMF and Reduction of Soil Erosion and Nutrient Leaching

In addition to soil structure enhancement, AMF also minimize the leaching of nutrients from the soil (He et al., 2021). Nutrient leaching and soil erosion are critical issues as they lead to the reduction of soil fertility and contamination of underground and surface water. Natural and intact ecosystems have great nutrient sequestration ability due to good adsorption and retention of nutrients, particularly through roots and AMF. However, agricultural soils are perturbed by agricultural activities, particularly tillage, and receive high quantities of fertilizers, especially N, P, and K. These, especially the very mobile nitrate, are vulnerable to leaching from the soil due to the absence of a strong system of nutrient holding capacity (Chen et al., 2018). The positive effects of AMF against nutrient leaching are evident at various levels. In the first place, improved soil structure allows nutrient retention improvement in soil, secondly, AMF take up nutrients from the soil solution (Evelin et al., 2019), and in the last place, AMF help soil have a greater solution retention capacity (Chen et al., 2018), which at the same time promotes the availability of nutrients and water for the plant (Zhang et al., 2018; Symanczik et al., 2020).

AMF and Reduction of Gaseous Emissions

Apart from leaching, significant nutrients amounts are also lost from the soil through gaseous emissions. These losses are harmful to the environment and costly in terms of agricultural production. AMF have the ability to reduce soil nutrient loss by widening the area of nutrient interception and preventing nutrient loss after rain-induced leaching episodes (Bhantana et al., 2021). AMF influence on greenhouse gas (GHG) emissions has been studied (Liang et al., 2019). Shen and Zhu (2021) have shown that AMF participate in the reduction of N_2O emissions, which is an important GHG, implying that they could play a role in climate change mitigation.

AMF could control N_2O emissions by reducing N loss from the soil and increasing plant uptake and assimilation of N, leading to a decrease in soluble N in the soil and, consequently, a reduction in denitrification (Shen and Zhu, 2021). Additionally, the correlation between key genes involved in N_2O production and consumption and the abundance of AMF suggests that the regulation of N_2O emissions is due to changes caused by AMF in soil microbial biomass and community composition (Balestrini et al., 2020). Furthermore, AMF could indirectly influence GHG emissions, by changing soil physical conditions, e.g. moisture, aggregation and aeration, all of which influence the GHG production and transport in the soil. Teutscherova et al. (2019) indicated that AMF symbiosis helps regulating N_2O emissions when soil moisture levels are high and suggested that the control of N_2O emissions by AMF plants could be motivated by increased soil water use rather than increased N uptake.

AMF Impacts on Soil Biological Properties

AMF inoculation can affect microbial processes and increase the number and activity of soil microorganisms (Qin et al., 2020). Several studies demonstrated AMF positive influences on soil enzymes activities, which is considered as an index of microbial activity (Ben-Laouane et al., 2020a; Qin et al., 2020). Enzymes activities in the soil are biologically interesting because they contribute to the cycle of mineral elements and influence their accessibility to plants. A meta-analysis of 56 studies showed that AMF generally increase soil enzymes activity such as urease, esterase and chitinase (Qin et al., 2020). The same authors showed that low available phosphorus and neutral pH lead to optimal AMF influence on enzymes activity in the soil .

Glomalin related soil protein (GRSP), a soil quality indicator, have positive effects on AMF interactions with other beneficial bacteria, which may cause a synergistic beneficial effect for the host plant (Gałązka et al., 2020). The same authors reported that the soils characterized by the highest GRSP content belonged to a soils group of a very good physical quality and characterized by a high biological activity, such as dehydrogenase activity, microbial biomass content, microbial nitrogen content and total number of bacteria.

AMF IMPACTS ON PLANT GROWTH AND CROP YIELD

AMF Impacts on Plant Growth

AMF are known as biofertilizers that can improve plant growth, yield and performance. Several studies revealed that AMF inoculation can effectively improve plants growth performance and development as well as crops productivity (Renaut et al., 2020).

In addition, previous studies showed that native and exogenous AMF inoculation increased plants growth principally leaf area, root length, and shoot and root biomass of date palm (Ait-El-Mokhtar et al., 2020a; Anli et al., 2020a), carob tree (Figure 2) (Boutasknit et al., 2020b), alfalfa (Ben-Laouane et al., 2020a, 2019), tomato (El Amerany et al., 2020) and maize (Renaut et al., 2020). Hu et al. (2009) have highlighted the positive effect of *Glomus caledonium* on the performance of maize in terms of biomass in extremely P-limited soils. Furthermore, a recent study conducted under open-field conditions demonstrated an improvement of saffron flower production by the application of a mixture of AMF strains (*Rhizophagus intraradices* and *Funneliformis mosseae*) (Caser et al., 2019). Also, plant growth promotion could be explained by the increase of mineral uptake, namely N and P nutrition (Symanczik et al., 2018) and water uptake by hyphae and root length (Zhang et al., 2018).

Figure 2. Shoot dry weight in non-mycorrhizal (NM) and arbuscular mycorrhizal (AM) carob. The values with different letters are significantly different as determined by Tukey's HSD test, after performing one-way ANOVA (P < 0.05). (Boutasknit et al., 2020b)

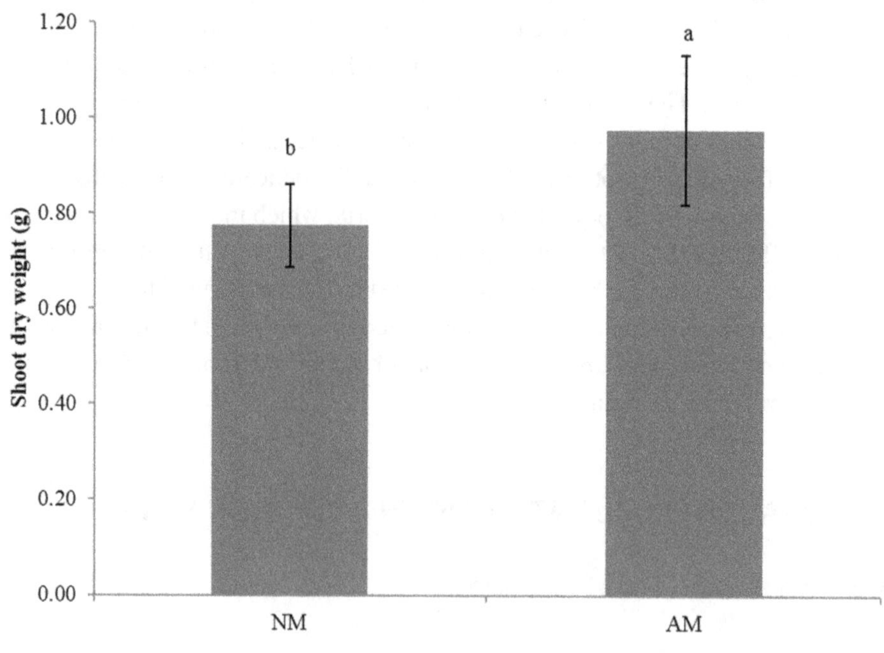

AMF Impacts on Crop Yield

The crop productivity improvement through biostimulants such as AMF is a promising biological alternative in agriculture. Indeed, potato (Alaux et al., 2018), corn, wheat, and soybean (Renaut et al., 2020) have shown a better yield owing to inoculation with *Rhizophagus irregularis*. Moreover, a significant yield improvement of saffron plants inoculated with *R. intraradices* and *F. mosseae* was recorded (Caser et al., 2019). Gao et al. (2020) reported the positive effect of *R. irregularis* on cotton productivity, seed production and development, and fiber quality. Improvement in growth and crops productivity could be due to the enhancement in the mineral nutrition by AMF. Several studies have shown the beneficial effect of AMF to improve crop productivity via direct absorption of mineral elements and water (Symanczik et al., 2018) and/or indirectly via fungal mycelia under drastic conditions (Harris-Valle et al., 2018). This improvement can also be explained by an improvement of the plant photosynthetic apparatus and gas exchanges as well as the exchange of nutrients (photosynthetic products and mineral elements) between shoot and root tissues. Furthermore, the stimulation of AMF symbiosis by root exudates, constitute an important source of organic carbon in the rhizosphere and a way of chemical communication between plant roots and fungi (Evelin et al., 2019). During symbiosis, the plants produced molecules such as strigolactones, which are considered as stimulators of fungal metabolism, hyphal branching, plant growth and root architecture (Nasir et al., 2019). This improvement could be due to the mobilization of nutrients by AMF or by the changes in the level of phytohormones involved in the symbiosis (Zhang et al., 2020).

AMF IMPACTS ON PLANT WATER AND NUTRIENTS UPTAKE

AMF Impacts on Plant Water

AMF have been widely used to improve plant performance under conditions of low soil water availability in both natural and agricultural ecosystems (Mariotte et al., 2018). Zhang et al. (2018) have shown that plants water content is increased via an extraradical hyphae. Hyphae can effectively improve water uptake from the soil through their capacity to explore micropores inaccessible to the plant root system (Boutasknit et al., 2020b). As a result, it is possible to reduce the risk of water movement in the soil even in areas that are not in contact with the roots to facilitate plants water absorption (Zhang et al., 2018). In fact, the major advantage of extraradical hyphae is their ability to reach areas beyond depletion in order to absorb and transport nutrients to the intracellular arbuscules of colonized cortical roots cells (Wipf et al., 2019). Furthermore, the establishment of AMF symbiosis leads to important changes in root physiology, and these changes are likely to be controlled by a specific gene expression pattern in the host plant (Feng et al., 2022). In addition, Symanczik et al. (2020) have shown that certain aquaporin genes are stimulated by mycorrhizal colonization. Aquaporins are ubiquitous membrane proteins that are involved in the maintenance of plant water homeostasis by facilitating the water and solutes flow (Chen et al., 2018). However, AMF plant aquaporins are not limited to the regulation of plant water status. AMF-regulated aquaporins are involved in the mobilization of nitrogen forms in plants, such as ammonium and urea (Quiroga et al., 2020). Increased hydraulic conductivity regulated by root aquaporins specifically in AMF plants could explain the maintenance of higher stomatal conductance, transpiration rate, relative water content and water potential observed in mycorrhizal plants compared to non-mycorrhizal plants (Symanczik et al., 2018). Additionally, the association with AMF alters plant water regulation by triggered hormonal signals, such as abscisic acid (ABA), which mediates stomatal conductance, or by osmolyte production (Hashem et al., 2018). Several studies have also shown that the development of AMF-induced mechanisms, including the modification of plant hormones content such as strigolactones, jasmonic acid and ABA, and the improvement of plant water status by increasing hydraulic conductivity (Bahadur et al., 2019). In addition to phytohormones, improving osmotic adjustment is considered an important element in mycorrhizal plants. AMF decreased the osmotic potential of leaves at both the point of complete turgidity and the point of loss of turgidity, accompanied by an increase in pressure potential at complete turgidity, thus maintaining a greater leaves water status of the host plant (Boutasknit et al., 2022). In addition, plants in symbiosis with AMF have a higher content of soluble sugar and total non-structural carbohydrates in leaves and roots, as well as K^+ and Ca^{2+} in leaves and roots (Boutasknit et al., 2020b). These positive osmolyte responses induced by AMF application protect macromolecules and structures from damage through increased osmotic adjustment capacity, helping to maintain a water potential gradient and water uptake from soil to roots (Symanczik et al., 2018; Zhang et al., 2018).

AMF Impacts on Nutrients Uptake

AMF improve plant growth performance and several major nutrients uptake such as nitrogen and phosphorus under non-stressed (Figure 3) and stressed environments (Boutasknit et al., 2020b; Sharma et al., 2020). However, fungal hyphae are much thinner than roots and have the capacity to enter smaller pores to uptake more mineral nutrients. AMF have been shown to be more effective to improve the mineral nutrients composition in plants namely poor mobility nutrients enhancing and/or selective nutrients uptake (Zhang et al., 2018). Mycorrhizal dependency increases with increasing stress conditions. Nutrient uptake is improved by AMF mainly as a function of the development of soil exploration by the external network of hyphae of these fungi (Symanczik et al., 2018; Zhang et al., 2018). As a result, mycorrhizal roots are often more effective than non-mycorrhizal roots in providing nutrients from soil to host plants (Kaur and Suseela, 2020). Mineral elements absorbed by AMF hyphae are easily translocated to fungal structures within the roots, where they are then exchanged for photosynthate from the host (Zhang et al., 2020).

Figure 3. Nutrient content in non-mycorrhizal (NM) and arbuscular mycorrhizal (AM) carob. P: phosphorus, K: potassium and Ca: calcium. Bars with different letters are significantly different as determined by Tukey's HSD test, after performing one-way ANOVA (P < 0.05). (Boutasknit et al., 2020b)

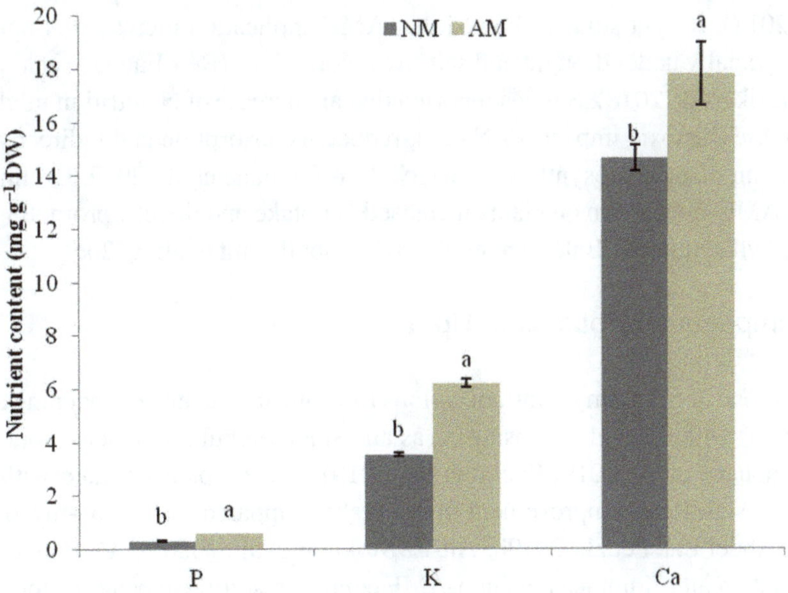

AMF Impacts on Phosphorus Uptake

AMF inoculation allows plants to increase the P concentration by improving its uptake, which is facilitated by the extensive fungal hyphae that allow them to reach a larger volume in the soil than non-mycorrhizal plants (Zhang et al., 2018; Zhang et al., 2020). Recently, Ait-El-Mokhtar et al. (2020a) recorded that P concentration in plants was reduced with increasing abiotic stress levels. Higher P concentrations in plants treated by AMF under stressed conditions suggest that AMF improved plants P uptake under drastic environments. Improved P nutrition in mycorrhizal plants may improve their growth performance and increase antioxidant activities as well as enhanced nitrogen fixation and nodulation in legumes (Ben-Laouane et al., 2019, 2020a).

AMF Impacts on Nitrogen Uptake

Climate change-related abiotic stresses negatively affect plant N acquisition and utilization. However, AMF inoculation improves N uptake in the host plant (Symanczik et al., 2018). Recent studies showed that AMF application increased N uptake in plant especially under drought and salt stress conditions (Ben-Laouane et al., 2019; Symanczik et al., 2018). Studies reported that an increase of N nutrition might help to decrease the toxic impacts of Na ion, reduce its absorption and indirectly helps to maintain plant photosynthesis activity (Ben-Laouane et al., 2019, Evelin et al., 2019). AMF inoculation on plants increased N uptake and thereby promoting plant chlorophyll synthesis (Balestrini et al., 2020; Boutasknit et al., 2020c).

AMF Impacts on Potassium Uptake

K^+ ion is one of the main important cations for stomatal closure and opening as well as protein synthesis and is considered as an osmotic regulator under abiotic stress (Ben-Laouane et al., 2019; Evelin et al., 2019). Indeed, plants treated with AMF showed a significant improvement in K uptake compared to the non-mycorrhizal plants (Abdel Latef et al., 2019b; Ait-El-Mokhtar et al., 2020a). AMF inoculation promoted greater K uptake than non-AMF control, resulting in better K uptake into the root xylem of mycorrhizal plants (Chandrasekaran, 2020).

AMF IMPACTS ON PLANT PHYSIOLOGICAL AND BIOCHEMICAL TRAITS

AMF Impacts on Plant Physiological Traits

AMF's influence on physiology and biochemistry was reported in many mycorrhizal plants growing under different levels of stress (Ait-El-Mokhtar et al., 2020a, 2019; Ben-Laouane et al., 2019; Meddich et al., 2020, 2021). AMF-inoculated date palm (Anli et al., 2020a) and carob (Boutasknit et al., 2020b) plants showed an increased photosynthetic performance by improving stomatal conductance and the efficiency of photosystem II (Figure 4) (Boutasknit et al., 2020b) and by regulating the energy flow between photochemical and non-photochemical reactions in stressful environment. The improvement in net assimilation rates through protection of PS II machinery and the increase in stomatal conductance under abiotic stress would be attributed to improved stress tolerance in AMF-inoculated plants (Symanczik et al., 2018).

Figure 4. Stomatal conductance (A) and chlorophyll fluorescence (Fv/Fm) (B) in non-mycorrhizal (NM) and arbuscular mycorrhizal (AM) carob. Bars with different letters are significantly different as determined by Tukey's HSD test, after performing one-way ANOVA (P < 0.05. (Boutasknit et al., 2020b)

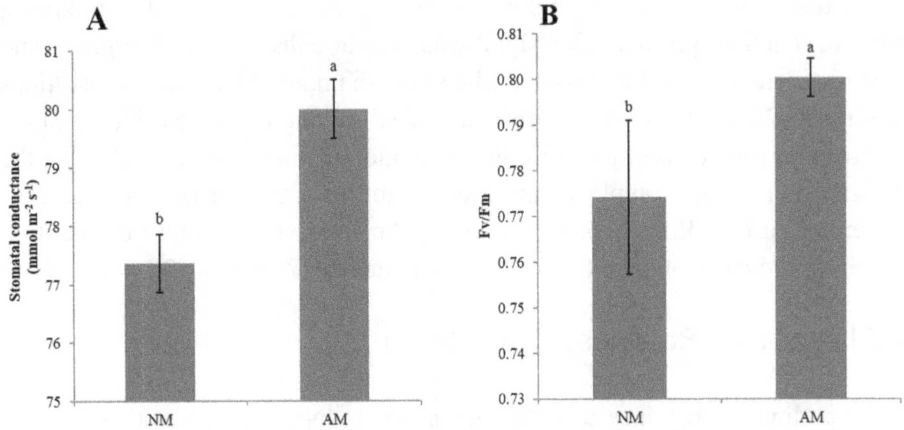

The improvement of nitrogen and phosphorus uptake due to the application of AMF could improve the photosynthetic attributes since these nutrients are considered to be determining components of the RuBisCO enzyme (Chen et al., 2017). Similarly, Begum et al. (2020) and Chang et al. (2018) reported that the improved uptake of K^+, Ca^{2+}, and Mg^{2+} due to the application of AMF could be possible reasons for increased photosynthetic capacity. This could also be attributed to the glycine betaine and proline accumulation in mycorrhizal plants that protects PSII pigment-protein complexes and CO_2 binding enzymes like RuBisCO and RuBisCO activase (Evelin et al., 2019) and without neglecting their contribution in osmotic adjustment mechanisms. Photosynthesis activity preserves and improves plant growth and development and afterward enhances plant tolerance capacity to overcome environmental stresses (Symanczik et al., 2018; Zhang et al., 2018). Abiotic stress significantly declined the chlorophyll pigments, photosynthesis activity (Kaur and Suseela, 2020). However, AMF inoculation alleviated this decline in mycorrhizal plants. Drought and salinity stresses induce photosynthetic pigments degradation by down-regulating the chlorophyll enzymes involved in the biosynthesis and up-regulating chlorophyllase activity (Evelin et al., 2019; Pan et al., 2020).

AMF Impacts on Plant Biochemical Traits

AMF Impacts on Proline Accumulation

Proline is involved in various plant developing processes and could be accumulated in significant concentrations under various environmental stresses (Ghosh et al., 2022). In addition, AMF inoculation caused an increase in proline concentrations in plant tissues under abiotic stresses (Hashem et al., 2018). Besides its known protective function, proline has a significant role in enhancing cell turgidity and performs several crucial functions for plant growth under abiotic stress conditions (Ghosh et al., 2022). These various roles include the reduction of possible oxidative membrane damage caused by ROS, the promotion of transduction pathways, the stabilization of protein complexes, and the availability of an alternative resource for nitrogen and carbon (Rumyantsev et al., 2021). An increase in proline concentration is a reliable indicator of salinity and drought tolerance (Zhu et al., 2020).

AMF Impacts on Betaines Accumulation

Besides proline, the betaines accumulation in plants subjected to abiotic stress is a crucial phenomenon for their protection (Sharma et al., 2019). Therefore, betaines can be used as an effective indicator for alleviation of environmental stresses (Sharma et al., 2019). They can also stabilize enzyme structures and activities and maintain membrane integrity against the harmful effects of water and salt stress (Shafi et al., 2019). These positive effects in combating environmental stresses are very pronounced when plants are inoculated with AMF (Chun et al., 2018). It was found that under severe stress conditions, the betaine content in plants inoculated with AMF was significantly higher than that in non-inoculated plants (Diagne et al., 2020).

AMF Impacts on Polyamines Synthesis

Free polyamines are osmolytes that are necessary for plant cell growth. They play an important role in plant responses to environmental stress like drought, salinity and antioxidative stress as well as high osmolarity (Evelin et al., 2019; Kaur and Suseela, 2020). They can also regulate root development under abiotic stress. Nevertheless, AMF inoculation increased polyamine concentrations in host plants. Several authors reported a significant increase in total free polyamine in plants inoculated by AMF (Zhang et al., 2020). The modulation of the polyamine pools could be a potential mechanism used by AMF to improve the adaptation of plants to saline and drought conditions (Evelin et al., 2019; Zhang et al., 2020).

AMF Impacts on Total Soluble Sugars Synthesis

Host plant mycorrhization level was positively correlated with high total carbohydrates accumulation as reported by Anli et al. (2020a) and Ben-Laouane et al. (2020b). In addition, Boutasknit et al. (2021) reported that plants inoculated with AMF had higher soluble sugars concentration compared to non-mycorrhizal plants. The positive relationship between sugar concentration and root mycorrhization could be due to the sink impact of the AMF demanding sugars from the shoot tissues (Wang et al., 2019). The increase in sugar accumulation could likewise be due to starch hydrolysis to sugars in the plants inoculated with AMF (Bona et al., 2018). Trehalose is the main storage carbohydrate in AMF, which plays a significant role as a stress protector by alleviating dehydrated membranes and enzymes and by protecting biological structures from desiccation damage (Sharma et al., 2020).

AMF Impacts on Antioxidant Activities

Plant cells contain an array of protection mechanisms and repairing systems that can minimize the occurrence of oxidative damage caused by ROS (Ait-El-Mokhtar et al., 2020a). The induction of ROS-scavenging enzymes, such as SOD, CAT, POD, APX, and glutathione reductase (GR) represent the most communal mechanism to prevent ROS accumulation in plant during stress response (Begum et al., 2020; Golubkina et al., 2020). Several studies demonstrated that antioxidant enzymes are commonly used as effective mechanisms to protect photosynthetic reaction centers from this damage in situations of biotic and/or abiotic stress (Meddich et al., 2018; Ait-El-Mokhtar et al., 2019b). However, plants treated with AMF may promote growth performances under abiotic stress, which linked with the antioxidant activity. Reports on plants treated with AMF and stress factors are contradictory; increase, no change, or even decrease in the SOD, CAT, POD and APX activities were reported in mycorrhizal plants subjected to drought and salinity (Abdel Latef et al., 2019b; Ghanbarzadeh et al., 2019; Begum et al., 2020).

AN OVERVIEW OF INDUCED ARBUSCULAR MYCORRHIZAL FUNGI EFFECTS LEADING TO CROP YIELD ENHANCEMENT UNDER CHANGING CLIMATE

AMF involvement in improving crop production and yield under changing environment conditions present many aspects and strategies (Figure 5). AMF can enhance photosynthesis functionality through mediating stomata opening and closure and improving chlorophyll content and photosystem II efficiency. Besides, the synthesis of aquaporins, AMF can enhance water uptake and hydraulic conductivity. In the other hand, AMF colonized roots produce glomalin and allowed extension of hyphae, providing availability and storage of nutrients for the plant. Furthermore, soil aggregation due to production of glomalin helps in increased water holding capacity and immobilization of nutrients. As a consequence, soil nutrient loss will be reduced. Therefore, the use of AMF regulates plant physiological and biochemical processes and improve soil physico-chemical parameters. In addition, AMF can enhance drought and salinity tolerance of host plants through mechanisms such as improvement of soil properties in rhizosphere, enlargement of root areas, enhancement of water use efficiency, N and P and other nutrient elements uptake (Symanczik et al., 2018; Zhang et al., 2018), soil organic carbon, efficient activation of defense system and protection against oxidative damages created by abiotic stress and thereby improve plants yield (Jiang et al., 2018; Ben-Laouane et al., 2020a). The use of AMF as bioinoculants in agriculture may increase the available P in soil, decrease the P-fertilizer application, minimize environmental pollution, and finally promote sustainable agriculture under a changing climate (Agnolucci et al., 2020).

Figure 5. Some potential mechanisms boosted by arbuscular mycorrhizal fungi application on soil and plant under a changing climate

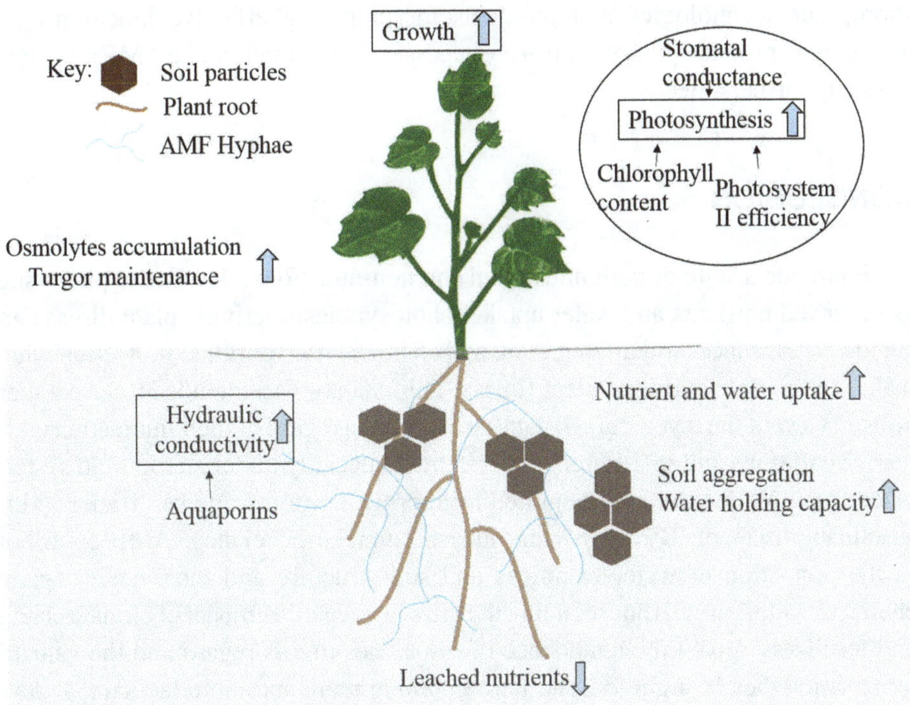

LIMITS OF AMF APPLICATION UNDER CHANGING CLIMATE TO BOOST CROP PRODUCTION

AMF can also have negative effects on crops. When colonizing plant roots, AMF may result in the suppression of important defense-related compounds in the roots of plants grown in nematode-infected soil. However, this root colonization is reduced by competition between different genera of AMF at the plant rhizosphere. Recently, a meta-analysis reported by Shi et al, (2002) showed that the application of a single mycorrhizal fungus also produced greater effects in terms of productivity than mixed mycorrhizal fungi. This is because the competition of mycorrhizal fungi for limited colonization areas in the root prevents root colonization affinity and therefore, the function of mycorrhizal fungi is negatively affected which also affects plant health and yield. In addition to these aspects, the use of AMF on a large scale by agriculture currently faces many challenges due to inconsistent results, problems with monitoring the success of inoculants in the field, gaps in knowledge and lack of methodology for large-scale production of inoculants, and quality control (Madawala, 2021). Thus, more research needs to be performed on the classification of the various AMF species based on their hosts and environmental preferences, with the purpose of defining the best inoculum for different crops and climatic conditions. Furthermore, appropriate technologies and protocols for preparing effective inoculums and maintaining product quality control are necessary to commercialize AMF inoculants to satisfy farmers' needs.

CONCLUSION

AMF provide a suite of both nutritional and non-nutritional benefits to plants such as improved nutrients and water uptake, photosynthesis activity, plant disease and chemical resistance, soil aggregation, and chemical transport and protection which makes them contributing to plant fitness. This chapter was intended to give a more holistic view of the mycorrhizal plant association in regard of their interaction under many conditions and in different ecosystems, especially under a crop field system, where multiple factors are integrated to improve nutrients uptake efficiency in a sustainable manner. By improving nutrients uptake efficiency, AMF contribute to the protection of water resources and soil structure and biodiversity against chemical pollution and thus reducing the growing negative impact of climate change and fertilizers. Given the abundance of researches in this regard and the valuable information they brought to light, it is becoming more and more factual that AMF are indispensable for agricultural sustainability and development especially under conditions dictated by climate change.

REFERENCES

Abdel Latef, A. A. H., Kordrostami, M., Zakir, A., Zaki, H., & Saleh, O. (2019a). Eustress with H_2O_2 facilitates plant growth by improving tolerance to salt stress in two wheat cultivars. *Plants*, *8*(9), 303. doi:10.3390/plants8090303 PMID:31461870

Abdel Latef, A. A. H., Mostofa, M. G., Rahman, M. M., Abdel-Farid, I. B., & Tran, L.-S. P. (2019b). Extracts from yeast and carrot roots enhance maize performance under seawater-induced salt stress by altering physio-biochemical characteristics of stressed plants. *Journal of Plant Growth Regulation*, *38*(3), 966–979. doi:10.100700344-018-9906-8

Agnihotri, R., Sahni, S., Sharma, M. P., & Gupta, M. M. (2022). Facets of AM fungi in sequestering soil carbon and improving soil health. In Fungal diversity, ecology, and control management, pp. 327-344. doi:10.1007/978-981-16-8877-5_15

Agnolucci, M., Avio, L., Palla, M., Sbrana, C., Turrini, A., & Giovannetti, M. (2020). Health-promoting properties of plant products: The role of mycorrhizal fungi and associated bacteria. *Agronomy (Basel)*, *10*(12), 1864. doi:10.3390/agronomy10121864

Ait-El-Mokhtar, M., Anli, M., Ben Laouane, R., Boutasknit, A., Boutaj, H., Draoui, A., Zarik, L., & Fakhech, A. (2019a). Food security and climate change. In Handbook of Research on Global Environmental Changes and Human Health, pp. 53–73. doi:10.4018/978-1-5225-7775-1.ch004

Ait-El-Mokhtar, M., Baslam, M., Ben-Laouane, R., Anli, M., Boutasknit, A., Mitsui, T., Wahbi, S., & Meddich, A. (2020a). Alleviation of detrimental effects of salt stress on date palm (*Phoenix dactylifera* L.) by the application of arbuscular mycorrhizal fungi and/or compost. *Frontiers in Sustainable Food Systems*, *4*, 131. doi:10.3389/fsufs.2020.00131

Ait-El-Mokhtar, M., Ben-Laouane, R., Anli, M., Boutasknit, A., Wahbi, S., & Meddich, A. (2019b). Use of mycorrhizal fungi in improving tolerance of the date palm (*Phoenix dactylifera* L.) seedlings to salt stress. *Scientia Horticulturae*, *253*, 429–438. doi:10.1016/j.scienta.2019.04.066

Ait-El-Mokhtar, M., Boutasknit, A., Ben-Laouane, R., Anli, M., El Amerany, F., Toubali, S., Lahbouki, S., Wahbi, S., & Meddich, A. (2020b). Vulnerability of Oasis Agriculture to Climate Change in Morocco In Impacts of Climate Change on Agriculture and Aquaculture, pp. 76–106. doi:10.4018/978-1-7998-3343-7.ch004

Ait-El-Mokhtar, M., Fakhech, A., Anli, M., Ben-Laouane, R., Boutasknit, A., Wahbi, S., & Meddich, A. (2020c). Infectivity of the palm groves arbuscular mycorrhizal fungi under arid and semi-arid climate and its edaphic determinants towards efficient ecological restoration. *Rhizosphere*, *15*, 1–5. doi:10.1016/j.rhisph.2020.100220

Ait-Rahou, Y., Ait-El-Mokhtar, M., Anli, M., Boutasknit, A., Ben-Laouane, R., Douira, A., Benkirane, R., El Modafar, C., & Meddich, A. (2020). Use of mycorrhizal fungi and compost for improving the growth and yield of tomato and its resistance to *Verticillium dahliae*. *Archiv für Phytopathologie und Pflanzenschutz*, 1–26. doi :10.1080/03235408.2020.1854938

Al Kharusi, L., Al Yahyai, R., & Yaish, M. W. (2019). Antioxidant response to salinity in salt-tolerant and salt-susceptible cultivars of date palm. *Agriculture*, *9*(1), 1–17. doi:10.3390/agriculture9010008

Alaux, P. L., César, V., Naveau, F., Cranenbrouck, S., & Declerck, S. (2018). Impact of Rhizophagus irregularis MUCL 41833 on disease symptoms caused by *Phytophthora infestans* in potato grown under field conditions. *Crop Protection (Guildford, Surrey), 107*, 26–33. doi:10.1016/j.cropro.2018.01.003

Andrino, A., Guggenberger, G., Kernchen, S., Mikutta, R., Sauheitl, L., & Boy, J. (2021). Production of organic acids by arbuscular mycorrhizal fungi and their contribution in the mobilization of phosphorus bound to iron oxides. *Frontiers in Plant Science*, *12*, 661842. doi:10.3389/fpls.2021.661842 PMID:34335645

Anli, M., Baslam, M., Tahiri, A., Raklami, A., Symanczik, S., Boutasknit, A., Ait-El-Mokhtar, M., Ben-Laouane, R., Toubali, S., Ait-Rahou, Y., Ait Chitt, M., Oufdou, K., Mitsui, T., Hafidi, M., & Meddich, A. (2020a). Biofertilizers as strategies to improve photosynthetic apparatus, growth, and drought stress tolerance in the date palm. *Frontiers in Plant Science*, *11*, 1–21. doi:10.3389/fpls.2020.516818 PMID:33193464

Anli, M., El Kaoua, M., Ait-El-Mokhtar, M., Boutasknit, A., Ben-Laouane, R., Toubali, S., Baslam, M., Lyamlouli, K., Hafidi, M., & Meddich, A. (2020b). Seaweed extract application and arbuscular mycorrhizal fungal inoculation: A tool for promoting growth and development of date palm (*Phoenix dactylifera* L.) cv «Boufgous». *South African Journal of Botany*, *132*, 15–21. doi:10.1016/j.sajb.2020.04.004

Anli, M., Symanczik, S., El Abbassi, A., Ait-El-Mokhtar, M., Boutasknit, A., Ben-Laouane, R., Toubali, S., Baslam, M., Mäder, P., Hafidi, M., & Meddich, A. (2020c). Use of arbuscular mycorrhizal fungus *Rhizoglomus irregulare* and compost to improve growth and physiological responses of *Phoenix dactylifera* "Boufgouss.". *Plant Biosystems*, 1–14. doi:10.1080/11263504.2020.1779848

Bahadur, A., Batool, A., Nasir, F., Jiang, S., Mingsen, Q., Zhang, Q., Pan, J., Liu, Y., & Feng, H. (2019). Mechanistic insights into arbuscular mycorrhizal fungi-mediated drought stress tolerance in plants. *International Journal of Molecular Sciences*, *20*(17), 1–18. doi:10.3390/ijms20174199 PMID:31461957

Balestrini, R., Brunetti, C., Chitarra, W., & Nerva, L. (2020). Photosynthetic Traits and Nitrogen Uptake in Crops: Which Is the Role of Arbuscular Mycorrhizal Fungi? *Plants*, *9*(9), 1105. doi:10.3390/plants9091105 PMID:32867243

Bárzana, G., Aroca, R., & Ruiz-Lozano, J. M. (2015). Localized and non-localized effects of arbuscular mycorrhizal symbiosis on accumulation of osmolytes and aquaporins and on antioxidant systems in maize plants subjected to total or partial root drying. *Plant, Cell & Environment*, *38*(8), 1613–1627. doi:10.1111/pce.12507 PMID:25630435

Basu, S., Rabara, R. C., & Negi, S. (2018). AMF: The future prospect for sustainable agriculture. *Physiological and Molecular Plant Pathology*, *102*, 36–45. doi:10.1016/j.pmpp.2017.11.007

Begum, N., Ahanger, M. A., & Zhang, L. (2020). AMF inoculation and phosphorus supplementation alleviates drought induced growth and photosynthetic decline in *Nicotiana tabacum* by up-regulating antioxidant metabolism and osmolyte accumulation. *Environmental and Experimental Botany*, *176*, 104088. doi:10.1016/j.envexpbot.2020.104088

Ben-Laouane, R., Ait-El-Mokhtar, M., Anli, M., Boutasknit, A., Ait Rahou, Y., Raklami, A., Oufdou, K., Wahbi, S., & Meddich, A. (2020a). Green compost combined with mycorrhizae and rhizobia: A strategy for improving alfalfa growth and yield under field conditions. *Gesunde Pflanzen*, *73*(2), 193–207. doi:10.100710343-020-00537-z

Ben-Laouane, R., Baslam, M., Ait-El-Mokhtar, M., Anli, M., Boutasknit, A., Ait-Rahou, Y., Toubali, S., Mitsui, T., Oufdou, K., Wahbi, S., & Meddich, A. (2020b). Potential of native arbuscular mycorrhizal fungi, rhizobia, and/or green compost as alfalfa (*Medicago sativa*) enhancers under salinity. *Microorganisms*, *8*(11), 1–27. doi:10.3390/microorganisms8111695 PMID:33143245

Ben-Laouane, R., Meddich, A., Bechtaoui, N., Oufdou, K., & Wahbi, S. (2019). Effects of arbuscular mycorrhizal fungi and rhizobia symbiosis on the tolerance of *Medicago sativa* to salt stress. *Gesunde Pflanzen*, *71*(2), 135–146. doi:10.100710343-019-00461-x

Bernardo, L., Carletti, P., Badeck, F. W., Rizza, F., Morcia, C., Ghizzoni, R., Rouphael, Y., Colla, G., Terzi, V., & Lucini, L. (2019). Metabolomic responses triggered by arbuscular mycorrhiza enhance tolerance to water stress in wheat cultivars. *Plant Physiology and Biochemistry*, *137*, 203–212. doi:10.1016/j.plaphy.2019.02.007 PMID:30802803

Bhantana, P., Malla, R., Vista, S. P., Rana, M. S., Mohamed, G., Joshi, B. D., Shah, S., Khadka, D., Timsina, G. P., Poudel, K., Zhihao, D., & Hu, C. X. (2021). Use of arbuscular mycorrhizal fungi (AMF) and zinc fertilizers in an adaptation of plant from drought and heat stress. *Biomedical Journal of Scientific & Technical Research*, *38*(3), 30357–30373. doi:10.26717/BJSTR.2021.38.006152

Bona, E., Todeschini, V., Cantamessa, S., Cesaro, P., Copetta, A., Lingua, G., Gamalero, E., Berta, G., & Massa, N. (2018). Combined bacterial and mycorrhizal inocula improve tomato quality at reduced fertilization. *Scientia Horticulturae*, *234*, 160–165. doi:10.1016/j.scienta.2018.02.026

Boutasknit, A., Ait-Rahou, Y., Anli, M., Ait-El-Mokhtar, M., Ben-Laouane, R., & Meddich, A. (2020a). Improvement of garlic growth, physiology, biochemical traits and soil fertility by *rhizophagus irregularis* and compost. *Gesunde Pflanzen*, *73*(2), 149–160. doi:10.100710343-020-00533-3

Boutasknit, A., Baslam, M., Ait-El-Mokhtar, M., Anli, M., Ben-Laouane, R., Ait-Rahou, Y., Mitsui, T., Douira, A., El Modafar, C., Wahbi, S., & Meddich, A. (2021). Assemblage of indigenous arbuscular mycorrhizal fungi and green waste compost enhance drought stress tolerance in carob (*Ceratonia siliqua* L.) trees. *Scientific Reports*, *11*(1), 1–23. doi:10.103841598-021-02018-3 PMID:34819547

Boutasknit, A., Baslam, M., Ait-El-Mokhtar, M., Anli, M., Ben-Laouane, R., Douira, A., El Modafar, C., Mitsui, T., Wahbi, S., & Meddich, A. (2020b). Arbuscular mycorrhizal fungi mediate drought tolerance and recovery in two contrasting carob (*Ceratonia siliqua* L.) ecotypes by regulating stomatal, water relations, and (In) organic adjustments. *Plants*, *9*(1), 80. doi:10.3390/plants9010080 PMID:31936327

Buras, A., Rammig, A. S., & Zang, C. (2020). Quantifying impacts of the 2018 drought on European ecosystems in comparison to 2003. *Biogeosciences*, *17*(6), 1655–1672. doi:10.5194/bg-17-1655-2020

Caser, M., Victorino, Í. M. M., Demasi, S., Berruti, A., Donno, D., Lumini, E., Bianciotto, V., & Scariot, V. (2019). Saffron cultivation in marginal alpine environments: How AMF inoculation modulates yield and bioactive compounds. *Agronomy (Basel)*, *9*(1), 12. doi:10.3390/agronomy9010012

Chandrasekaran, M. (2020). A meta-analytical approach on arbuscular mycorrhizal fungi inoculation efficiency on plant growth and nutrient uptake. *Agriculture*, *10*(9), 1–12. doi:10.3390/agriculture10090370

Chang, W., Sui, X., Fan, X. X., Jia, T. T., & Song, F. Q. (2018). Arbuscular mycorrhizal symbiosis modulates antioxidant response and ion distribution in salt-stressed *Elaeagnus angustifolia* seedlings. *Frontiers in Microbiology*, *9*, 1–14. doi:10.3389/fmicb.2018.00652 PMID:29675008

Chen, B., Hu, Y., Li, T., Xu, L., Zhang, X., & Hao, Z. (2018). The role of AM symbiosis in plant adaptation to drought stress. *Journal of Integrated Field Science*, *15*, 22–25.

Chen, S., Zhao, H., Zou, C., Li, Y., Chen, Y., Wang, Z., Jiang, Y., Liu, A., Zhao, P., Wang, M., & Ahammed, G. J. (2017). Combined inoculation with multiple arbuscular mycorrhizal fungi improves growth, nutrient uptake and photosynthesis in cucumber seedlings. *Frontiers in Microbiology*, *8*, 1–11. doi:10.3389/fmicb.2017.02516 PMID:29312217

Chun, S. C., Paramasivan, M., & Chandrasekaran, M. (2018). Proline accumulation influenced by osmotic stress in arbuscular mycorrhizal symbiotic plants. *Frontiers in Microbiology*, *9*, 2525. doi:10.3389/fmicb.2018.02525 PMID:30459731

Diagne, N., Ngom, M., Djighaly, P. I., Fall, D., Hocher, V., & Svistoonoff, S. (2020). Roles of arbuscular mycorrhizal fungi on plant growth and performance: Importance in biotic and abiotic stressed regulation. *Diversity (Basel)*, *12*(10), 1–25. doi:10.3390/d12100370

El Amerany, F., Rhazi, M., Wahbi, S., Taourirte, M., & Meddich, A. (2020). The effect of chitosan, arbuscular mycorrhizal fungi, and compost applied individually or in combination on growth, nutrient uptake, and stem anatomy of tomato. *Scientia Horticulturae*, *261*, 109015. doi:10.1016/j.scienta.2019.109015

Etesami, H., Jeong, B. R., & Glick, B. R. (2021). Contribution of arbuscular mycorrhizal fungi, phosphate–solubilizing bacteria, and silicon to P uptake by plant. *Frontiers in Plant Science*, *12*, 699618. doi:10.3389/fpls.2021.699618 PMID:34276750

Evelin, H., Devi, T. S., Gupta, S., & Kapoor, R. (2019). Mitigation of salinity stress in plants by arbuscular mycorrhizal symbiosis: Current understanding and new challenges. *Frontiers in Plant Science*, *10*, 1–21. doi:10.3389/fpls.2019.00470 PMID:31031793

Feng, J., Lv, W., Xu, J., Huang, Z., Rui, W., Lei, X., Ju, X., & Li, Z. (2022). Overlapping root architecture and gene expression of nitrogen transporters for nitrogen acquisition of tomato plants colonized with isolates of *Funneliformis mosseae* in hydroponic production. *Plants*, *11*(9), 1176. doi:10.3390/plants11091176 PMID:35567176

Field, K. J., & Pressel, S. (2018). Unity in diversity: Structural and functional insights into the ancient partnerships between plants and fungi. *The New Phytologist*, *220*(4), 996–1011. doi:10.1111/nph.15158 PMID:29696662

Gałązka, A., Gawryjołek, K., Grządziel, J., & Księżak, J. (2017). Effect of different agricultural management practices on soil biological parameters including glomalin fraction. *Plant, Soil and Environment*, *63*(7), 300–306. doi:10.17221/207/2017-PSE

Gałązka, A., Niedźwiecki, J., Grządziel, J., & Gawryjołek, K. (2020). Evaluation of changes in Glomalin-Related Soil Proteins (GRSP) content, microbial diversity and physical properties depending on the type of soil as the important biotic determinants of soil quality. *Agronomy (Basel)*, *10*(9), 1279. doi:10.3390/agronomy10091279

Gao, X., Guo, H., Zhang, Q., Guo, H., Zhang, L., Zhang, C., Gou, Z., Liu, Y., Wei, J., Chen, A., Chu, Z., & Zeng, F. (2020). Arbuscular mycorrhizal fungi (AMF) enhanced the growth, yield, fiber quality and phosphorus regulation in upland cotton (*Gossypium hirsutum* L.). *Scientific Reports*, *10*(1), 1–13. doi:10.103841598-020-59180-3 PMID:32034269

Garcia, C. L., Dattamudi, S., Chanda, S., & Jayachandran, K. (2019). Effect of Salinity Stress and Microbial Inoculations on Glomalin Production and Plant Growth Parameters of Snap Bean (*Phaseolus vulgaris*). *Agronomy (Basel)*, *9*(9), 545. doi:10.3390/agronomy9090545

Ghanbarzadeh, Z., Mohsenzadeh, S., Rowshan, V., & Moradshahi, A. (2019). Evaluation of the growth, essential oil composition and antioxidant activity of *Dracocephalum moldavica* under water deficit stress and symbiosis with *Claroideoglomus etunicatum* and *Micrococcus yunnanensis*. *Scientia Horticulturae*, *256*, 108652. doi:10.1016/j.scienta.2019.108652

Ghosh, U. K., Islam, M. N., Siddiqui, M. N., Cao, X., & Khan, M. A. R. (2022). Proline, a multifaceted signalling molecule in plant responses to abiotic stress: Understanding the physiological mechanisms. *Plant Biology*, *24*(2), 227–239. doi:10.1111/plb.13363 PMID:34796604

Giovannini, L., Palla, M., Agnolucci, M., Avio, L., Sbrana, C., Turrini, A., & Giovannetti, M. (2020). Arbuscular mycorrhizal fungi and associated microbiota as plant biostimulants: Research strategies for the selection of the best performing inocula. *Agronomy (Basel)*, *10*(1), 106. doi:10.3390/agronomy10010106

Goicoechea, N., Baslam, M., Erice, G., & Irigoyen, J. J. (2014). Increased photosynthetic acclimation in alfalfa associated with arbuscular mycorrhizal fungi (AMF) and cultivated in greenhouse under elevated CO_2. *Journal of Plant Physiology*, *171*(18), 1774–1781. doi:10.1016/j.jplph.2014.07.027 PMID:25240322

Golubkina, N., Amagova, Z., Matsadze, V., Zamana, S., Tallarita, A., & Caruso, G. (2020). Effects of arbuscular mycorrhizal fungi on yield, biochemical characteristics, and elemental composition of garlic and onion under selenium supply. *Plants*, *9*(1), 84. doi:10.3390/plants9010084 PMID:31936528

Harris-Valle, C., Esqueda, M., Gutiérrez, A., Castellanos, A. E., Gardea, A. A., & Berbara, R. (2018). Physiological response of *Cucurbita pepo* var. pepo mycorrhized by *Sonoran desert* native arbuscular fungi to drought and salinity stresses. *Brazilian Journal of Microbiology*, *49*(1), 45–53. doi:10.1016/j.bjm.2017.04.005 PMID:28887008

Hashem, A., Alqarawi, A. A., Radhakrishnan, R., Al-Arjani, A. B. F., Aldehaish, H. A., Egamberdieva, D., & Abd Allah, E. F. (2018). Arbuscular mycorrhizal fungi regulate the oxidative system, hormones and ionic equilibrium to trigger salt stress tolerance in *Cucumis sativus* L. *Saudi Journal of Biological Sciences*, *25*, 1102–1114. doi:10.1016/j.sjbs.2018.03.009 PMID:30174509

He, Y., Yang, R., Lei, G., Li, M., Li, T., Zhan, F., & Li, Y. (2021). Arbuscular mycorrhizal fungus–induced decrease in phosphorus loss due to leaching in red soils under simulated heavy rainfall. *Journal of Soils and Sediments*, *21*(2), 881–889. doi:10.100711368-020-02849-z

Hu, J., Lin, X., Wang, J., Dai, J., Cui, X., Chen, R., & Zhang, J. (2009). Arbuscular mycorrhizal fungus enhances crop yield and P-uptake of maize (*Zea mays* L.): A field case study on a sandy loam soil as affected by long-term P-deficiency fertilization. *Soil Biology & Biochemistry*, *41*(12), 2460–2465. doi:10.1016/j.soilbio.2009.09.002

Jiang, Y., Zhou, H., Chen, L., Yuan, Y., Fang, H., Luan, L., Chen, Y., Wang, X., Liu, M., Li, H., Peng, X., & Sun, B. (2018). Nematodes and Microorganisms Interactively Stimulate Soil Organic Carbon Turnover in the Macroaggregates. *Frontiers in Microbiology*, *9*, 2803. doi:10.3389/fmicb.2018.02803 PMID:30534114

Kaur, S., & Suseela, V. (2020). Unraveling Arbuscular Mycorrhiza-Induced Changes in Plant Primary and Secondary Metabolome. *Metabolites*, *10*(8), 335. doi:10.3390/metabo10080335 PMID:32824704

Kour, D., Rana, K. L., Yadav, A. N., Yadav, N., Kumar, M., Kumar, V., Vyas, P., Dhaliwal, H. S., & Saxena, A. K. (2020). Microbial biofertilizers: Bioresources and eco-friendly technologies for agricultural and environmental sustainability. *Biocatalysis and Agricultural Biotechnology*, *23*, 101487. doi:10.1016/j.bcab.2019.101487

Lanfranco, L., Fiorilli, V., Venice, F., & Bonfante, P. (2018). Strigolactones cross the kingdoms: Plants, fungi, and bacteria in the arbuscular mycorrhizal symbiosis. *Journal of Experimental Botany*, *69*(9), 2175–2188. doi:10.1093/jxb/erx432 PMID:29309622

Liang, J. F., An, J., Gao, J. Q., Zhang, X. Y., Song, M. H., & Yu, F. H. (2019). Interactive effects of biochar and AMF on plant growth and greenhouse gas emissions from wetland microcosms. *Geoderma*, *346*, 11–17. doi:10.1016/j.geoderma.2019.03.033

Madawala, H. M. S. P. (2021). Arbuscular mycorrhizal fungi as biofertilizers: Current trends, challenges, and future prospects. *Biofertilizers*, 83-93. doi:10.1016/B978-0-12-821667-5.00029-4

Mahfudz, M., Saleh, S., Antara, M., Anshary, A., Bachri, S., Made, U., Hasanah, U., & Rauf, R. A. (2019). Adoption and advantages of eco-friendly technology application at the shallot farming system in Indonesia. *Agronomy Research (Tartu)*, *17*, 1679–1687. doi:10.15159/AR.19.188

Mariotte, P., Mehrabi, Z., Bezemer, T. M., De Deyn, G. B., Kulmatiski, A., Drigo, B., Veen, G. F. C., van der Heijden, M. G. A., & Kardol, P. (2018). Plant–soil feedback: Bridging natural and agricultural sciences. *Trends in Ecology & Evolution*, *33*(2), 129–142. doi:10.1016/j.tree.2017.11.005 PMID:29241940

Mateus, I. D., Masclaux, F. G., Aletti, C., Rojas, E. C., Savary, R., Dupuis, C., & Sanders, I. R. (2019). Dual RNA-seq reveals large-scale non-conserved genotype × genotype-specific genetic reprograming and molecular crosstalk in the mycorrhizal symbiosis. *Multidisciplinary Journal of Microbial Ecology*, *13*(5), 1226–1238. doi:10.103841396-018-0342-3 PMID:30647457

Meddich, A., Ait El Mokhtar, M., Bourzik, W., Mitsui, T., Baslam, M., & Hafidi, M. (2018). Optimizing growth and tolerance of date palm (*Phoenix dactylifera* L.) to drought, salinity, and vascular fusarium-induced wilt (*Fusarium oxysporum*) by application of arbuscular mycorrhizal fungi (AMF). In *Root biology* (pp. 239–258). Springer. doi:10.1007/978-3-319-75910-4_9

Meddich, A., Ait El Mokhtar, M., Wahbi, S., & Boumezzough, A. (2017). Evaluation of the mycorrhizal potential in relation with the physico-chemical properties of soils in Moroccan palm groves (Marrakech and Tafilalet). *Cahiers Agricultures*, *26*, 45012. doi:10.1051/cagri/2017044

Meddich, A., Ait Rahou, Y., Boutasknit, A., Ait-El-Mokhtar, M., Fakhech, A., Lahbouki, S., Benaffari, W., Ben-laouanen R., & Wahbi, S. (2021). Role of mycorrhizal fungi in improving the tolerance of melon (*Cucumus melo*) under two water deficit partial root drying and regulated deficit irrigation. *Plant Biosystems-An International Journal Dealing with all Aspects of Plant Biology*, 1-22.

Meddich, A., Oufdou, K., Boutasknit, A., Raklami, A., Tahiri, A., Ben-Laouane, R., Ait-El-Mokhtar, M., Anli, M., Mitsui, T., Wahbi, S., & Baslam, M. (2020). Use of Organic and Biological Fertilizers as Strategies to Improve Crop Biomass, Yields and Physicochemical Parameters of Soil. In Nutrient Dynamics for Sustainable Crop Production, pp. 247–288. Springer. doi:10.1007/978-981-13-8660-2_9

Meglouli, H., Fontaine, J., Verdin, A., Magnin-Robert, M., Tisserant, B., Hijri, M., & Sahraoui, A. L. H. (2019). Aided phytoremediation to clean up dioxins/furans-aged contaminated soil: Correlation between microbial communities and pollutant dissipation. *Microorganisms*, *7*(11), 1–18. doi:10.3390/microorganisms7110523 PMID:31684182

Nasir, F., Shi, S., Tian, L., Chang, C., Ma, L., Li, X., Gao, Y., & Tian, C. (2019). Strigolactones shape the rhizomicrobiome in rice (*Oryza sativa*). *Plant Science*, *286*, 118–133. doi:10.1016/j.plantsci.2019.05.016 PMID:31300137

Olivares, O. E. A., Torres, S. S., Jiménez, B. S. I., Enríquez, C. J. O., Zignol, F., Reygadas, Y., & Tiefenbacher, J. P. (2019). Climate change, land use/land cover change, and population growth as drivers of groundwater depletion in the Central Valleys, Oaxaca, Mexico. *Remote Sensing*, *11*(11), 1290. doi:10.3390/rs11111290

Pan, J., Huang, C., Peng, F., Zhang, W., Luo, J., Ma, S., & Xue, X. (2020). Effect of arbuscular mycorrhizal fungi (AMF) and plant growth-promoting bacteria (PGPR) inoculations on *Elaeagnus angustifolia* L. in saline Soil. *Applied Sciences (Basel, Switzerland)*, *10*(3), 945. doi:10.3390/app10030945

Pylak, M., Oszust, K., & Frąc, M. (2019). Review report on the role of bioproducts, biopreparations, biostimulants and microbial inoculants in organic production of fruit. *Reviews in Environmental Science and Biotechnology*, *18*(3), 597–616. doi:10.100711157-019-09500-5

Qin, M., Zhang, Q., Pan, J., Jiang, S., Liu, Y., Bahadur, A., Peng, Z., Yang, Y., & Feng, H. (2020). Effect of arbuscular mycorrhizal fungi on soil enzyme activity is coupled with increased plant biomass. *European Journal of Soil Science*, *71*, 84–92. doi:10.1111/ejss.12815

Quiroga, G., Erice, G., Aroca, R., & Ruiz-Lozano, J. M. (2020). Elucidating the possible involvement of maize aquaporins in the plant boron transport and homeostasis mediated by *Rhizophagus irregularis* under drought stress conditions. *International Journal of Molecular Sciences*, *21*(5), 148. doi:10.3390/ijms21051748 PMID:32143345

Renaut, S., Daoud, R., Masse, J., Vialle, A., & Hijri, M. (2020). Inoculation with *Rhizophagus irregularis* does not alter arbuscular mycorrhizal fungal community structure within the roots of corn, wheat, and soybean crops. *Microorganisms*, *8*(1), 83. doi:10.3390/microorganisms8010083 PMID:31936180

Riaz, M., Kamran, M., Fang, Y., Wang, Q., Cao, H., Yang, G., Deng, L., Wang, Y., Zhou, Y., Anastopoulos, I., & Wang, X. (2021). Arbuscular mycorrhizal fungi-induced mitigation of heavy metal phytotoxicity in metal contaminated soils: A critical review. *Journal of Hazardous Materials*, *402*, 123919. doi:10.1016/j.jhazmat.2020.123919 PMID:33254825

Rimington, W. R., Pressel, S., Duckett, J. G., Field, K. J., Read, D. J., & Bidartondo, M. I. (2018). Ancient plants with ancient fungi: Liverworts associate with early-diverging arbuscular mycorrhizal fungi. *Proceedings. Biological Sciences*, *285*(1888), 20181600. doi:10.1098/rspb.2018.1600 PMID:30305437

Rivera-Becerril, F., van Tuinen, D., Chatagnier, O., Rouard, N., Béguet, J., Kuszala, C., Soulas, G., Gianinazzi-Pearson, V., & Martin-Laurent, F. (2017). Impact of a pesticide cocktail (fenhexamid, folpel, deltamethrin) on the abundance of Glomeromycota in two agricultural soils. *The Science of the Total Environment*, *577*, 84–93. doi:10.1016/j.scitotenv.2016.10.098 PMID:27817923

Rumyantsev, A., Sidorin, A., Volkov, A., Al Shanaa, O., Sambuk, E., & Padkina, M. (2021). Transcriptome analysis unveils the effects of proline on gene expression in the Yeast *Komagataella phaffii*. *Microorganisms*, *10*(1), 67. doi:10.3390/microorganisms10010067 PMID:35056516

Sangwan, S., & Prasanna, R. (2021). Mycorrhizae helper bacteria: Unlocking their potential as bioenhancers of plant–arbuscular mycorrhizal fungal associations. *Microbial Ecology*, *84*(1), 1–10. doi:10.100700248-021-01831-7 PMID:34417849

Shafi, A., Zahoor, I., & Mushtaq, U. (2019). Proline accumulation and oxidative stress: Diverse roles and mechanism of tolerance and adaptation under salinity stress. In *Salt Stress, Microbes, and Plant Interactions: Mechanisms and Molecular Approaches* (pp. 269–300). Springer. doi:10.1007/978-981-13-8805-7_13

Sharma, A., Shahzad, B., Kumar, V., Kohli, S. K., Sidhu, G. P. S., Bali, A. S., Handa, N., Kapoor, D., Bhardwaj, R., & Zheng, B. (2019). Phytohormones regulate accumulation of osmolytes under abiotic stress. *Biomolecules*, *9*(7), 285. doi:10.3390/biom9070285 PMID:31319576

Sharma, K., Gupta, S., Thokchom, S. D., & Kapoor, R. (2021). The Potential Roles of arbuscular mycorrhizal fungi in soil health and conservation. *International journal of plant and environment*, *7*, 39-48.

Sharma, M. P., Grover, M., Chourasiya, D., Bharti, A., Agnihotri, R., Maheshwari, H. S., Pareek, A., Buyer, J. S., Sharma, S. K., Schütz, L., Mathimaran, N., Singla-Pareek, S. L., Grossman, J. M., & Bagyaraj, D. J. (2020). Deciphering the role of trehalose in tripartite symbiosis among rhizobia, arbuscular mycorrhizal fungi, and legumes for enhancing abiotic stress tolerance in crop plants. *Frontiers in Microbiology*, *11*, 509919. doi:10.3389/fmicb.2020.509919 PMID:33042042

Shi, Y., Wang, S., Guo, J., Xu, Z., Wang, S., & Sang, Y. (2022). Effects of arbuscular mycorrhizal inoculation on the phytoremediation of PAH-contaminated soil: A meta-analysis. *Chemosphere*, *136033*, 136033. doi:10.1016/j.chemosphere.2022.136033 PMID:35981621

Soh, M. C., Mitchell, N. J., Ridley, A. R., Butler, C. W., Puan, C. L., & Peh, K. S. H. (2019). Impacts of habitat degradation on tropical montane biodiversity and ecosystem services: A systematic map for identifying future research priorities. *Frontiers in Forests and Global Change*, *2*, 83. doi:10.3389/ffgc.2019.00083

Subhashini, D. V. (2016). Effect of NPK fertilizers and co-inoculation with phosphate-solubilizing arbuscular mycorrhizal fungus and potassium-mobilizing bacteria on growth, yield, nutrient acquisition, and quality of tobacco (*Nicotiana tabacum* L.). *Communications in Soil Science and Plant Analysis*, *47*(3), 328–337. doi:10.1080/00103624.2015.1123724

Symanczik, S., Krützmann, J., Nehls, U., Boller, T., & Courty, P. E. (2020). Expression of major intrinsic protein genes in Sorghum bicolor roots under water deficit depends on arbuscular mycorrhizal fungal species. *Soil Biology & Biochemistry*, *140*, 107643. doi:10.1016/j.soilbio.2019.107643

Symanczik, S., Lehmann, M. F., Wiemken, A., Boller, T., & Courty, P. E. (2018). Effects of two contrasted arbuscular mycorrhizal fungal isolates on nutrient uptake by *Sorghum bicolor* under drought. *Mycorrhiza*, *28*(8), 779–785. doi:10.100700572-018-0853-9 PMID:30006910

Teutscherova, N., Vazquez, E., Arango, J., Arevalo, A., Benito, M., & Pulleman, M. (2019). Native arbuscular mycorrhizal fungi increase the abundance of ammonia-oxidizing bacteria, but suppress nitrous oxide emissions shortly after urea application. *Geoderma*, *338*, 493–501. doi:10.1016/j.geoderma.2018.09.023

Toubali, S., Tahiri, A., Anli, M., Symanczik, S., Boutasknit, A., Ait-El-Mokhtar, M., Ben-Laouane, R., Oufdou, K., Ait-Rahou, Y., Ben-Ahmed, H., Jemo, M., Hafidi, M., & Meddich, A. (2020). Physiological and biochemical behaviors of date palm vitroplants treated with microbial consortia and compost in response to salt stress. *Applied Sciences (Basel, Switzerland)*, *10*(23), 8665. doi:10.3390/app10238665

Umar, A. (2017). Phytohyperaccumulator-AMF (arbuscular mycorrhizal fungi) interaction in heavy metals detoxification of soil. *Acta Biológica Paranaense*, *46*, 123–148. doi:10.5380/abpr.v46i0.57147

Wang, F., Sun, Y., & Shi, Z. (2019). Arbuscular mycorrhiza enhances biomass production and salt tolerance of sweet sorghum. *Microorganisms*, *7*(9), 289. doi:10.3390/microorganisms7090289 PMID:31450847

Wipf, D., Krajinski, F., van Tuinen, D., Recorbet, G., & Courty, P. E. (2019). Trading on the arbuscular mycorrhiza market: From arbuscules to common mycorrhizal networks. *The New Phytologist*, *223*(3), 1127–1142. doi:10.1111/nph.15775 PMID:30843207

Zhang, F., Zou, Y. N., & Wu, Q. S. (2018). Quantitative estimation of water uptake by mycorrhizal extraradical hyphae in citrus under drought stress. *Scientia Horticulturae*, *229*, 132–136. doi:10.1016/j.scienta.2017.10.038

Zhang, F., Zou, Y. N., Wu, Q. S., & Kuča, K. (2020). Arbuscular mycorrhizas modulate root polyamine metabolism to enhance drought tolerance of trifoliate orange. *Environmental and Experimental Botany*, *171*, 103926. doi:10.1016/j.envexpbot.2019.103926

Zhang, L., Xu, M., Liu, Y., Zhang, F., Hodge, A., & Feng, G. (2016). Carbon and phosphorus exchange may enable cooperation between an arbuscular mycorrhizal fungus and a phosphate-solubilizing bacterium. *The New Phytologist*, *210*(3), 1022–1032. doi:10.1111/nph.13838 PMID:27074400

Zhang, Z., Mallik, A., Zhang, J., Huang, Y., & Zhou, L. (2019). Effects of arbuscular mycorrhizal fungi on inoculated seedling growth and rhizosphere soil aggregates. *Soil & Tillage Research*, *194*, 104340. doi:10.1016/j.still.2019.104340

Zhongming, Z., Linong, L., Xiaona, Y., Wangqiang, Z., & Wei, L. (2018). 'Soil probiotics' promise bigger, healthier crops, but there's a downside. *Global S&T Development Trend Analysis Platform of Resources and Environment*. http://resp.llas.ac.cn/C666/handle/2XK7JSWQ/124877

Zhu, H., Zhou, Y., Zhai, H., He, S., Zhao, N., & Liu, Q. (2020). A novel sweetpotato *WRKY* transcription factor, *IBWRKY2*, positively regulates drought and salt tolerance in transgenic *Arabidopsis*. *Biomolecules*, *10*(4), 506. doi:10.3390/biom10040506 PMID:32230780

KEY TERMS AND DEFINITIONS

Agriculture: Is the process of producing food, forage, fiber, and various other products by the cultivation the soil and the rearing of domesticated animals.

Arbuscular Mycorrhizal Fungi: Type of fungi which form symbiosis with roots of vascular plants.

Climate Change: Refers to all variations in climatic parameters of the global climate of the Earth or its various regional climates over time.

Sustainability: The ability to be maintained at a certain rate or level over a period of time.

Soil fertility: refers to the ability of soil to sustain agricultural plant growth by providing necessary resources.

Salinity: refers to the concentrations of salts in water or soils.

Drought: refers to a prolonged dry period in the natural climate cycle that can occur anywhere in the world.

Chapter 6
Affordable Methods and Techniques to Optimize Farming Productivity in a Context of Climate Change in Arid Areas

Ahmed Karmaoui
https://orcid.org/0000-0003-3881-4029
Université Moulay Ismail, Meknès, Morocco

Adil Moumane
https://orcid.org/0000-0003-0296-2679
Université Ibn-Tofail, Morocco

Guido Minucci
Department of Architecture and Urban Studies, Politecnico di Milano, Italy

Samir El Jaafari
Moulay Ismail University, Morocco

Hassan Chaachouay
Moulay Ismail University, Morocco

Lhoussain Hajji
https://orcid.org/0000-0002-7785-8991
Moulay Ismail University, Morocco

ABSTRACT

Worldwide, the arid lands occupy one of the largest areas on Earth, covering more than 30% of the total area. In these lands, scarcity of resources has emerged as a consequence of climate change (CC) and anthropogenic pressure. The reduced resources require sustainable management, including the use of efficient methods and techniques to adapt to CC. The purpose of this chapter is to present a collection of

DOI: 10.4018/978-1-6684-4649-2.ch006

affordable techniques and practices to optimize farming productivity in the context of CC. This chapter also presents the strengths and weaknesses and highlights the replication of these methods where water supply and food security are far from being achieved. Each technic must be studied and adapted with caution in context of the defined arid area, taking into account the local socio-economic and ecological characteristics. The proposed techniques may be used in governmental planning to support the adaptation and mitigation strategies.

INTRODUCTION

The decrease in precipitation and the increase in temperatures and extreme weather events are affecting food security (Ayinu et al., 2022), due to declining of crop productivity and the emergence of plants and farm animal diseases. The crops and farm animals are exposed to the risks of water scarcity (Leal Filho et al., 2022), soil deterioration and moisture reduction, increased water and soil salinity, increased energy costs, and demand for products. The control of the parameters of these services and resources may help increase productivity, which provides more food for more people even if in vulnerable areas such as arid environments. In arid areas, sustainable agricultural development is essential, affected by water resources (Zhou et al., 2021). In addition to water, soil quality has always played an important role in food security and economic stability in these harsh environments. Unfortunately, climate change and human pressure is affecting these fragile resources. They are threatened by several phenomena such as salinity and desertification. Moumane et al. (2022) found a significant advancement in desertification from 1991 to 2021 and observed a considerable reduction of cultivated lands (−29.6%) in the pre-Sahara of Morocco. Otherwise, according to the last IPCC special report on "climate change, desertification, land degradation…", desertification will reduce the provision of ecosystem services including losses in biodiversity in the crop, and livestock productivity in drylands (Mirzabaev et al., 2019). In cultivated land, for example, soil erosion and degradation cause a loss of 5 to 7 Mha every year (Sivanappan, 1995). Water resources are also threatened in these areas. Dryland zones have serious problems meeting demands due to the population, infrastructure growth, and climate change (Koch & Missimer, 2016). Climate change (Lee et al., 2021), soil salinity (Yin et al., 2021), groundwater reduction and salinity (Moumane et al. 2021), water overuse, droughts, land degradation and desertification (Karmaoui et al., 2022; Lee et al., 2021) are the main causes of degradation in arid ecosystems and agro-ecosystems. These risks push for better management of agricultural systems and natural resources using sustainable solutions. Food security in arid regions depends on the resilience of agricultural systems to the problems of climate change.

Sustainable management of the water, soil, and biodiversity resources can provide key services and meet the development needs in these vulnerable regions. Alternative techniques that have proven to be effective in several areas can be used to increase productivity, diversify crops, and resist insect pests in desert areas. Optimizing the productivity of farms in the context of climate change is a significant step toward meeting the needs of a growing global population. The techniques are numerous and depend on the type of ecosystem services such as water, soil, or biodiversity, and the use of renewable energies. These techniques support the efficient use of vulnerable resources and assure the sustainability of the farms. The purpose of the current chapter is to present an inventory of simple and affordable methods, techniques, and practices to optimize farming productivity in the context of climate change. The chapter was organized into two axes: Axis 1 assists in identifying examples of some recent techniques and practices used in the context of the land, water, and energy nexus while the second axis deals with traditional practices and equipment used in the arid areas.

LAND, WATER, AND ENERGY NEXUS TO OPTIMIZE AGRICULTURE AND TACKLE CLIMATE CHANGE

Water is the most important resource that supports all kind of ecosystems in arid areas. The techniques used as solutions to optimize the use of water in these zones are numerous. The conversion of agricultural land to fallow land is reported as a technique that may help reduce the overuse of water in these vulnerable regions. In arid regions, groundwater is the main source of irrigation and the overuse extraction of groundwater is a major issue facing these environments (Pourmohamad et al., 2019). Pourmohamad et al., (2019) considered the conversion of some agricultural land to fallow land as a solution to reduce groundwater overuse. Rainwater harvesting is another useful method recommended in areas with scarce water resources. The technique aims to harvest rainwater from a surface or a structure to use it later. Flat surfaces can be designed or operated to collect rainwater and store it in cisterns or direct it to aquifers by infiltration. This highlights the importance of increasing soil infiltration to enhance the potential of groundwater resources. Usually, it is used by digging regular holes and filling them with loose soil to increase the infiltration of rainwater. In this context, the vegetation also allows better infiltration through the roots. In some arid areas, the dew collection technique is highly suggested for harvesting water. The dew is strong at night favored by the temperature range (between day and night). During low temperatures, the plants and the soil cool down, which induces the condensation of part of the water in the air. From air to soil, the buried diffuser can be used to save irrigation water and maintain soil quality. The buried diffuser is an innovative method for underground irrigation. Subsurface drip irrigation technique that leads water to a depth depending on soil nature and crop roots distribution (Zemni et al., (2022). In a Saharan oasis, Zemni et al., (2022) investigated the Chahbani buried diffuser for date palm recommending that this diffuser must be combined with infrequent surface irrigation and be placed at X = 0.5 m (positions from the trunk of the tree) and Z = 0.6 m (soil depths). The technique was tested for plants in arid and semi-arid climates in Tunisia (Gasmi et al., 2019) where soil and water are affected by high salinity rates. To fight against the salinity of water, magnetic water irrigation is recommended to reduce the concentration of salinity. The Magnetizer is an innovative physical water treatment technique used for the physical treatment of water. Several studies reveal the importance of the physical treatment of water by a static magnetic field (Surendran et al., 2016; Hilal and Hilal, 2000). This supports plant growth with recycled water. It is through the application of a magnetic or electrostatic field on the water without removing the limestone and protecting the pipes against scaling. Magnetic water irrigation is an important technique mainly for slightly saline soil than saline soils; it improves the water holding capacity, redistributes water and salt in salinized soils, and decreases the soil salt contents (Zhou et al., 2021). Land and soil are among the

Figure 1. The intervention of the local government to fight against desertification and dunes advancement toward the oasis agroecosystem Oases and the temporary river in the Pre-Sahara of Morocco showing the stabilizing physical and biological systems. Source: author processing using images from Google Earth

Figure 2 Adoption of measures by the government and local people to combat this desertification and sand dunes. A, A village threatened by large sand dunes; B, A farm invaded by sand; C, Physical fixation; and D, Photo 4: Biological fixation; E, Mixt fixation. Source: Authors

principal support of the ecosystem services provided by arid areas. Unfortunately, this supporting service is threatened mainly by desertification as above-mentioned. The fight against desertification is the most common strategy used in these areas that include physical and biological stabilizing systems (Figure 1 and 2).

Desertification is frequent in arid lands and its impact is significant on the socioecological system. The use of shelterbelts is an important technique to fight desertification and grow crop productivity in arid areas (Li et al., 2020). A farmland shelterbelt may provide maximum protection with a minimum of water and land resources (Li et al., 2020). The yield of cotton was significantly correlated with the spatial distribution of shelterbelts (Li et al., 2020).

Energy is considerable pillar in the agriculture sector and renewable energy systems are considered an ecologic solution. Industrials and academics considered renewable energy systems potential, feasible, reliable, and sustainable electrical energy sources (Nassar et al., 2021). The pumped hydroelectric storage technique is an example of a solution used in agriculture. Pumped Hydroelectric Storage (PHS) is a hydroelectric energy storage technique that produces energy from two water reservoirs at different elevations. Commercially PHS is a viable technology and eligible to be associated with the renewable energy systems (Nassar et al., 2021). These clean and economical energies poses a challenge to the overuse of water resources primarily groundwater, which requires thinking about the regulation of extraction. The overuse may be controlled in many arid areas by the introduction of water pricing regulations. Water pricing was investigated in the study of Shannak (2022) in Saudi Arabia on crop cultivation. This author found that the increase of prices would reduce crop cultivation advancing that prices in agriculture must be determined with caution.

The above-mentioned methods, techniques, and practices may be adopted and adapted to local socio-economic and ecologic conditions to optimize farming productivity in the context of climate change. Karmaoui et al., (2021) suggested that the human response should be concentrated on adaptation strategies including both traditional and new technologies to protect soil and water resources such as small dam construction, composting, and recycling solid and liquid wastes.

TRADITIONAL OR INDIGENOUS KNOWLEDGE

The indigenous knowledge gained global recognition by UNESCO, United Nations Conference on Environment and Development (UNCED) in 1992 (Rio de Janeiro), and by the International Union for the Conservation of Nature and Natural Resources (IUCN), due to its cultural, ecological, and socio-economic values. By definition, indigenous knowledge is the local knowledge that is unique to a given culture or society (Risiro, 2013). After Warren (1991) it is the basis for local decision-making in agriculture, health, food education, and environmental management. It covers local, traditional, and non-western beliefs, practices, and customs and usually refers to informal forms of knowledge (Horsthemke, 2004). It is the knowledge of people of a particular geographical area that has survived for a long period (Mapara, 2009). In many cases, traditional knowledge has been orally passed for generations from one person to another and this can be expressed through stories, songs art, and even laws (Risiro, 2013). Indigenous knowledge can be called by various names such as indigenous knowledge of knowing (Mapara, 2009), traditional knowledge, rural knowledge as well as ethno-science (Altieri, 1995). UNESCO gives the following definition [1]:

«*Intangible Cultural Heritage includes traditions or living expressions inherited from our ancestors and passed on to our descendants, such as* oral traditions, performing arts, social practices, rituals, festive events, knowledge, and practices concerning nature and the universe *or the* knowledge and skills to produce traditional crafts.. ».

Indigenous knowledge is therefore generated by a particular society within a geographical area and transmitted from one generation to another to provide solutions to the existing problems of that time (Risiro, 2013). Among the most serious problems, the climate change. The indigenous population is vulnerable to the impacts of climate change due to its connection with the environment. Degraded land is more susceptible to climate change and variability impacts such as droughts and floods (Henry et al., 2006), which could cause regression of land productivity. This lack of land productivity has led to declines in the carrying capacity of the land and the emergence of a chronically poor group of people, even during good weather (Glanz, 1999). In the recent past, some 25% or 320 million hectares of the already fragile resource base in African drylands have further been degraded by unsustainable land uses, such as over-grazing, over-cultivation, illegal and excessive fuel wood collection compounded by poorly conceived policies, and ineffective governance (UNEP, 2002). In order to cope with this issue, indigenous population of African drylands has developed a rich arsenal of experiences relating to the environmental challenges. The Middle Draa Valley community has accumulated also enormous indigenous knowledge using a variety of techniques and materials in all field types. This knowledge is threatened. This is due to the fact that this knowledge is not well documented and it stands in danger of being lost as its custodians are passing away (UNEP, 2008). The traditional practices are various and differ by their geographical origin and socio-cultural. Khettaras, pendulum well, composting and human waste recycling are some examples described in this axis.

Khettaras and Pendulum well

Traditional methods are adopted in arid areas such as *Khettaras* (qarez or foggara), present in China and Sothern Morocco. A sloping gallery dug down to the piezometric level of the water table. Water flows from the ground and trickles. The soil on which the water flows must be impermeable to limit infiltration to the aquifer; the vertical wells are used for the maintenance of the Khettara galleries (Figure 3).

The pendulum well is a traditional shallow well used in the oasis of Ktaoua whose drawing system is carried out using a counterweight. The water elevation system consists of 2 uprights, a pole, a rope, and a dlou (l'gnina) (Figure 4). This system is used to extract groundwater for irrigation of oasis farms, and it serves as a watering hole for animals.

Composting and Waste Recycling

The oasis offers a range of materials used by the oasis population including organic wastes that can be used in some clean industries, such as the production of organic fertilizer. This is used in soil fertilization and reduces the deterioration of quality. This practice may also help to adapt to droughts, water scarcity, and poor quality, soil erosion, desertification and sand, soil salinity, and poverty. This technique

Figure 3. Khettaras system used to irrigate the oases in the Pre-Saharan area (dryland). A large landscape showing the expansion of wells and galleries of the Khettaras in Errachidia province, the arid area of Draa Tafilalet region of Morocco, and B, enlargement of picture A. Source: author processing using images from Google Earth

Figure 4. Pendulum well "L'gnina" in Oasis of Ktaoua. a-d, the Pendulum; e-f, the well, and the drainage system. Source: Authors

aims at producing compost using organic residues, vegetable, and fruit residues as a solution to preserve the environment as well as create jobs at the local scale. The production of compost depends on a biological process called aerobic fermentation of palm leaves from leaves and leaves to add the residues of vegetables and fruits. We can summarize the

The production can be summarized as follows:

- Collection and grinding of dry palm residues,
- Collecting vegetable residues from the weekly markets for green vegetables and leaves,
- Crushing organic waste
- Addition of spray water free from salinity,
- The material is buried and covered for one month flipping periodically with the addition of water when needed,
- Leave the mixture for another month and then fill in the bags
- Tracking the quality of the final product.
- Planting various plantations by conducting experiments to compare the final product with other media such as soil wet and then imported organic fertilizer.

Figure 5. Composting experience conducted in the pre-Sahara of Morocco. Source: Authors

Composting has significant impacts on agroecological and socio-economic systems. This technique improves the physicochemical and biological properties of the soil to improve their productivity, promoting sustainable socio-economic development and also creating jobs.

Despite the severe climate, the farmer oasis has significant advantages, particularly plant and animal material quality with good adaptation to difficult conditions (Sabbari & Larbi, 1988):

Figure 6. Types of traditional fertilizers. 1, ash; 2, manure; 3 and 4, solid compost. Source: Authors

- High productivity of certain local products (Cereals: *Fartas, Cheguira, forage alfalfa, Filalie*, etc.).
- Precocity of production (fruits, vegetables, and cereals).
- Exceptional characteristics (high fecundity) of the ovine race (*D'man* race).

Biodiversity, floors cultures practice is a means of adaptation to climate change and better use of natural resources. The soil valuation is done using all possible organic waste (manure and animal manure from sanitary tanks). Legumes are predominant and highly complementary major contributors to the increased fertility of soils, favorable to other cultures, and restrict the use of nitrogen fertilizers, which pollute groundwater (Ferry & Toutain, 1990).

Ecological Sanitation

Ecological sanitation is another popular technique that plays several roles in the preservation of water, soil, biodiversity, and human health in drylands. This ecological method is recommended in regions threatened by water scarcity, land degradation, and desertification (Karmaoui et al., 2021Ecosan). This system is called locally "*Chikh eddar" (*Figure 7) that allows to collection solid human waste and transport it to the fields to expose it to the sun for some time for disinfection.

The reuse of human excreta as a fertilizer is very old since it has been used for more than 2,500 years ago in China, and Nepal (Lüthi et al., 2010). In the Pre-Saharan oases, this traditional practice is called locally "Canib" or "Chikh Eddar". This practice must be manipulated with precaution because of the risks of the present pathogenic species such as bacteria, protozoa, and viruses.

Climate change and droughts influence the vegetation cover and the forage availability which reduces the volume of livestock and then food security. Floods affect most the communities and causes loss of people, livestock, crops, and household property. The indigenous knowledge of disaster prediction is based on

Figure 7. The traditional system of ecological sanitation called locally "Chikh eddar". Source: the authors

the behavior of some animals like birds (including the cry), insects, air, and water temperatures, clouds… The birds make more nests before a wet season and fewer nests in case of a dry season and migrate to new nesting places. Other species are frogs and toads. The absence of frogs and toads indicated the coming of the dry season. Other survival practices included the observation of the wind, temperatures, and clouds forms. Based on these observations, the local population starts preparing for the planting season or preparing to adapt to floods.

Appropriate Equipment and working tools in the Oasis Farm

Several working tools were observed in the pre-Saharan area suitable for work (clearing, cultivation, harvesting, and transportation (Figure 8) such as Tamaskart, Menjel, Atla, Hajamya, etc (Figure 9).

Figure 9 provides the main tools used in the agricultural field.

At the farm, the tools are manufactured manually and with local material.

Figure 8. Farm activities work including harvesting and transportation. Source: the authors

Figure 9. The main tools used by traditional communities in the agrarian field. 1, El'Fes; 2, Tamaskert ; 3, El'atla ; 4, Manjel ; 5, Makyat ; 6, Outed ; 7, Gadoum ; 8, Makhyat. Source: the authors

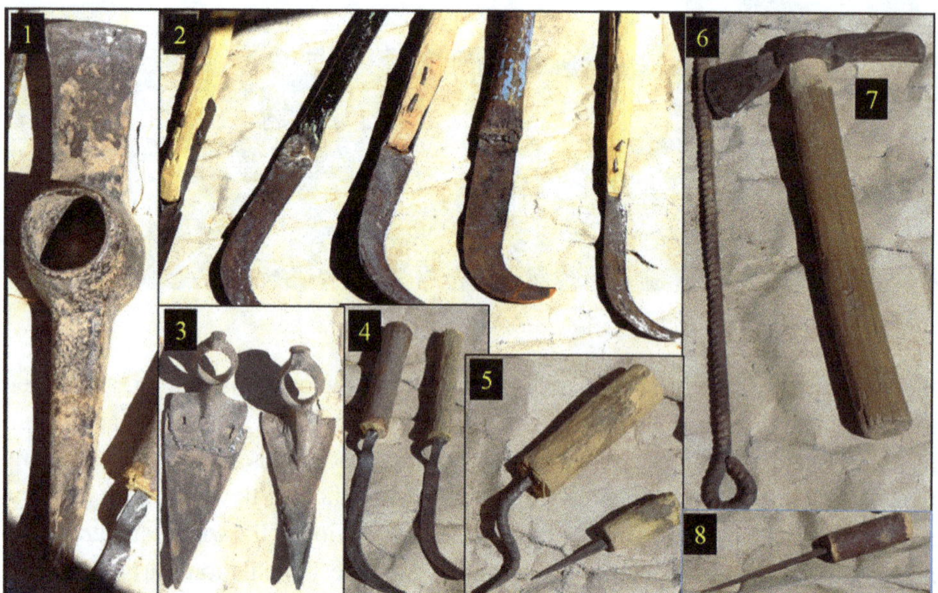

Many of the traditional practices and techniques were harmonious with the environment. The most important was and are for the present the mixed cropping or intercropping cereals with alfalfa and with vegetables. Several other practices allow optimizing land production under harsh environmental conditions:

Figure 10. Construction materials. Source: the authors

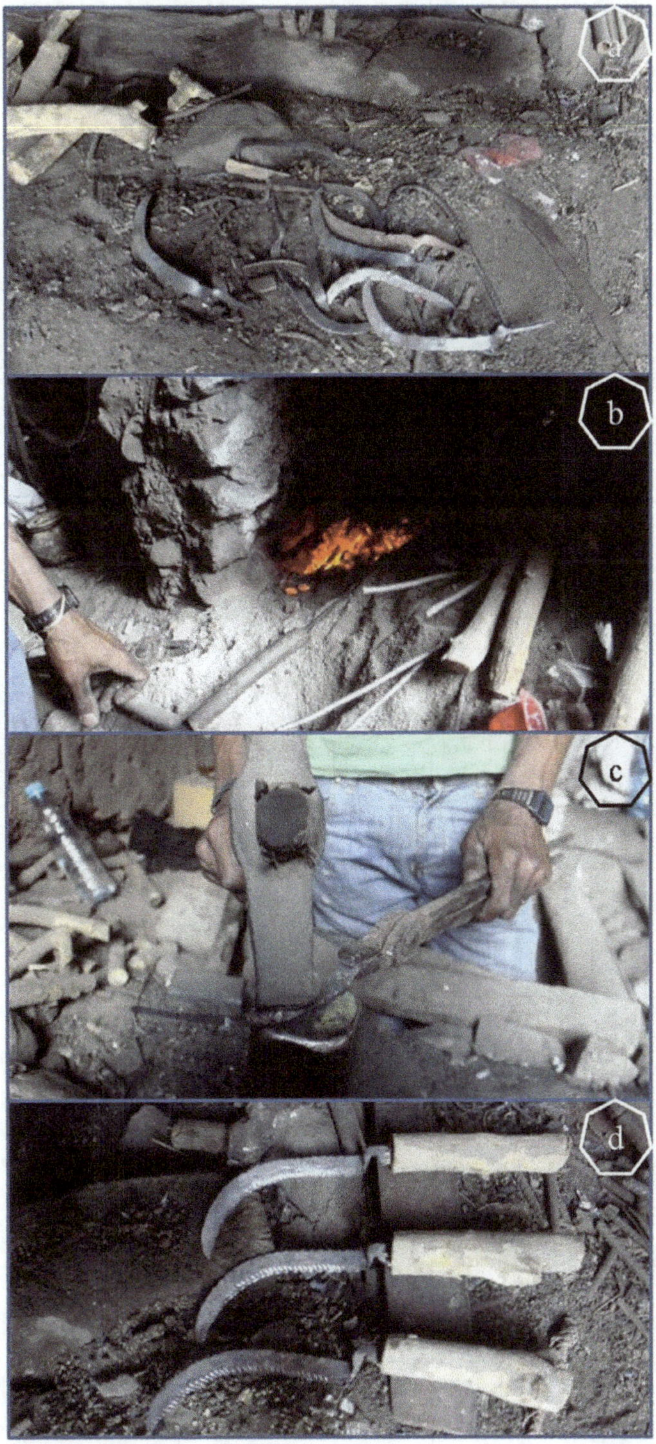

Table 1. Strengths and Weaknesses of Techniques /practices for optimizing farming in era of climate change

Technique /practice	Description	Strengths/advantages	Weaknesses	Recommendations/ Reference
The Buried Diffuser	An innovate method for underground irrigation characterized by absence of loss by evaporation and use less water compared to the drip technique	§ Reduces losses due to evaporation § Used for different crops § Better than drip irrigation § Saves irrigation water, fertilizers, and energy § Increased productivity § Easy installation and use	§ Strange Techniques § Expensive and technical § Risk of clogging	§ Training and sensitization of farmers § At school
Magnetic water irrigation	The Magnetizer is an innovative physical water treatment technique used for physical treatment of water avoiding the formation of deposits of limescale, salt, and corrosion	§ Easy installation § Water softener § Improve energy qualities and oxygen absorption capacity § The device does not require maintenance	The effectiveness is not confirmed scientifically/	For more information about the technique: Taimourya et al., 2015
The polyter hydro-retentive	The polyter is a revolutionary technique, a water-retentive, invented by French horticulturist Philippe Ouaki Di Giorno. Formed by grains that can replace water.	• Innovative and high performance. • Hydro-retentive, • Fertilizer, • Naturally degradable		
Vermicompost or lombricompost	It is a compost using earthworms through their droppings. Transforming domestic organic waste or manure by worms.	§ More rapid, break down organic wastes more quickly § Scalable and easy to do § Product is rich in nutrients (potassium, phosphorus, nitrogen) § The population of worms is growing rapidly § Aeration of the wastes	§ Attraction of fruit flies and presence of pathogens § Defined temperature and selective waste	
Biocompost Organic waste composting	The compost is produced using domestic solid waste, animal manure, and farm waste. It is a natural organic fertilizer that support and improve the structure of soil of farms or green areas of cities.	§ Technically easy to produce § Organic and natural § Recycling solid wastes § An important alternative of chemical fertilizers § Improve of the soil quality	Long time to produce Large quantities of solid wastes to produce small quantities	See the the work of: Bonoli et al., 2019
Smart water management	The use of digital technology and intelligent systems to improve sustainability and reduce costs. The use of smart meters, sensors, remote monitoring and control, data recording and acquisition, geographic information systems, and remote sensing technologies.	§ Sense and measure § Interconnections and instrumentation of water systems (supply, treatments) § Detect and monitor the change § Quick analyze of a situation, and fast responses, § Save money	§ Expensive and complicated technology § Experts and specialized engineers	

continues on following page

Table 1. Continued

Technique /practice	Description	Strengths/advantages	Weaknesses	Recommendations/ Reference
Hydroponic	The technique allows the growing of plants without soil using aqueous nutrients, with their roots supported by a media including gravel. A monitoring of the some variables such as pH are crucial to success the productivity.	§ Reduced area, saves of water, pH and nutrients control, § Rapid growth, § Yield	§ Technicity, § system failure, § Relatively expensive to install the system § Disease spread	
Sprouting *of* cereal	Fodder production Category: Sprouted grains are thought as having exceptional nutritive capacity (protein and vitamin) for dairy cattle	§ Immunity and growth § Good energy/protein balance, § Reduction of seed consumption § Decrease the risk of acidosis § More minerals and vitamins § Cheaper material for making sprouts.	§ Risk of mold,	
Heat tolerant varieties	Many crops are sensitive to high heat during different growth stages, which influences both the yield and quality. Researchers and scholars are developing new Heat Tolerant Varieties to combat the heath or thermal stress.	§ Support food security § Adapt to climate change and drought § Costs and time consuming	§ Health problems associated to genetically modified organism § Sometimes, bigger size and poor nutrients § Technicity	
Ladybugs (beetles or ladybird beetles)	Is a very beneficial group of beetles, insects widely used in biological control because they are considered predators of aphids and plant pests. Breeding can be done by purchasing ladybug larvae, a terrarium, and breeding kits and preparing a growth medium.	§ Beneficial predatory insects used frequently in biological control § Vorace: devours hundreds of aphids every day.	§ In absence of pests, ladybugs attack some fruits with little damage, § Fly away from the farms where they are released.	
Black Soldier Fly (BSF)	Category: Anima food production The larva of the insect is rich in proteins interesting alternative food for different livestock sectors such as chickens and fish.	§ Larvae rich in protein and fat. § High nutritional quality for animals § The larvae neutralize pathogens that transmit disease, which protects animals and humans. § Inexpensive and simple breeding § Break down organic waste and produce a kind of rich compost.	§ Hygiene and unknown risks.	

Source: the authors

- Conserving soil fertility (symbiotic relationships such as nitrogen fixation);
- Minimizes the failure through hazards;
- Optimize the use of soil nutrients and gives high yields;
- Help keeping livestock (improve fertility by manure);
- Diversify crops;
- Resistance to pests and diseases

INVENTORY OF THE TECHNIQUES USED TO OPTIMIZE THE FARM PRODUCTIVITY

An increasing interest in techniques to optimize farming productivity in the context of climate change is recorded in the literature. Several practices were classified into three main components water, land, and energy (Table 1). Certainly, these techniques have advantages but also are associated with weaknesses. Each method or practice is explored as much as possible in light of its advantages including reliability, feasibility, energy consumption, cost, security, utility and productivity, transferability, and ease of use.

DISCUSSION

This chapter aims to review affordable methods, techniques, and practices to optimize farming productivity in arid areas in the context of climate change. In these environments, the responses to climate change must be focused on the reduction of impacts and risks of climate change, vulnerability, and the adoption of mitigation and adaptation strategies. The population of these areas has always been able to adapt to climate change, and also to anything that causes socioeconomic consequences. These areas offer an important example of areas to apply adaptation. Adaptations are both individually and collectively (Benoît, 2008).

Solutions to adapt vary in terms of the type of measures, cost, time, and effort of implementation. Several practices were found in the literature applied in the agriculture sector in physical components (water, soil, and energy), or biological component (the use of biological material including animals). The physical component grouped techniques such as the conversion of agricultural land to fallow land that help reduce the overuse of water (Pourmohamad et al., 2019) and the use of rainwater harvesting to cope with water scarcity in arid regions with the challenge of public acceptance and motivation to use it (Liu et al., 2022). The dew collection is another technique for harvesting water while buried diffusers can be used to both save irrigation water and maintain soil quality. In arid areas, the soil quality including soil salinity is a critical issue, which suggests using of techniques such as magnetic water irrigation. This method was used to fight against the salinity of water, which supports plant growth with recycled water. Among the effects of water salinity, land degradation and desertification can emerge. The fight against desertification is the most common strategy used in these areas that include physical and biological stabilizing systems. The use of farmland shelterbelts protects with a minimum of water and land resources (Li et al., 2020). In terms of energy, many techniques were used in arid areas including the solar and eolian energy systems and pumped hydroelectric. These techniques may be adopted to optimize farming productivity in the context of climate change. These recent techniques can be supported by traditional technologies to protect soil and water resources such as Khattaras, composting, and waste recycling.

It is therefore necessary to diversify the means of living, especially in the event of a bad harvest.

The agriculture sector is marked by growing challenges such as droughts soil and water degradation. The increasing associated issues are worsened by years of a continuously increasing population. Water and soil ecosystem services are threatened by scarcity and salinity. Although arid areas have a significant traditional potential for adaptation to harsh environments, the technologies are not well-conserved by the current generation. In Morocco, many studies and projects were implemented to incorporate these techniques. An inventory of the most relevant techniques used to optimize sustainable farm productivity was performed. In case of water scarcity, and unfavorable climatic and soil conditions, these solutions are important. For the replication of these techniques, it is necessary to evaluate the relevance of a technique through the study of the investment/production ratio.

CONCLUSION

In arid areas, scarcity of resources is considered a consequence of the conjugated effect of CC and human pressure. The shortage of resources requires sustainable management including the use of efficient methods and techniques to adapt to CC in many fields including the agriculture sector. In this context, this chapter presents a set of affordable methods and practices to optimize farming productivity in arid areas. The presented methods should be used taking into account the local socio-economic and ecological characteristics. The study suggests for policymakers accept and replicate these methods to address the challenges of CC in arid areas.

REFERENCES

Ayinu, Y. T., Ayal, D. Y., Zeleke, T. T., & Beketie, K. T. (2022). Impact of climate variability on household food security in Godere District, Gambella Region, Ethiopia. *Climate Services, 27*, 100307. doi:10.1016/j.cliser.2022.100307

Benoît, É. (2008). Les changements climatiques: vulnérabilité, impacts et adaptation dans le monde de la médecine traditionnelle au Burkina Faso. *VertigO-la revue électronique en sciences de l'environnement, 8*(1).

Bonoli, A., Zanni, S., & Awere, E. (2019). Organic waste composting and sustainability in low-income communities in Palestine: Lessons from a pilot project in the village of Al Jalameh, Jenin. *International Journal of Recycling of Organic Waste in Agriculture, 8*(3), 253–262. doi:10.100740093-019-0264-8

Gasmi, I., Aljoumani, B., Sànchez-Espigares, J. A., Mechergui, M., & Moussa, M. (2019). A linear mixed effect (LME) model for soil nutrients and soil salinity changes based on two localized irrigation techniques (drip irrigation and buried diffuser). *Arabian Journal of Geosciences, 12*(11), 1–12. doi:10.100712517-019-4400-6

Hilal, M. H., & Hilal, M. M. (2000). Application of magnetic technologies in desert agriculture. I-Seed germination and seedling emergence of some crops in a saline calcareous soil. *Egyptian Journal of Soil Science, 40*(3), 413–422.

Karmaoui, A., El Jaafari, S., Chaachouay, H., & Hajji, L. (2021). The socio-ecological system of the pre-Sahara zone of Morocco: A conceptual framework to analyse the impact of drought and desertification. *GeoJournal*, 1–14. doi:10.100710708-021-10546-8

Koch, M., & Missimer, T. M. (2016). Water resources assessment and management in drylands. *Water (Basel), 8*(6), 239. doi:10.3390/w8060239

Leal Filho, W., Totin, E., Franke, J. A., Andrew, S. M., Abubakar, I. R., Azadi, H., & Team, T. G. A. M. I. (2022). Understanding responses to climate-related water scarcity in Africa. *The Science of the Total Environment, 806*, 150420. doi:10.1016/j.scitotenv.2021.150420 PMID:34571220

Lee, R. H., Navarro-Navarro, L. A., Ley, A. L., Hartfield, K., Tolleson, D. R., & Scott, C. A. (2021). Spatio-temporal dynamics of climate change, land degradation, and water insecurity in an arid rangeland: The Río San Miguel watershed, Sonora, Mexico. *Journal of Arid Environments, 193*, 104539. doi:10.1016/j.jaridenv.2021.104539

Li, X., Liu, L., Xie, J., Wang, Z., Yang, S., Zhang, Z., Qi, S., & Li, Y. (2020). Optimizing the quantity and spatial patterns of farmland shelter forests increases cotton productivity in arid lands. *Agriculture, Ecosystems & Environment, 292*, 106832. doi:10.1016/j.agee.2020.106832

Liu, Y., Li, G., Zeng, P., Zhang, X., Tian, T., Feng, H., & Che, Y. (2022). Challenge of rainwater harvesting in Shanghai, China: A public psychological perspective. *Journal of Environmental Management, 318*, 115584. doi:10.1016/j.jenvman.2022.115584 PMID:35753131

Lüthi, C., Panasar, A., Schütze, T., Norström, A., McConville, J., & Saywell, D. (2010). Sustainable Sanitation in Cities: A Framework for Action. *Sustainable Sanitation Alliance.* https://publications.lib.chalmers.se/records/fulltext/192469/local_192469.pdf

Mattingly, D. J., Bokbot, Y., Sterry, M., Cuénod, A., Fenwick, C., Gatto, M. C., & Mugnai, N. (2017). Long-term history in a Moroccan oasis zone: The Middle Draa Project 2015. *Journal of African Archaeology, 15*(2), 141–172. doi:10.1163/21915784-12340009

Mirzabaev, A., Wu, J., Evans, J., Garcia-Oliva, F., Hussein, I. A. G., Iqbal, M. H., & Tena, F. (2019). Desertification. In Climate Change and Land: an IPCC special report on climate change, desertification, land degradation, sustainable land management, food security, and greenhouse gas fluxes in terrestrial ecosystems.

Moumane, A., Al Karkouri, J., Benmansour, A., El Ghazali, F. E., Fico, J., Karmaoui, A., & Batchi, M. (2022). Monitoring long-term land use, land cover change, and desertification in the Ternata oasis, Middle Draa Valley, Morocco. *Remote Sensing Applications: Society and Environment, 100745*, 100745. doi:10.1016/j.rsase.2022.100745

Moumane, A.; El Ghazali, F.E.; Al Karkouri, J.; Delorme, J.; Batchi, M.; Chafiki, D.; & Karmaoui, A. (2021). Monitoring spatiotemporal variation of groundwater level and salinity under land use change using integrated field measurements, GIS, geostatistical, and remote-sensing approach: case study of the Feija aquifer, Middle Draa watershed, Moroccan Sahara. *Environ Monit Assess, 193*, 769. doi:10.1007/s10661-021-09581-2

Nassar, Y. F., Abdunnabi, M. J., Sbeta, M. N., Hafez, A. A., Amer, K. A., Ahmed, A. Y., & Belgasim, B. (2021). Dynamic analysis and sizing optimization of a pumped hydroelectric storage-integrated hybrid PV/Wind system: A case study. *Energy Conversion and Management, 229*, 113744. doi:10.1016/j.enconman.2020.113744

Pourmohamad, Y., Alizadeh, A., Baygi, M. M., Gebremichael, M., Ziaei, A. N., & Bannayan, M. (2019). Optimizing cropping area by proposing a combined water-energy productivity function for Neyshabur Basin, Iran. *Agricultural Water Management*, *217*, 131–140. doi:10.1016/j.agwat.2019.02.040

Shannak, S. (2022). Optimizing dynamics of water-energy-food nexus in a desert climate. *Energy Policy*, *164*(C), 112884. doi:10.1016/j.enpol.2022.112884

Sivanappan, R. K. (1995). Soil and water management in the dry lands of India. *Land Use Policy*, *12*(2), 165–175. doi:10.1016/0264-8377(94)00001-G

Surendran, U., Sandeep, O., & Joseph, E. J. (2016). impacts of magnetic treatment of irrigation water on plant, water and soil characteristics. *Agricultural Water Management*, *178*, 21–29. doi:10.1016/j.agwat.2016.08.016

Taimourya, H., Oussible, M., Bourarach, E., El Harif, A., Hassanain, N., Masmoudi, L., & BAAMAL, L. (2015). Évaluation de la productivité de la pomme de terre (Solanum Tuberosum L.), sous l'effet de l'irrigation avec une eau traitée magnétiquement dans la région de Chaouia (Maroc). *Revue Marocaine des Sciences Agronomiques et Vétérinaires, 3*(3), 30-36. https://agromaroc.com/index.php/Actes_IAVH2/article/view/406

Yin, X., Feng, Q., Li, Y., Liu, W., Zhu, M., Xu, G., Zheng, X., & Sindikubwabo, C. (2021). Induced soil degradation risks and plant responses by salinity and sodicity in intensive irrigated agroecosystems of seasonally-frozen arid regions. *Journal of Hydrology (Amsterdam)*, *603*, 127036. doi:10.1016/j.jhydrol.2021.127036

Zemni, N., Slama, F., Bouksila, F., & Bouhlila, R. (2022). Simulating and monitoring water flow, salinity distribution and yield production under buried diffuser irrigation for date palm tree in Saharan Jemna oasis (North Africa). *Agriculture, Ecosystems & Environment*, *325*, 107772. doi:10.1016/j.agee.2021.107772

Zhou, B., Yang, L., Chen, X., Ye, S., Peng, Y., & Liang, C. (2021). Effect of magnetic water irrigation on the improvement of salinized soil and cotton growth in Xinjiang. *Agricultural Water Management*, *248*, 106784. doi:10.1016/j.agwat.2021.106784

ENDNOTE

1 http://www.unesco.org/culture/ich/index.php?lg=en&pg=00003

Chapter 7
Population Projections, Food Consumption, and Agricultural Production are Used to Optimize Agriculture Under Climatic Constraints

Nassima AMIRI
Faculty of Sciences Semlalia, Cadi Ayyad University, Morocco

Mohammed Yacoubi
https://orcid.org/0000-0001-6579-1884
Cadi Ayyad University, Morocco

Mohammed Messouli
Cadi Ayyad University, Morocco

ABSTRACT

Climate change is expected to have a significant impact on agricultural production both locally and globally. High temperatures and irregular precipitation projected by the IPCC could lead to a decline in agricultural production, which would require significant changes in agricultural practices. A concomitant factor to climate change will be the increase in population and its distribution and level of consumption, which also influences agricultural production strategies, the conversion of agricultural land, and the type of irrigation. Determining the impacts of these forcings on land use will depend on our understanding of past changes in the agricultural sector (cereal) and how future climate change will occur. It is proposed to develop an innovative methodology that will integrate agricultural production, socio-economic statistics, demographic trends, and patterns of agricultural land use caused by climate change and human development.

DOI: 10.4018/978-1-6684-4649-2.ch007

INTRODUCTION

Food has the power to give us vital energy and nutrients. It helps us stay healthy. Cereals, vegetables and fruits provide this vital energy for human survival and daily activities and are also the basis of the food industry. These agricultural products come from arable land. To have these products, we need fertile fields, water and all favorable climatic conditions for a profitable agricultural yield. Unfortunately, climate change has worsened agricultural productivity through adverse conditions reflected in erratic rainfall and high temperatures that have lowered yields to the limit. Under projected emission scenarios, the Intergovernmental Panel on Climate Change (IPCC) predicts global warming at least until the middle of the century.

The global food problem has become a significant issue in the long run, with substantial environmental implications affecting agriculture. Therefore in Morocco, since 1956, priority has been given to agricultural development (Faysse, 2015), and agriculture has become the country's leading economic sector, generating about 14% of gross domestic product (GDPGDP) (MAPMDREF, 2014).

The cereal sector is the basis of human food in the world and Morocco. Its availability and inadequacy are the two factors that influence human food security. Furthermore, most arable land depends on rain, and according to the IPCC report (IPCC, 2007), Morocco is among the countries vulnerable to climate change, as a result, it is directly confronted with the impacts of climate change. So all its conditions would lead to a decline in ceramic agricultural production in Morocco.

Agricultural yield is always dependent on climatic conditions, especially in the dry zones, a rainy year leads to satisfactory production, and a dry year leads to an increased decrease in yield.

Agronomic drought, primarily caused by a water deficit, is a significant natural hazard for agricultural rainfall systems (Kumar, 1998; Ribeiro et al., 2019), especially in countries dependent on rainfed crops, the situation will become more serious; according to scientific studies, it has been shown that in the semi-arid regions of the southern Mediterranean, 80% of the arable land is rainfed, there will be an increase in the vulnerability of agriculture to drought events (Schilling et al., 2012; Tigkas and Tsakiris, 2015).

Flooding is also one of the causes of yield loss, and other studies have shown that increased droughts and floods lead to agricultural production (Battisti and Naylor, 2009).

Unfortunately, irrigated land occupies only 15% of cultivated land (Maazouz, 2016), and in the case of rainfed land, the non-irrigated areas are the most cultivated. However, most of them are less productive and severely affected by drought during the peak of the summer (MAPMDREF, 2014).

For all these reasons and constraints, this study proposes a simulation of the projections of changes in agricultural land use and covers environmental and socio-economic linkages over the next 20 years.

Understanding the links between these various factors and their implications for agricultural production is essential to determine the likelihood of significant future trends in land use variation concerning agriculture.

Through this chapter, we will model these links between the environment and the socio-economic level and project their consequences on the current and potential land cover and the change in land use over the following years.

BACKGROUND

Agricultural yield is always dependent on climatic conditions, especially in the dry zones, a rainy year leads to satisfactory production, and a dry year leads to an increased decrease in yield.

In countries vulnerable to climate change and dependent on rainfed crops, the situation will become more serious (Kumar, 1998; Ribeiro et al., 2019, Schilling et al., 2012; Tigkas and Tsakiris, 2015).

At the beginning of our era, the world population was 188 million, however, thanks to the industrial revolution and the development in the health field, a significant change has occurred (Frόna et al., 2019). This demographic expansion is expected to continue for several decades before peaking at nearly 10 billion people in the 21st century (Boretti and Rosa, 2019; Bongaarts, 2009). Today, the world's population is estimated to have reached 7.8 billion by March 2020 and is expected to reach 8.5 billion by 2030 (Boretti and Rosa, 2019).

This population, which is constantly growing, needs to be fed, but since agriculture is negatively impacted by climate change, this will pose a significant challenge for feeding this global population.

In line with global warming trends, the observations of the National Directorate of Meteorology of Morocco show an increase in temperatures, less precipitation and an increase in drought, widening the gap between water supply and demand. Average temperatures are expected to rise between 2 and 5°C by the end of this century, in contrast, precipitation is expected to fall by 20 to 30% (UNDP, 2009), so the Moroccan population is also concerned about being part of the challenge of ensuring food security.

The work presented in this chapter clarifies some aspects of food security and what initiatives are needed to address this constraint.

MAIN FOCUS OF THE CHAPTER[1]

The main objective of this study is to determine the impact of climate change on land use, and It will depend on our understanding of past changes in the wheat sector in Morocco and how future climate change, irrigation, population size and distribution affect cereal production and its relationship to the expansion of agricultural arable land.

Issues, Controversies, Problems

Morocco is declared among the countries vulnerable to climate change, as a result, it is directly confronted with the impacts of climate change, so all its conditions would lead to a decline in ceramic agricultural production in Morocco.

Interventions in the agriculture sector have already assessed Morocco's potential vulnerability to climate change at the regional level. Is it necessary to prioritize certain regions since they will be the most deteriorated, and there will be an opportunity to grow cereals only in certain regions which will be able to satisfy the Moroccan population or to grow varieties appropriate to the field conditions?

Despite the constraints of climate change, the work given in this chapter will clarify how we will be able to meet cereal needs for all of Morocco and even in the future.

MATERIALS AND METHODS

Study Area

Morocco is located in northwest Africa, overlooking the Mediterranean Sea to the north and the Atlantic Ocean to the west, between latitudes 21°N to 36°N and longitudes 1°W to 17°W. It is separated from the European continent only by the Strait of Gibraltar.

Morocco has an administrative division of 12 regions which are: (Figure 1)

Region 1: Tanger-Tetouan-Al Hoceïma
Region 2: Oriental
Region 3: Fes-Meknes
Region 4: Rabat-Sale-Kenitra
Region 5: Beni Mellal-Khenifra
Region 6: Casablanca-Settat
Region 7: Marrakech-Safi
Region 8: Draa-Tafilalet

Region 9: Souss-Massa
Region 10: Guelmim-Oued Noun
Region 11: Laayoune-Sakia El Hamra
Region 12: Dakhla-Oued Ed-Dahab

Figure 1. Map of the 12 administrative regions of Morocco.
Source: (Portail national des collectivités territoriales, 2022)

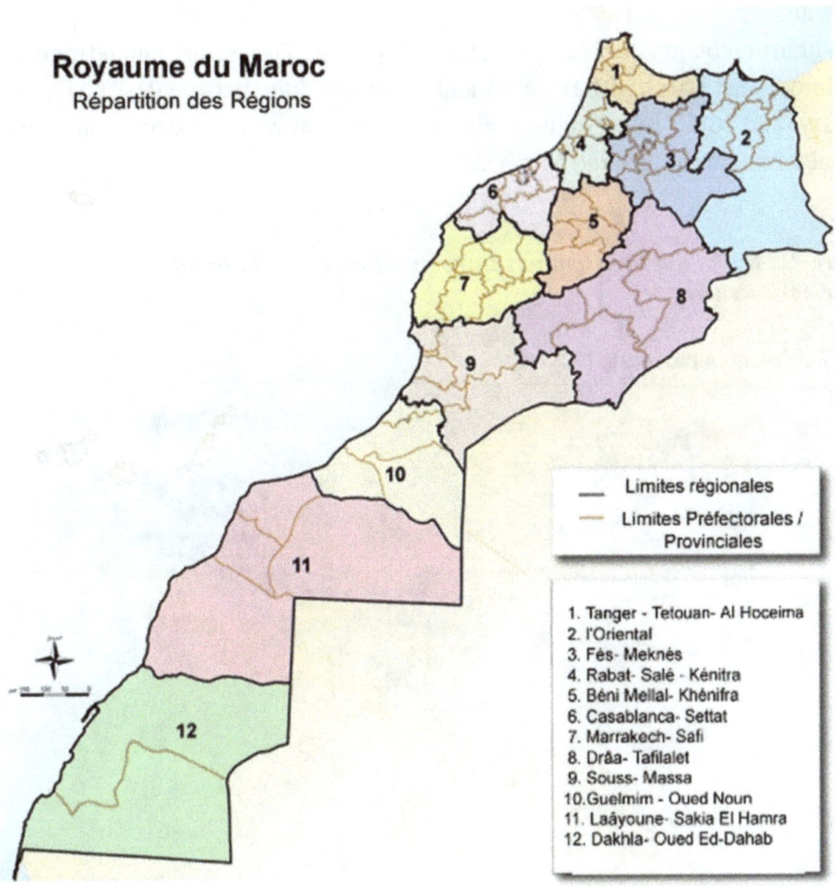

The territory of Morocco covers an area of 710,850 km. It is characterized by well-diversified geography: the plains, high mountain ranges where many massifs exceed 3000 meters (Rif and Atlas ranges) and the desert.

Morocco's climate is very nuanced, Mediterranean in the north, oceanic in the west, continental in the interior and Saharan in the south (Aridoculture), with mild and humid winters and hot and dry summers (World Bank Climate Change Knowledge Portal, 2021).

The Atlantic Ocean strongly influences precipitation and temperatures to the west. Within the country, variations in precipitation and temperatures are primarily regulated by the geographical diversity of the territory. The presence of the sea attenuates temperature variations, tempers the seasons and increases the humidity of the air.

Within the country, the climate varies with altitude. The average annual temperature for Morocco is 17.5°C, with average monthly temperatures between 9.4°C (December, January) and 26°C (July, August). Figure 2 below shows the distribution of average annual temperatures around Morocco.

Figure 2. Map of average temperatures in Morocco by Balaghi.
Source: (Aridoculture)

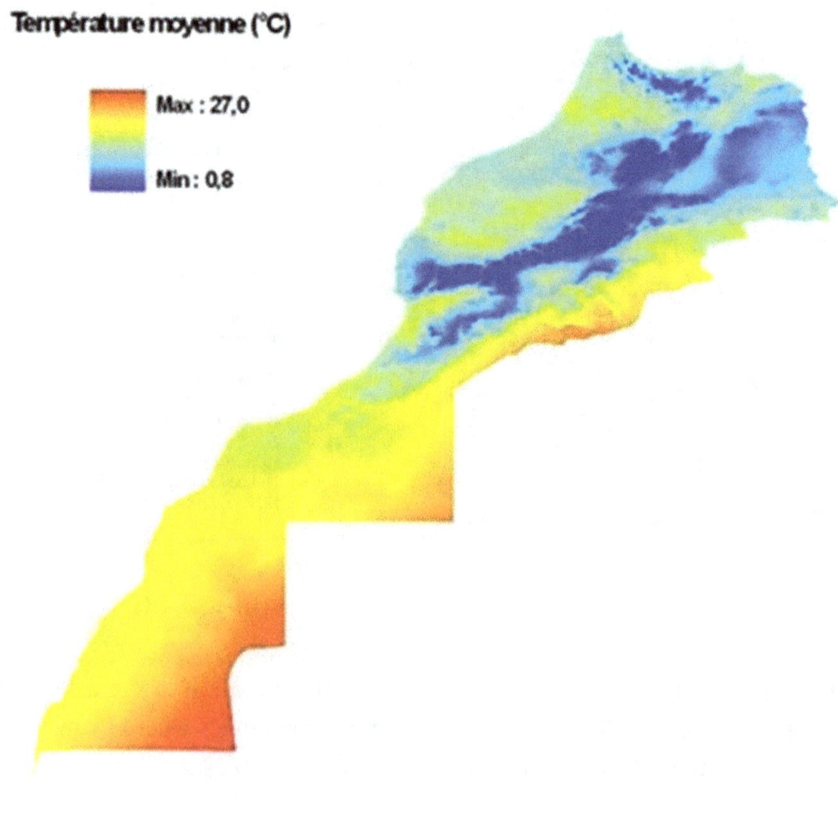

Average annual precipitation is 318.8 mm, the highest occurring from October to April, while meagre precipitation occurs between June and August (World Bank Climate Change Knowledge Portal, 2021). Figure 3 shows the distribution of annual rainfall in Morocco.

Figure 3. Annual rainfall map in Morocco (Evolution of rainfall from 1960 to 2015 in Morocco).
Source: (EL AJHAR et al., 2018)

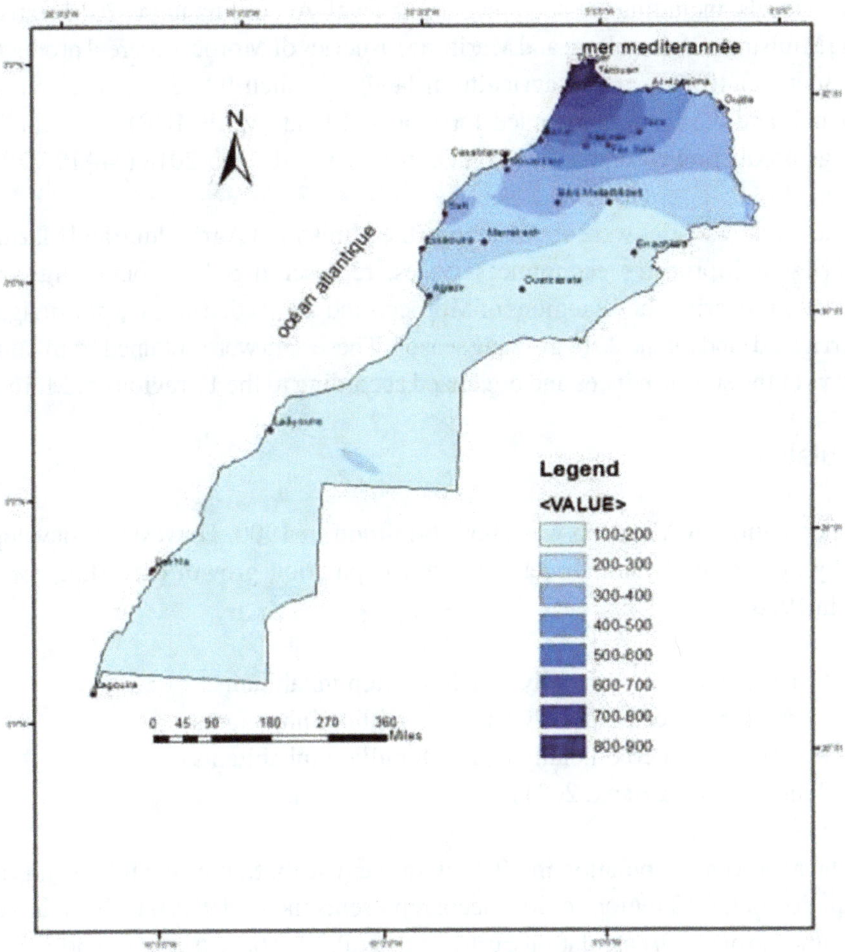

Data

Cereal Production

Cereals are mainly grown in soft wheat, barley, durum wheat and maize. Sorghum and rice are also practised but with marginal importance. Nevertheless, they are a primary source for human consumption and animal feed, which makes it possible to develop cereal feed resources (barley, straw, stubble...)

The cereal sector plays a crucial role in Morocco's agricultural economy at various levels, including the socio-economic level. According to the 2014 statistics of the Ministry of Agriculture and Maritime Fisheries of Morocco, cereal production represents 63.40% of arable agricultural land, of which 92.62% of the area was cropland, and 7.38% was intended for irrigated land (MAPMDREF, 2021). The average production rose to 86 million Qx between 2015 and 2018 (MAPMDREF, 2019).

The cereal statistics were obtained from the Ministry of Agriculture and Maritime Fisheries of Morocco's economic services, representing data containing cereal production covering the 12 regions of Morocco and the seeded areas in non-irrigated and irrigated land for the 2014 growing season. These data were obtained from sample surveys of the sub-provinces and organized according to the 12 regions of Morocco.

Population

The population of Morocco was about 6 million in 1900. However, it developed rapidly during the twentieth century, with population growth exceeding 3% per year in 1950.

- In 1981, Morocco officially had 20 million inhabitants;
- In 2004, Morocco officially had 29.6 million inhabitants;
- In 2014, Morocco officially had 33.8 million inhabitants.
  Source: (World Bank, 2021).

The Moroccan population in 2014, with the distribution of the urban and rural population of the 12 regions of Morocco, represents the statistics on which the rural population projection calculations, urban and total, have been triggered and will also be used for the calculation of actual and projected consumption (HCP)

Cereal Consumption

The HCP carries out consumption surveys on all agricultural products. We are interested in those cereals of the year 2014, the consumption of the urban population was 181 kg/person, and that of the rural was 191 kg/person. These values were used to calculate Moroccan urban consumption, Moroccan rural consumption and total Moroccan consumption (urban and rural).

The fraction of the urban, rural and total population is calculated to estimate the consumption of each region.

Climate Change: Growing Season and Temperature

According to the work of Asseng et al. (Asseng et al., 2015) on crop modelling to assess the threat of climate change, to local and global food production, by predicting wheat grain yield by simulating how crops react to temperature, it is inferred that wheat production falls by 6% for each additional temperature increase°C, this result is used to calculate the decrease in cereal production in Morocco taking into account its temperatures at the scale of its 12 regions.

Based on research with farmers and agricultural engineers in each region based on their experiences and direct contact with the agricultural community, cereal growing seasons are determined based on rainfed and irrigated land.

Can be determined two types of growing periods that depend on semi-early and semi late

- **Rainfed land:**
  The factor responsible for the semi is the rain, the growing season is from October to May.
- **Irrigated land:**
  The factor responsible for the semi is the crop rotation that designates the distribution of crops in the soil, the growing season begins from September or October to May, and since most starts in September, the growing season is chosen for irrigated land from September to May. The following table 1 shows the cereal growing season in Morocco by land type.

Table 1. Morocco's growing season of cereals on rainfed and irrigated land.

		Early Cereal Growing Season	
Land	**Factor Responsible for semi**	**Semi early**	**Harvest after semi early**
Rainfed	Rain	October	May
Irrigated	crop rotation	September or October	May

Then, the average monthly temperatures related to the growing season, which depends on rainfed and irrigated land, are determined according to its growing season.

For 2014, this is the 2013/2014 crop year, so it starts on 1 September 2013 and ends on 31 August 2014, and then identifies the growing season of the 2013/2014 crop year.

- **Rainfed land**
  For rainfed land, It starts in October 2013 and ends in May 2014
- **Irrigated land**
  For irrigated land, it starts in September 2013 and ends in May 2014.

Therefore, the monthly mean temperatures for the growing seasons for 2013/2014, 2023/2024 and 2033/2034 crop years are determined based on the Future RCPRCP Projection 4.5 and the Canadian Earth System Model (CANESM) (MOSAICC Morocco).

Mosaicc is a site developed by the Ministry of Agriculture and Maritime Fisheries, Rural Development and Water and Forests), with technical support from FAO and in partnership with the National Institute of Agricultural Research (INRA) and the Directorate of National Meteorology (DMNDMN), which provides the projected climate data in Morocco at various scales by decades, For the years 2013, 2014, 2023,2024, 2033 and 2034, the monthly mean temperatures have been deduced from the trend line equations of the years 2010, 2020, 2030 and 2040, then the months that determine the rainland and irrigated growing seasons previously explained in Table1.

The following table 2 shows the results of calculating the mean temperatures of the growing seasons.

The study shows that each 1°C increase contributes to the 6% decrease in wheat production. According to Asseng and colleagues (Asseng et al., 2015), the difference in temperature averages related to growing seasons between crop years 2023/2024 and 2013/2014 is first calculated, followed by 2033/2034 and 2013/2014, respectively.

The following equations can be used to determine the decrease in production in each region according to the following equations:

Percentage decrease in production (2023/2024)= 6% × X °C
Percentage decrease in production (2033/2034)= 6% × Y °C

Which:

X °C: is the temperature difference between 2023/2024 and 2013/2014
Y °C: the temperature difference between 2033/2034 and 2013/2014.

Table 2. Projected mean temperatures of the cereal growing seasons in Morocco.

Regions	Temperature related to growing season °C					
	2013/2014		2023/2024		2033/2034	
	Tp_raifed land	Tp_irrigated land	Tp_raifed land	Tp_irrigated land	Tp_raifed land	Tp_irrigated land
Region 1	14,0	14,9	14,4	15,2	14,9	15,6
Region 2	14,6	15,6	14,9	15,9	15,4	16,4
Region 3	14,4	15,4	14,8	15,8	15,4	16,3
Region 4	16,1	16,9	16,4	17,1	16,7	17,5
Region 5	15,2	16,2	15,6	16,6	16,2	17,1
Region 6	16,0	16,8	16,3	17,0	16,7	17,4
Region 7	16,1	16,9	16,5	17,2	17,0	17,7
Region 8	15,5	16,7	16,1	17,2	17,0	18,0
Region 9	16,6	17,3	16,9	17,6	17,4	18,0
Region 10	18,2	18,8	18,5	19,1	18,9	19,4
Region 11	19,7	20,3	20,0	20,6	20,4	20,9
Region 12	21,5	22,0	21,7	22,2	22,1	22,5

After determining the percentage of production, the 2024 and 2034 production is calculated by the following equations:

- **Rainfed Production 2024 (Kg)**

Rainfed Production(2024) = Rainfed Production2014 – (Rainfed Production2014 ×% decrease in production by temperature related to the rainfed growing season)

- **Irrigated Production 2024 (kg)**

Irrigated Production(2024) = irrigated Production2014 – (Irrigated Production2014 ×% decrease in production in 2024 by temperature related to the irrigated growing season)

- **Total Production 2024**

Total Production(2024) = Rainfed Production2024 + Irrigated Production2024

- **Rainfed Production 2034 (Kg)**

Rainfed Production(2034) = Rainfed Production2014 – (Rainfed Production2014 ×% decrease in production by temperature related to the rainfed growing season)

- **Irrigated Production 2034 (kg)**

Irrigated Production(2034) = irrigated Production2014 – (Irrigated Production2014 ×% decrease in production in 2024 by temperature related to the irrigated growing season)

- **Total Production 2034**

Total Production(2034) = Rainfed Production2034 + Irrigated Production2034

Methodology

The modelling discussed in this chapter will be used to identify future projections and their current and potential impacts on land use change over the next 20 years. Our approach aimed at a terrestrial system model linked to the environment at the socio-economic level through feeding population data in urban and rural areas, consumption around Morocco, cereal production and area in rainland and irrigated, taking into account all 12 regions of Morocco.

Scenarios will be studied to determine how variations in climate change factors, socio-economic level and technological development will affect the future of agricultural land use over the next 20 years, which could have significant consequences for human well-being regarding food satisfaction.

The HCP carries out consumption surveys on all agricultural products. We are interested in those cereals of the year 2014, the consumption of the urban population was 181 kg/person, and that of the rural was 191 kg/person, these values were used to calculate Moroccan urban consumption, Moroccan rural consumption and total Moroccan consumption (urban and rural). Then the fraction of the urban, rural and total population is calculated to estimate the consumption of each region.

The growth rate of the Moroccan population is around 1.22% for the urban population, 1.22% for the rural population and 1.22% for the total population (2021 World Population by Country). We use this rate to calculate projections of Morocco's urban, rural and total population.

RESULTS

Moroccan Cereal Production and Population

The results show in figure 4 that the balance of cereal production in 2014 remains positive, despite eight regions with a production deficit, of which two non-productive regions in southern Morocco are R11 and R12, respectively Laayoune-Sakia Hamra and Dakhla-Ouad Deheb. The country has four major cereal production regions, Fes-Meknes (R3), Rabat-Sale-Kenitra (R4), Beni Mellal-Khenifra (R5) and Casablanca-Settat (R6), the size of the deficit was established based on local production and population size.

Figure 4. Total production, consumption and Deficit/gaint.

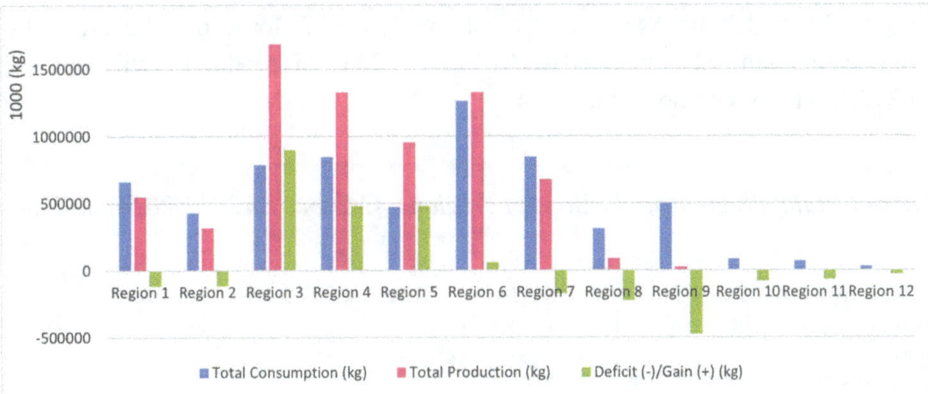

According to the population projection calculations in Figure 5, the Moroccan population changes linearly, which requires more food.

Climate Change by Temperature

The results of the projected decrease in production due to the temperature increase are presented in figures 6 and 7, showing the decrease in kg in the 2024 and 2034 scenarios.

First, cereal production was zero in 2014 in regions 11 and 12 and is also considered zero in 2024 and 2034.

Figure 5. Population growth in Morocco from 2014 to 2034.

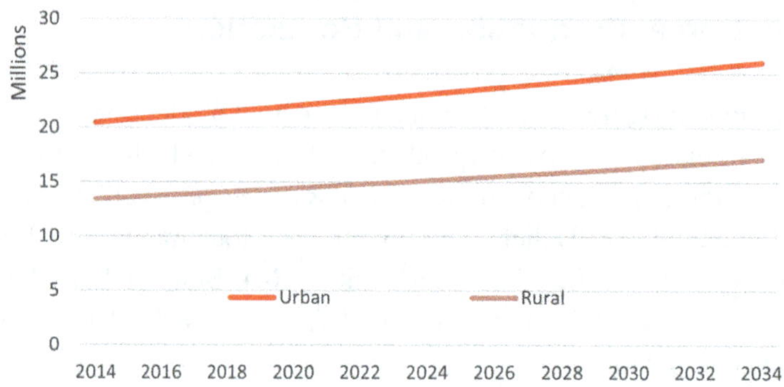

In 2014, cereal production was zero in Regions 11 and 12 and was considered zero in 2024 and 2034. As a result, cereal production in Morocco will decrease by 146,666 tonnes in 2024 compared to 2014, and by 2034 this production will decrease by 372,153 tonnes compared to 2014.

Figure 6. Rainfall cereal production in 12 regions of Morocco in 2014, 2024 and 2034.

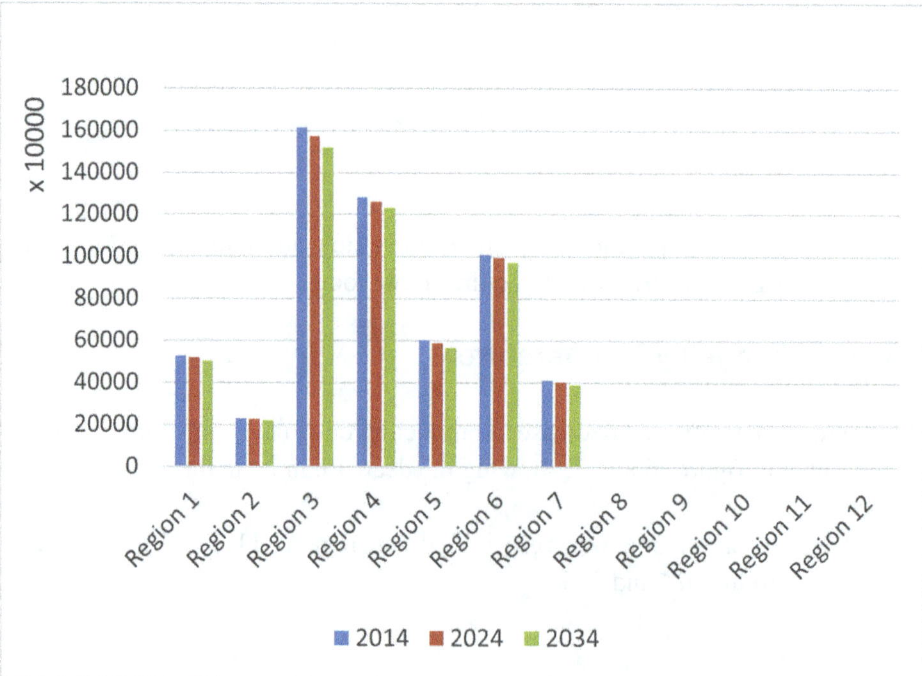

Figure 7. Irrigated cereal production in 12 regions of Morocco in 2014, 2024 and 2034.

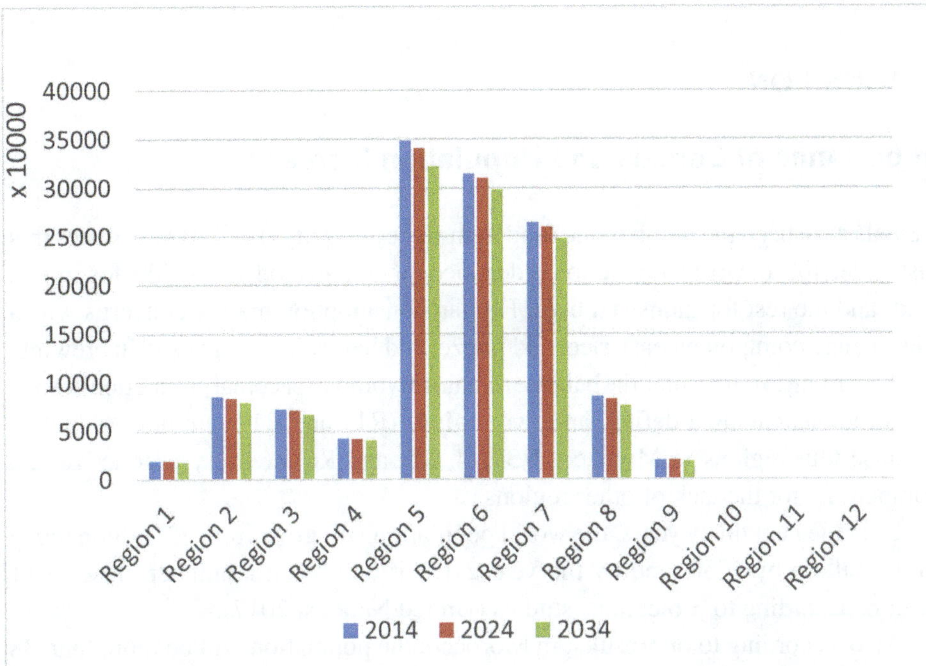

Most of the production comes from this rainfed land, as rainfed production represents 82% and irrigated production 18%. Hence, the cultivation of cereals in Morocco depends first on the climate. According to our study of the increase in temperature according to the two scenarios, 2024 and 2034, this climate constraint worsens the situation by decreasing more and more cereal production.

Suppose we focus on Morocco's three major agricultural regions, R3 (Fes-Meknes), R4 (Rabat-Sale-Kenitra) and R6 (Casablanca-Settat), the overall decreases in cereal production in these regions compared to 2014 will be a decrease in rainfed production of 89,766.71 tonnes in 2024 and 210,756.63 tonnes in 2034. The decrease in irrigated production in these regions will be 14,830.52 tonnes in 2024 and 34,295.57 tonnes in 2034.

Moreover, speaking about the most productive regions in Morocco being R3 (Fes-Meknes), R4 (Rabat-Sale-Kenitra) and R6 (Casablanca-Settat), in 2024, the total decreases in these three regions compared to 2014 in Bour land will be 77658,12 Tons and that of irrigated land will be 17851.48 Tons. Moreover, for 2034 compared to 2014, the total decrease in cereal production in these three major regions in terms of cereal production in dry land will be 181253.53 Tonnes, and that of irrigated land will be 58970.20 Tonnes.

DISCUSSION

Importance of Cereals and Population Increase

Cereals have been the number one food item since antiquity. The FAO estimates that just under 40% of world production is destined for human food, about 50% for animal feed, and the rest for industrial uses. Human consumption mainly concerns wheat (durum and common wheat), rice, and maize, and barley is mainly used in brewing.

According to our results, the balance remains favorable for cereal production despite eight regions having a deficit, and two regions (R11 and R12) are not productive because four regions in Morocco (R3, R4, R5, and R6) are very productive and compensate for the lack of other regions.

Over the next thirty years, the world population will grow from 7.7 billion today to 9.7 billion by 2050, and by the year 2100, it will reach a number close to 11 billion, according to projections studies (United Nations, 2017).

Also, according to our results, in Morocco, the population will be more than 38 million in 2024 and more than 43 million in 2034, and all the countries of the world will undergo this demographic development.

Based on the scenarios obtained from the population, consumption, and cereal production impacted by climate change, analyses have shown that the Moroccan population does not cease to increase over time like any population in the world, which requires its diet to satisfy its nutritional need so as not to fall in cases of cereal shortages, especially in certain more vulnerable regions, what aggravates the situation is the decrease in production due to climate change by treating the case of rising temperatures and the decrease in precipitation.

Morocco must establish to increase the production or import of cereals. It is also to improve eating habits by considering all the social categories. Lowering the annual average for cereals per inhabitant also seems practical, considering all social categories.

Impact of Climate Change on Crops

According to studies, the hot spots due to climate change are the regions of the Mediterranean and Northern Europe, followed by the regions of the northern hemisphere at high latitudes and Central America (Lionello & Scarascia, 2018; Barkhordarian et al., 2013). Other studies show that global warming in the future will occur during the year, but it will be more pronounced in summer than in winter and will be more critical on land than on the seas (Déqué et al., 2012).

Several scientific researchers studied the impact of climatic change on agricultural production (Ludwig and Asseng, 2006). Land use also influences agricultural production, according to studies on Mediterranean regions (Valverde et al., 2015). Other studies have shown that, apart from climatic conditions, local and individual agricultural practices also influence productivity (Gornall et al., 2010). For example, higher temperatures impact agricultural productivity, farm incomes and food security during the growing season (Battisti and Naylor, 2009). Other studies have shown that extreme climatic conditions such as heat waves, storms, drought, salinity and flooding influence cereal production (Farooq et al., 2015; Farooq et al., 2014; Dyer et al., 2014).

To date, considerable efforts have been made by scientists to understand better the causes and effects of climate change in recent years on a global scale (Tol, 2010). Climatic conditions mainly influence the agricultural use of water is the essential prerequisite for sustainable agriculture. The problem of water scarcity is even more significant in arid and semi-arid regions (Wrachien, 2013; Wrachien, 2014).

This study deals with climate change based on temperature related to the growing season of plants. 2 scenarios are developed based on 2014, which is the scenario of 2024 and another of 2034. The results show that cereal production will decrease mainly in rainfed areas.

In line with global warming trends, the observations of the National Directorate of Meteorology of Morocco show an increase in temperatures, less precipitation and an increase in drought. Average temperatures are expected to rise between 2 and 5°C by the end of this century. In contrast, precipitation is expected to fall by 20 to 30% (UNDP, 2009). Other studies on Morocco of the impact of climate change on agriculture carried out by INRA (National Institute of Agricultural Research) within the framework of the AGRIMAP (Adaptation to climate change in agriculture in the Maghreb), the results showed that compared to 2010, temperatures will increase by 1.3°C (0.2°C) by 2050 and by 2.3°C (0.9°C) by 2090.

Amiri and collaborators (Amiri et al., 2021) have shown that in Morocco, cereal production will decrease due to climate change by treating the case of rising temperatures and decreasing rainfall by developing scenarios taking into account rainfall and irrigated land. The decrease in rainfed production represents 87%, and that of irrigated production represents 13% because the land in Morocco is the most dominant. Moreover, other research on Morocco has shown that, on average, there will 10% is the percentage decrease in wheat grain and sugar yields in 2050 due to climate change (Asseng et al., 2015).

Variations will influence the growing season of crops in temperature and precipitation. In this context, scientific research has shown that this period will decrease by 30 days in 2050 and 90 days in 2090 compared to 2010 (Schilling et al., 2020).

In 2050, the growth period will be from November to April, and for 2090, it will last from January to March. This growing period reduction will have significant consequences on crop yield, and its determination is essential to predict crop yields and know how to react to ensure food security (Balaghi et al., 2008).

Climate change does not always have a detrimental effect on agriculture, as other regions like Canada, Europe, northern China and northern Russia are increasing their agricultural production because the effects of climate change on carbon fertilization have been taken into account (Cline, 2007).

Climate change may affect crops indirectly through effects on pests and diseases (Gornall et al., 2010). Some pests respond positively to a rise in CO_2, such as aphids (Newman, 2004), as their winter mortality is reduced by increasing temperature (Zhou et al., 1995). Other studies in this context have shown that sizeable daily temperature amplitudes positively influence insect development, survival, longevity and fecundity (*Sitobion avenae)* (Xing et al., 2021).

SOLUTIONS AND RECOMMENDATIONS

It is, therefore, necessary to increase cereal production under climatic constraints and demographic development. Consequently, the right and brilliant solution would be straining the agricultural sector, which plays a crucial role in the country's development. It is necessary to consider water-saving irrigation techniques such as drip. It is compulsory to sow agricultural areas with the most productive wheat varieties. In addition, more attention should be paid to small and medium-sized farms to integrate them into modern and industrial agriculture by providing them with the necessary support, particularly in areas with a cereal deficit.

FUTURE RESEARCH DIRECTIONS

Research on the impact of climate change on cereals as primary human food production and all elements of food security must be conducted and encouraged. This research can be contributed by yield forecasting modelling to identify and streamline food security strategies, especially in countries most vulnerable to climate change. Genetic improvement is also relevant, provided that it is subject to health and environmental protection policies.

CONCLUSION

Modelling and future projections of agricultural production, human development and consumption patterns are practical solutions that can be used to consider new actions for human well-being. According to the study, the balance of cereal production in Morocco remains positive, and the most productive regions are Fes-Meknes (R3), Rabat-Sale-Kenitra (R4), Beni Mellal-Khenifra (R5) and Casablanca-Settat (R6).

Under the climate constraint and demographic development, the Moroccan economy will also have to increase its production or import cereals.

The improvement of agricultural production may be achieved by actions being:

- Development in the irrigated sector plays a crucial role in the country's agricultural development case; therefore, it is necessary to consider water-saving irrigation techniques such as drip;
- Giving more attention to small and medium-sized farms is also a good initiative by integrating them into modern and industrial agriculture by providing them with the necessary support, especially those with a cereal deficit;
- Seeding agricultural areas with the most productive wheat varieties taking into account genetic engineering, which also seems a relevant solution;
- We must consider health and environmental protection policy promote healthier and more sustainable consumption patterns;
- We must educate the public through a healthy and balanced diet that meets the needs of all social groups;
- Research on the impact of climate change on cereals as primary human food production and all elements of food security must be conducted and encouraged.

ACKNOWLEDGMENT

This research was supported by the Laboratory of Hydrobiology Ecology and Ecotoxicology and Global Changes, Faculty of Sciences Semlalia, Cadi Ayyad University.

FUNDING: This research received no specific grant from any funding agency in the public, commercial, or not-for-profit sectors.

REFERENCES

AgriMaroc. (2014). *L' agriculture marocaine en chiffres* 2014. AgriMaroc.ma. https://www.agrimaroc.ma/lagriculture-marocaine-en-chiffres-2014/

Amiri, N., Lahlali, R., Amiri, S. E. L., Jarroudi, M., Khebiza, M. Y., & Messouli, M. (2021). Development of an Integrated Model to Assess the Impact of Agricultural Practices and Land Use on Agricultural Production in Morocco under Climate Stress over the Next Twenty Years. *Sustainability, 13*(21), 11943. doi:10.3390u132111943

Aridoculture. (n.d.) SAADA. *Climat.* https://sites.google.com/site/aridoculture/home/climat

Asseng, S., Ewert, F., Martre, P., Rötter, R. P., Lobell, D. B., Cammarano, D., Kimball, B. A., Ottman, M. J., Wall, G. W., White, J. W., Reynolds, M. P., Alderman, P. D., Prasad, P. V. V., Aggarwal, P. K., Anothai, J., Basso, B., Biernath, C., Challinor, A. J., De Sanctis, G., & Zhu, Y. (2015). Rising temperatures reduce global wheat production. *Nature Climate Change, 5*(2), 143–147. doi:10.1038/nclimate2470

Balaghi, R., Tychon, B., Eerens, H., & Jlibene, M. (2008). Empirical regression models using NDVI, rainfall and temperature data for the early prediction of wheat grain yields in Morocco. *International Journal of Applied Earth Observation and Geoinformation, 10*(4), 438–452. doi:10.1016/j.jag.2006.12.001

Barkhordarian, A., von Storch, H., & Bhend, J. (2013). The expectation of future precipitation change over the Mediterranean region is different from what we observe. *Climate Dynamics, 40*(1–2), 225–244. doi:10.100700382-012-1497-7

Battisti, D. S., & Naylor, R. L. (2009). Historical warnings of future food insecurity with unprecedented seasonal heat. *Science, 323*(5911), 240–244. doi:10.1126cience.1164363 PMID:19131626

Bongaarts, J. (2009). Human population growth and the demographic transition. *Philosophical Transactions of the Royal Society of London. Series B, Biological Sciences, 364*(1532), 2985–2990. doi:10.1098/rstb.2009.0137 PMID:19770150

Boretti, A., & Rosa, L. (2019). Reassessing the projections of the World Water Development Report. *Npj Clean Water, 2*(1), 1–6. doi:10.103841545-019-0039-9

Cline, W. R. (2007). *Global Warming and Agriculture: Impact Estimates by Country.* Peterson Institute.

de Wrachien, D. (2014). The Great Potential of Micro-Irrigation Technology for Poor-Rural Communities. *Irrigation & Drainage Systems Engineering, 3*(2). doi:10.4172/2168-9768.1000e124

Déqué, M., Somot, S., Sanchez-Gomez, E., Goodess, C. M., Jacob, D., Lenderink, G., & Christensen, O. B. (2012). The spread amongst ENSEMBLES regional scenarios: Regional climate models, driving general circulation models and interannual variability. *Climate Dynamics*, *38*(5), 951–964. doi:10.100700382-011-1053-x

Dyer, G. A., López-Feldman, A., Yúnez-Naude, A., & Taylor, J. E. (2014). Genetic erosion in maize's center of origin. *Proceedings of the National Academy of Sciences of the United States of America*, *111*(39), 14094–14099. doi:10.1073/pnas.1407033111

El Ajhar, E. A., Douae, E. K., Ahmed, E. B., Khadija, E. K., & Driss, B. (2015). *Evolution De La Pluviométrie De 1960 À 2015 Au Maroc,* 10. Filière céréalière, Ministère de l'agriculture. https://www.agriculture.gov.ma/fr/filiere/Cerealiere

Farooq, M., Hussain, M., & Siddique, K. H. M. (2014). Drought Stress in Wheat during Flowering and Grain-filling Periods. *Critical Reviews in Plant Sciences*, *33*(4), 331–349. doi:10.1080/07352689.2014.875291

Farooq, M., Hussain, M., Wakeel, A., & Siddique, K. H. M. (2015). Salt stress in maize: Effects, resistance mechanisms, and management. A review. *Agronomy for Sustainable Development*, *35*(2), 461–481. doi:10.100713593-015-0287-0

Faysse, N. (2015). The rationale of the Green Morocco Plan: Missing links between goals and implementation. *The Journal of North African Studies, 20*(4). https://www.tandfonline.com/doi/abs/10.1080/13629387.2015.1053112

Fróna, D., Szenderák, J., & Harangi-Rákos, M. (2019). The Challenge of Feeding the World. *Sustainability*, *11*(20), 5816. doi:10.3390u11205816

Gornall, J., Betts, R., Burke, E., Clark, R., Camp, J., Willett, K., & Wiltshire, A. (2010). Implications of climate change for agricultural productivity in the early twenty-first century. *Philosophical Transactions of the Royal Society of London. Series B, Biological Sciences*, *365*(1554), 2973–2989. doi:10.1098/rstb.2010.0158

HCP. (2022). *Site institutionnel du Haut-Commissariat au Plan du Royaume du Maroc*. Haut-Commissariat. https://www.hcp.ma/

Kumar, V. (1998). An early warning system for agricultural drought in an arid region using limited data. *Journal of Arid Environments*, *40*(2), 199–209. doi:10.1006/jare.1998.0437

Lionello, P., & Scarascia, L. (2018). The relation between climate change in the Mediterranean region and global warming. *Regional Environmental Change*, *18*(5), 1481–1493. https://doi.org/10.1007/s10113-018-1290-1

Ludwig, F., & Asseng, S. (2006). Climate change impacts on wheat production in a Mediterranean environment in Western Australia. *Agricultural Systems*, *90*(1), 159–179. https://doi.org/10.1016/j.agsy.2005.12.002

Maazouz, S. (2016). *Le développement de l'irrigation durable au Maroc*. AgriMaroc. https://www.agrimaroc.ma/le-developpement-de-lirrigation-durable-au-maroc/

MAPMDREF. (2019). Filière céréalière. Retrieved 27 December 2021, from https://www.agriculture.gov.ma/fr/filiere/Cerealiere

MAPMDREF. (2021). *Publication d'un rapport sur le bilan et impacts du Plan Maroc Vert*. Ministère de l'agriculture. https://www.agriculture.gov.ma/fr/actualites/publication-dun-rapport-sur-le-bilan-et-impacts-du-plan-maroc-vert

Morocco, M. O. S. A. I. C. C. Retrieved 30 March 2022, from http://www.changementclimatique.ma/indexe77d.html?q=en/node/20

Newman, J. A. (2004). Climate change and cereal aphids: The relative effects of increasing CO2 and temperature on aphid population dynamics. *Global Change Biology*, *10*(1), 5–15. doi:10.1111/j.1365-2486.2003.00709.x

Portail national des collectivités territoriales. (2022). *La Région*. Portail national des collectivités territoriales. https://www.collectivites-territoriales.gov.ma/fr/la-region

Ribeiro, A. F. S., Russo, A., Gouveia, C. M., & Páscoa, P. (2019). Modelling drought-related yield losses in Iberia using remote sensing and multiscalar indices. *Theoretical and Applied Climatology*, *136*(1), 203–220. https://doi.org/10.1007/s00704-018-2478-5

Schilling, J., Freier, K. P., Hertig, E., & Scheffran, J. (2012). Climate change, vulnerability and adaptation in North Africa with focus on Morocco. *Agriculture, Ecosystems & Environment*, *156*, 12–26. https://doi.org/10.1016/j.agee.2012.04.021

Schilling, J., Hertig, E., Tramblay, Y., & Scheffran, J. (2020). Climate change vulnerability, water resources and social implications in North Africa. *Regional Environmental Change*, *20*(1), 15. https://doi.org/10.1007/s10113-020-01597-7

Tigkas, D., & Tsakiris, G. (2015). Early Estimation of Drought Impacts on Rainfed Wheat Yield in Mediterranean Climate. *Environmental Processes, 2*(1), 97–114. doi:10.1007/s40710-014-0052-4

Tol, R. S. J. (2010). The Economic Impact of Climate Change. *Perspektiven der Wirtschaftspolitik*, *11*(Supplement), 13–37. https://doi.org/10.1111/1468-2516.0326

UNDP. (2009). Climate Change Adaptation. Morocco. Retrieved 30 March 2022, from https://www.adaptation-undp.org/explore/arab-states/morocco

United Nations. (2017). World population projected to reach 9.8 billion in 2050, and 11.2 billion in 2100. Retrieved 20 March 2022, from https://www.un.org/en/desa/world-population-projected-reach-98-billion-2050-and-112-billion-2100

Valverde, P., de Carvalho, M., Serralheiro, R., Maia, R., Ramos, V., & Oliveira, B. (2015). Climate change impacts on rainfed agriculture in the Guadiana river basin (Portugal). *Agricultural Water Management*, *150*, 35–45. doi:10.1016/j.agwat.2014.11.008

World Bank. (2021). Morocco. Total Population. Retrieved 7 January 2022, from https://data.worldbank.org/indicator/SP.POP.TOTL?locations=MA

World Bank Climate Change Knowledge Portal. (2021). Morocco. Retrieved December 2021, from https://climateknowledgeportal.worldbank.org/country/morocco

Wrachien, D. D. (2013). Food Production and Irrigation and Drainage Systems Development Perspective and Challenges. *Irrigation & Drainage Systems Engineering*, *2*(3). doi:10.4172/2168-9768.1000e122

Xing, K., Sun, D., Zhang, J., & Zhao, F. (2021). Wide Diurnal Temperature Amplitude and High Population Density Can Positively Affect the Life History of Sitobion avenae (Hemiptera: Aphididae). *Journal of Insect Science*, *21*(2), 6. doi:10.1093/jisesa/ieab011

Zhou, X., Harrington, R., Woiwod, I. P., Perry, J. N., Bale, J. S., & Clark, S. J. (1995). Effects of temperature on aphid phenology. *Global Change Biology*, *1*(4), 303–313. doi:10.1111/j.1365-2486.1995.tb00029.x

KEY TERMS AND DEFINITIONS

Climate Change: Refers to all the variations in climate characteristics in a given place over time: warming or cooling.

Consumption of Cereals: The quantity of cereals consumed per inhabitant in the year

Demography: The quantitative study of human populations and their dynamics based on their components: fertility, conjugality, migration, ageing and mortality

Growing Season: The time when weather conditions are conducive to plant growth

Food Security: A situation that guarantees population access to food qualitatively and quantitatively at all times. Eating habits must be sufficient to ensure a healthy and active lifestyle.

Irrigated Land: An area where water, other than rainwater, is artificially supplied to improve grazing or agricultural production.

Rainfed Land: An area where farming relies on rainfall for water.

Vulnerability to Climate Change: The degree to which a system is likely to suffer or be negatively affected by the adverse effects of climate change, including climate variability and extreme events.

Chapter 8
Global Warming:
A Reassessment of the Origin of the Problem, and an Effective Human-Scale Solution

Sebastian Cruft
Independent Researcher, UK

Ahmed Karmaoui
https://orcid.org/0000-0003-3881-4029
Moulay Ismail University, Morocco

ABSTRACT

A bibliometric analysis and a short discussion of some solutions to global warming. The current chapter attempts to reassess the origin of global warming and proposes an effective human-scale solution. The bibliometric review reveals an increasing trend of indexed publications from 1936 to 2020. The trend analysis shows research turning towards topics such as performance, life-cycle assessment, electric vehicles, construction, thermal comfort, mechanical properties, and quality. With regard to the origin of global warming, doubt is cast on the view that increased heat in the atmosphere is chiefly caused by increased carbon. From the point of view of chemistry and physics, CO2 is not a cause of heat, but a result of what creates heat (combustion). Moreover, the low-carbon strategies suggested to control the increase of heat seem slow, difficult to measure and control, and so far, ineffective. In industry, it is axiomatic that good technologies have measurable results, and always work in the same way if conditions are as foreseen.

DOI: 10.4018/978-1-6684-4649-2.ch008

INTRODUCTION

Global warming (GW) refers to a general rise in average temperatures that effect the weather balances and ecosystems. Currently, GW is among the most popular and debatable topics (Jian-Bin et al., 2012). It is strongly associated with the rise of greenhouse gases (RHG) in the atmosphere, which exacerbate the greenhouse effect. The natural greenhouse effect makes possible life on Earth and its absence induces a drastic decrease in surface temperature of about -19°C, instead of the current 14°C (Cassia et al., 2018; Le Treut et al., 2007). Greenhouse effect is caused by greenhouse gases including water vapor (<5%), CO2 (0.03–0.04%), methane, nitrous oxide (NxO), ozone, and chlorofluorocarbons (Inglezakis, 2016). These gases absorb and emit radiation in infrared interval (IPCC, 2014). The association between human pressure and RHD are found in the literature. Beenstock et al., (2012) reported no continuous effect of greenhouse gas forcings global temperature but a temporary effect. The Non-governmental International Panel on Climate Change (NIPCC) disagree about global warming that is man-made but state that it is a natural change. However, the studies dealing with the fact that the cause of the increase in greenhouse effect is an anthropogenic phenomenon are numerous including the reports of the most recent publication of the Intergovernmental Panel on Climate Change (IPCC), IPCC Sixth Assessment Report on Climate change 2022: Impacts, adaptation and vulnerability (Pörtner et al., 2022). Estrada et al., 2013 highlighted that "long-term trends in total radiative forcing and temperatures have largely been determined by atmospheric GHG concentrations" (stronger human effect).

So, scientists believe that the anthropogenic activities induce an increase in atmospheric concentrations of the greenhouse gases, which make the Earth much warmer than usual with a continuous record of temperature and atmospheric CO2 (Reichle, 2019). Experts reported that the global average temperature has increased and expected an additional increase in the current century if no effective measures are taken, which impact seriously the global population and ecosystems. Several human activities were reported to be the causes of this global warming including traditional sources of energy, such as coal, gas, and oil, deforestation, and intensive farming, etc. It is well established that this effect is provoked by human and affects many sectors including health, ecosystems, and agriculture. Governmental and non-governmental efforts to fight against this global issue are multiplied, proposing intelligent solutions in many domains including industry, transport, agriculture… The documents reporting innovative solutions that would limit GW are constantly increasing in number, including the use of renewable power-generation, transferring energy-saving technologies, imposing carbon taxes, carbon sequestration, and global cooling. The current chapter aims to explore the structure and trends of research and discuss the origin of the Global Warming, and proposes an effective human-scale solution.

BIBLIOMETRIC ANALYSIS ON SOLUTIONS OF GLOBAL WARMING

In this section, we explored the diffusion and evolution of the knowledge in the field of global warming. This approach allows us to reflect the research structure based on 2,000 papers using a search based on the key terms "solutions of global warming" conducted on Web of Sciences database, April 14, 2022. The bibliometric analysis highlighted five main clusters (Figure 1) with a focus on the keywords climate change, performance, CO_2, and energy (Figure 2). Cluster 1 comprises red items including 'climate change, temperature, adaptation, impact, vulnerability, environment, and policy'. The second cluster encompasses green items such as 'carbon-dioxide, CO_2, conversion, CO_2 capture and sequestration'… The yellow items (The third cluster of key terms) groups the keywords 'performance, optimization, energy, electric vehicles, solar energy, and hydrogen' while the blue cluster gathers and associates terms such as 'sustainable development, energy efficiency, systems, and thermal comfort'. Finally, the purple group of key terms includes 'biomass, emissions, and carbon'.

Figure 1. Occurrence and links Network using the keywords "Solutions of global warming" and VOSviewer tool

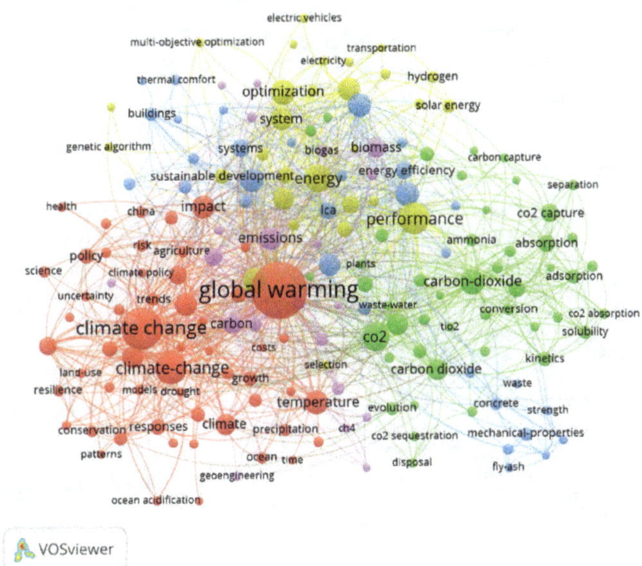

Figure 2. density-visualization using the keywords "Solutions of global warming" and VOSviewer tool

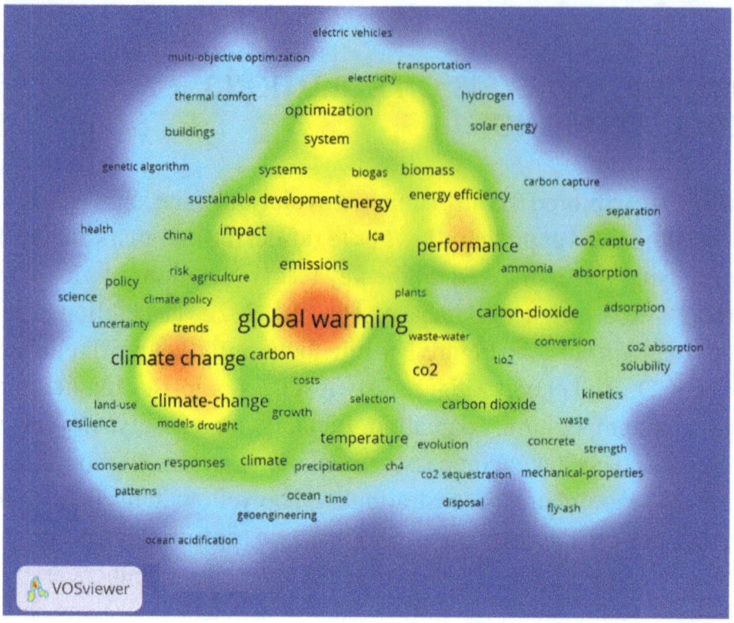

Figure 3. Trending analysis using the keywords "Solutions of global warming" and VOSviewer tool;

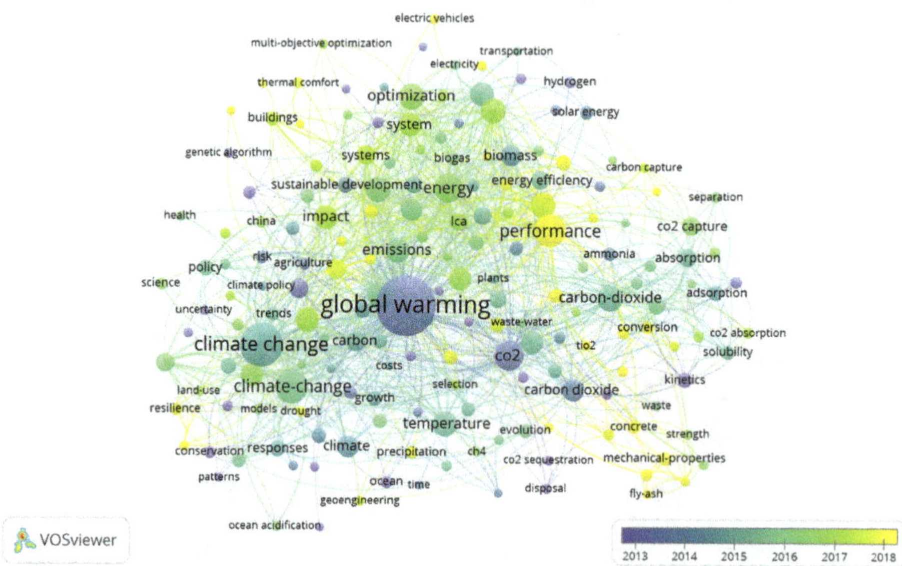

For the trend analysis, research is turning towards the topics such as performance, life-cycle assessment, electric vehicles, construction, thermal comfort, mechanical properties, and quality (Figure 2).

SOLUTIONS TO THE PROBLEM OF THE GLOBAL WARMING

Many aspects of the view that increased heat in the atmosphere is caused by increased carbon seem self-contradictory. In chemistry and physics, CO_2 is not a cause of heat, but a result of combustion, and thus a result of the process of the creation of heat. Additionally, we don't say: this carbon is intolerable – open the window. We say: this heat is intolerable – do whatever you can to cool down. We suggest that Global Warming is a result of:

- The heat from combustion (as created day-and-night in internal combustion engines, electricity generating plants, burning wood, coal and oil): plus
- **This heat being amplified by** the greenhouse gases which it creates. What the relationship between the two factors is, will be an important subject for future study.

Moreover, the low-carbon strategies suggested to control the increase of heat seem slow, difficult to measure and control, and so far, ineffective. In industry it is axiomatic that **good technologies have measurable results**, and always work in the same way if conditions are as foreseen.

Suspecting that there might be factors as yet unpublicised, we decided to investigate other possible reasons behind Anthropogenic Global Warming.

We started by looking at the 'Global Average Temperature Change' Graph 1850 – 2025' published by NASA GISS and 4 other international scientific organisations (Figure 4):

From this, it is clear that temperatures are rising fast (Figure 4). However, it is also clear that the opinion that Global Warming is the result of industrialisation, and the often-quoted aim of returning the Earth's temperature to 'preindustrial levels' are both too generalised to describe a graph which contains significant downward movements as well as upward spikes.

So, we analysed the temperature movements further (all temperatures are marked in Centigrade, as per the NASA-GISS graph (Figure 5). Firstly we placed a thick brown line at the approximate lowest point of the main temperature changes. This helped to show the correlation between temperature increases and decreases, and the timeline.

Figure 4. Global average temperature change from five different scientific organizations NASA GISS, HadCRUT.5.0.1.0, NCDC NOAA, Japan Meteorological Office, and Berkeley Earth. Source: Wikipedia

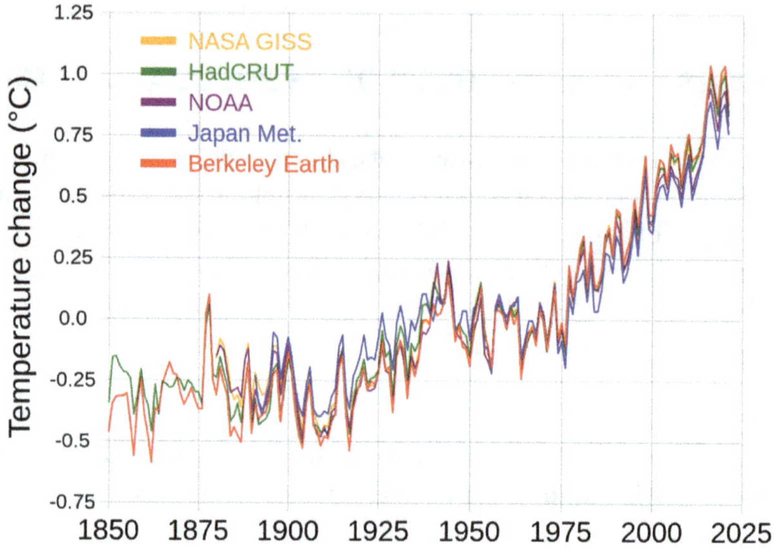

Figure 5. Global average temperature change from five different scientific organizations NASA GISS, HadCRUT.5.0.1.0, NCDC NOAA, Japan Meteorological Office, and Berkeley Earth as adapted in May 2022. Source: GlobalCoolingandIrrigationSystem.org

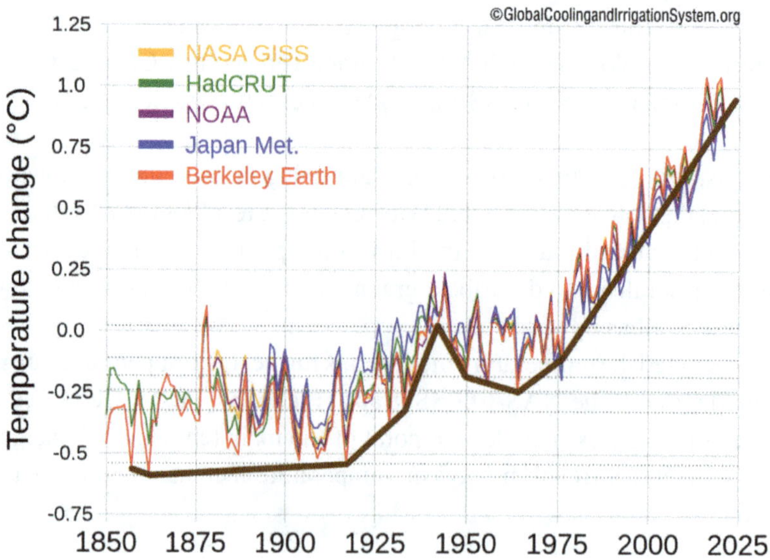

At first we were surprised to see any downward gradients at all, but when we compared the readings with the dates, we saw new explanations. We increased the precision of the dates and marked the increase or decrease by degrees of angle. Downward movements (periods of cooling) were shown by a 'minus' sign - e.g. -20º (as at 1860 and 1950), and upward movements (periods of heating) by a figure without the 'minus' sign – e.g. 3º and 38º (as at 1865 and 1915). Obviously, we are aware that the actual angle will depend on the format of the graph, but the relationship between the angles is still significant. We then started to add interpretations.

Our attention was immediately drawn to the remarkable upward spikes in the 1940s. To someone born in London in 1946, seeing gaps like missing teeth in the terraced streets from bombing in the 'Blitz', and reading somber histories of 1000-plane raids using incendiary bombs on Hamburg and Dresden, which created week-long temperatures of about 800ºC, and the massive fire-bombing of Japan, the spikes in the period of the Second World War started to become clear (Figure 6).

Add to this the testing of nuclear weapons and their use in war - no wonder there was a **62º downturn at the end of the war!** We now suggest that the enormous temperatures generated in the Second World War initiated the present problem of Global Warming.

Figure 6 Heating Effect of Hiroshima Bomb. Source: GlobalCoolingandIrrigationSystem.org

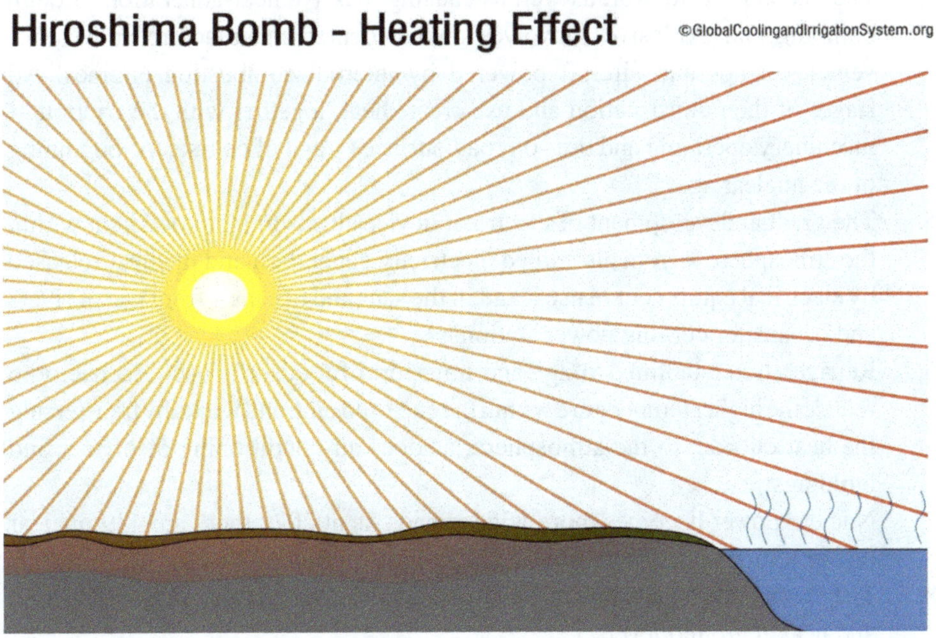

According to the Hiroshima Peace Memorial Museum: https://hpmmuseum.jp/modules/exhibition/index.php?action=ItemView&item_id=59&lang=eng

The nuclear bomb on Hiroshima created a temperature of 7,000°C, at a height of 580 m. Its heat will have remained in the land, buildings and sea until it moved over a period to cooler locations by the normal processes of radiation, convection, conduction and condensation.

As at July 2022, there have been over 2,000 recorded nuclear explosions, as well as accidents at power plants. We have analysed the effects of these and other 'extra-heating events' and further causes of heating at or near ground/sea level since the 1920s. These are set out in the following Graph, with approximations of dates. In a roughly chronological order, we list these significant heat-generating developments:

1. Coal-based industry from about 1865. Obviously the engineering breakthroughs were made well before this date, but that is the year when we see the start of a 3° upward temperature gradient, which lasts until 1915 (the First World War)… approx. 65 years.
2. After this we see an upward gradient of 38° until about the mid-1930s, with what looks like a downward spike reflecting the Great Depression, and the temporary collapse of activity in the industrialised nations.
3. We see a further increase in heat-generating activity in the 1930s, with the growth of the mass-market for cars and lorries and the gradual increase in air-travel;
4. The Second World War, as well as causing massive heat-generation through bombing and explosives, involves unprecedented manufacture and use of vehicles, ships and aircraft powered by internal combustion engines. All stages of their construction and use create heat, together with the friction of machinery operating and tires on road-surfaces – and of course, the beginning of the nuclear age;
5. The gradual development of electrical devices leads to increased heat within the atmosphere – from air-cooled electricity generation and the heat released by electrical equipment in use – and in the seas and oceans – from use of rivers and oceans for cooling power stations;
6. Refrigeration enabling storage and transport of foodstuffs uses electricity to reduce the higher temperature within the enclosed unit, and works by transferring the heat outside, to the atmosphere, as does air-conditioning of offices and buildings;
7. Nuclear power becomes more widespread. As at 2022 there are 439 nuclear power stations operating, and 62 under construction;
8. The use of rockets, whether military or scientific, leads to massive combustion and heat at ground level;

9. Anthropogenic deforestation removes longer-term evaporative cooling. Wildfires remove this in the short-term, and add the heat of their combustion;
10. In 1963-73, 388,000 tons of Napalm incendiary material were used in the Vietnam War;
11. The growth of computers and the Internet increases the need for electricity generation. Computers create heat, which is vented to the atmosphere.

Of course, there are many other normal activities which release heat into the air, sea or earth. Endothermic and exothermic reactions in manufacturing and construction all ultimately release heat into the partially-closed envelope of the Earth's atmosphere.

Between 1860 and 1915 the gradient for the overall temperature rise was about 3° (Figure 7). This indicates that the amount of heat received from the Sun and the amount of heat which the Earth could was lose naturally were approximately in equilibrium, with just a small heat-increment. How to reduce the modern rate of heat-increase of 53° or more is the present challenge.

How to Reduce the Modern Rate of heat-increase

The brown line in the graph above starts to curve downwards again after the Cop27 Conference in Egypt, when Global Cooling Technology is introduced. Here is how it will be done. The technology is based on two constant physical principles:

1. Heat moves from a warmer environment to a cooler one unless prevented by insulation;
2. The air cools by approximately 0.5 - 1°C for every 100 metres of increased altitude.

Temperature changes at altitude have been measured since the 1930s by Radiosonde balloons, at an increasing range of locations and frequency. Changes of pressure and other atmospheric data have been added to the records. A useful site to research weather data (past and present) is: http://weather.uwyo.edu/upperair/sounding.html. (An obvious demonstration of this is that snow and ice tend to form on the **upper** surfaces of mountains. What is called the Lapse Rate is routinely used by glider pilots: https://en.wikipedia.org/wiki/Lapse_rate

The graph 'Differential Temperature vs Height' demonstrates that the pattern of heat reduction with increased altitude is worldwide, with readings from both Hemispheres and the Equator (Figure 8). Counter-intuitively, the readings go **upwards** with **falling** temperatures – apologies for that…

Figure 7. Global average temperature change associated to the main anthropogenic impacts. Source: GlobalCoolingandIrrigationSystem.org

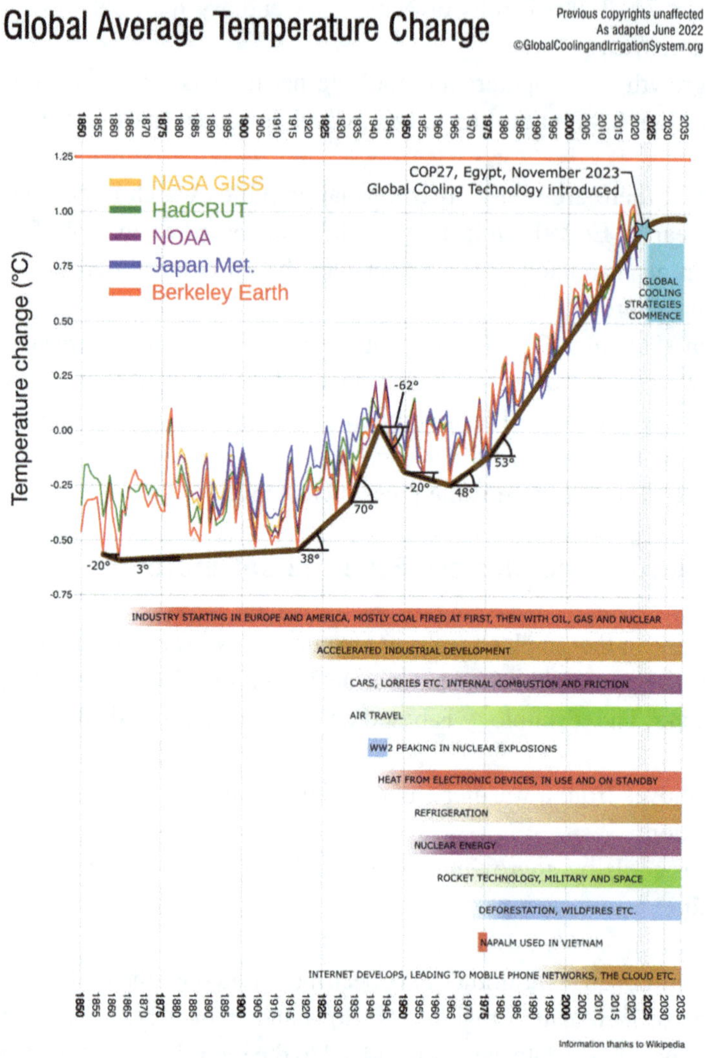

At altitudes of c. 800m there was a temperature reduction of 5-10°C except in the case of Melbourne, Australia, where the air at 800m was warmer by 2°C. (This may have been a freak reading caused by thermals). Over 1500m there is an absolutely clear pattern of temperature reduction.

At present there is widespread distress due to Global Warming (or the Climate Catastrophe, as it is sometimes called). As well as terrible fires and flooding, it is leading to despair and mental illness, particularly in young people.

Figure 8. Differential Temperature vs Height. Source: GlobalCoolingandIrrigationSystem.org

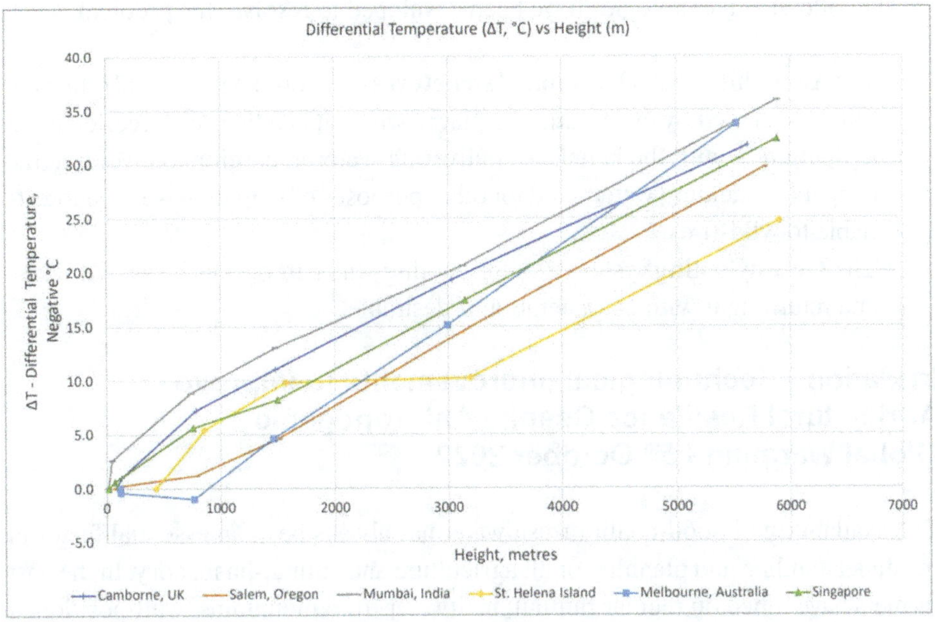

We suggest that the origin of high-speed Global Warming is the old destroyer of human happiness, civilizations, and cultures – namely WAR. It is terrible that this continues in the 21st Century. However, there is still a constant generation of heat – especially in power stations, to provide electricity, which <u>we really need!</u> – and there is still no technology to dispose of it.

Since the bombing raids of the 1940s it has been as if we were all sitting in a bath, and nobody has noticed that the hot tap is constantly running. Excess water is leaving through the overflow, but the temperature is rising faster all the time as new heat is being generated through combustion and friction of all sorts – in physics (in very general terms), work equals, or creates, heat. Our culture is based on the work of human beings and machines, and it would be very harmful to us all to try to deny this.

There is no need to blame individuals, governments or companies. Industrialisation has saved us from the need for physical slavery and the insecurity of subsistence farming – all we have to do is be alert to what is happening… and install and employ suitable technology.

A clearer definition of our aims than 'returning to pre-industrial levels' would also be useful. If we make the world much colder, we shall have to heat it again.

Looking at the NASA/GISS Graph as adapted, the Global Temperature of 1960-75 looks like a good temperature to aim for.

The following technologies have been developed to resolve this problem:

1. The **Armadillo™** and **Fontana™ structures** - for cooling the air and returning moisture cooled by evaporation to places where it can be used beneficially in agriculture – with the benefit of delivering water underground, avoiding the dangers of standing water - and for other purposes of irrigation – e.g. in forests liable to wild-fires.
2. The **OceanCoolingSystem™** - for cooling water in oceans, seas and lakes, and returning it with extra aeration if desirable.

Irrigation – Technological Improvements to Improve Agricultural Resilience Despite Anthropogenic Global Warming - 5th October 2022

The availability of good-quality fresh water has always been an essential factor in the understanding and planning of all agriculture and animal-husbandry. In the past farmers have relied on regular rainfall, meltwater from mountains, well-positioned stones to promote condensation, qanats and wells. Rain has the advantage that, because it is a product of evaporation, it does not contain salt or other contaminants.

However, the combination of Global Warming, damming and water-extraction has led to many previously well-watered areas becoming dry, which is a disaster for the inhabitants.

What is now proposed is the use of **Armadillo™** and **Fontana™ Air-and-Land Cooling structures.** These use the moisture-content of the air (Earth's atmosphere contains about 13,000 cubic kilometres of water) which they raise in well-insulated pipework to a height where the temperature is cold enough to cause the moisture in the air to condense. It is then carried down in pipes into underground irrigation systems, which can be directed – still underground - to fields, orchards or pastures.

In calculating the height needed the designers should note that the atmosphere loses on average about 1 degree Celsius (the 'lapse rate' - https://en.wikipedia.org/wiki/Lapse_rate) and 0.02 hpA of pressure for every 100 metres of height increase. Within the structures the moisture-content in the air is condensed by contact with metallic plates which are chilled by conduction in air at high level.

Advantages over rainwater are:

1. Rainwater frequently does not penetrate the dry upper surface of the soil, and evaporates before it can significantly moisten the soil beneath it;
2. Rainwater runs downhill, leaving slopes dry, and causing flooding below;

3. Rainwater sometimes collects in stagnant pools, which are favourable to vectors of disease, such as sand-flies, mosquitos etc.;
4. Rainwater can be blocked by mountains, causing 'Rain-Shadows', where no rain falls;
5. If the amount of water collected is not what is required, the structure can be modified (raised to collect more, shortened for less, or fitted with more drains to irrigate additional areas).

The structures are designed to be produced comparatively inexpensively in kit form.

The designs were originally intended for a patent-application, but they are now available freely to everyone.

The **Armadillo™** is has no fan and requires no attention other than periodic maintenance. The **Fontana™** has a fan. Both can be powered by Photovoltaic power.

Figure 9. The drawing of the Armadillo™. Source: GlobalCoolingandIrrigationSystem. org:

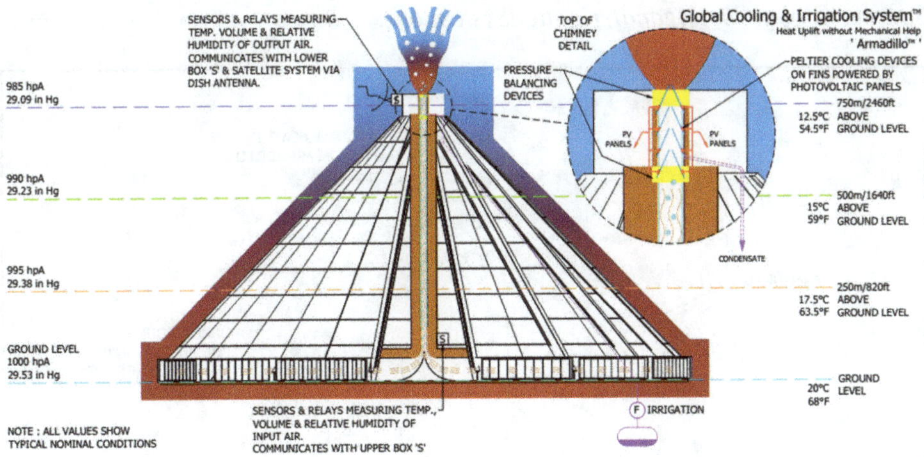

CONCLUSION

The chapter starts from a bibliometric analysis which was performed to show the most relevant key terms used in the field of global warming solutions. This chapter also provides an insight about the origin of the problem, and presents some effective human-scale solutions. The results reveal that research in the field of solutions of the global warming is turning toward topics such as performance, life-cycle assessment, electric vehicles, construction, thermal comfort, mechanical properties, and quality.

Figure 10. The drawing of the Fontana™. Source: GlobalCoolingandIrrigationSystem.org.

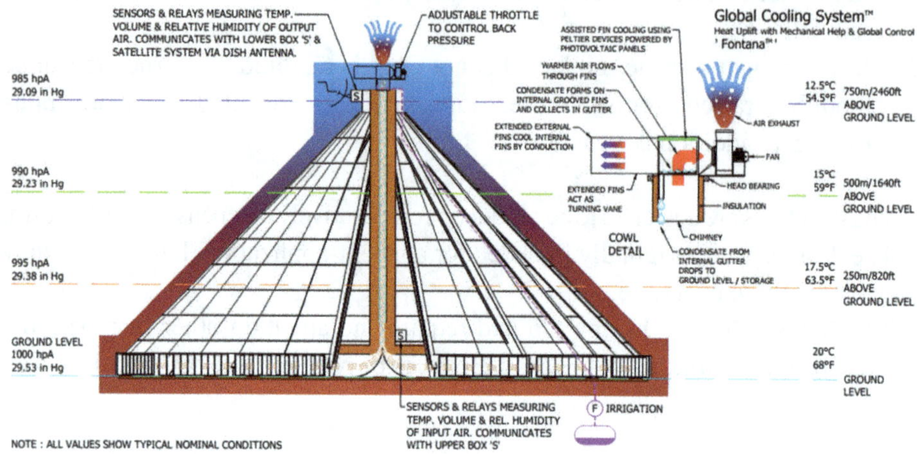

Figure 11. The drawing of the underground irrigation system used by both structures. Source: GlobalCoolingandIrrigationSystem.org.

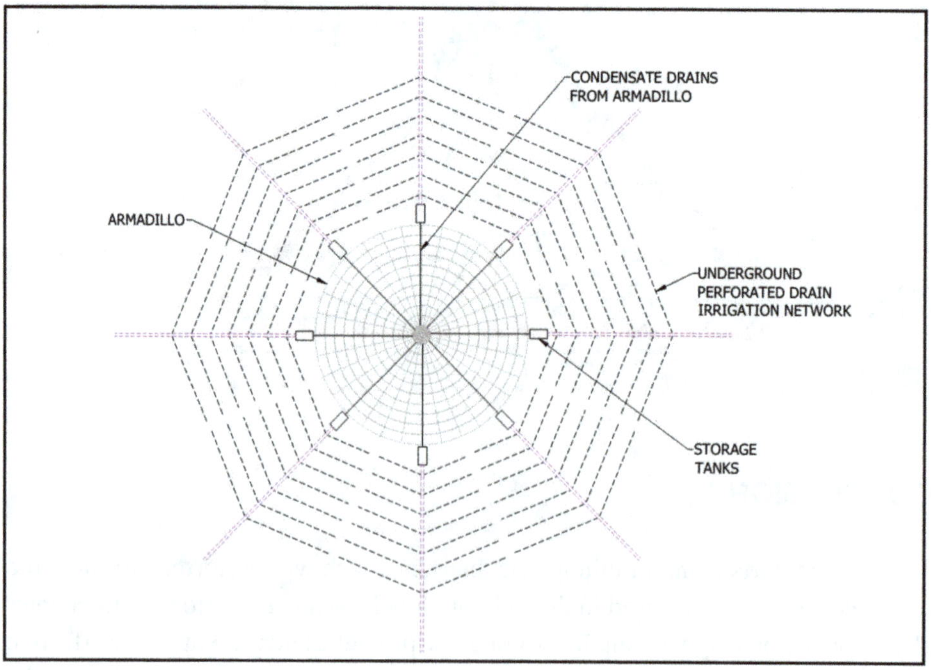

Figure 12. The drawing of a Fontana™ with an entrance to the underground irrigation system used by both structures. Source: GlobalCoolingandIrrigationSystem.org.

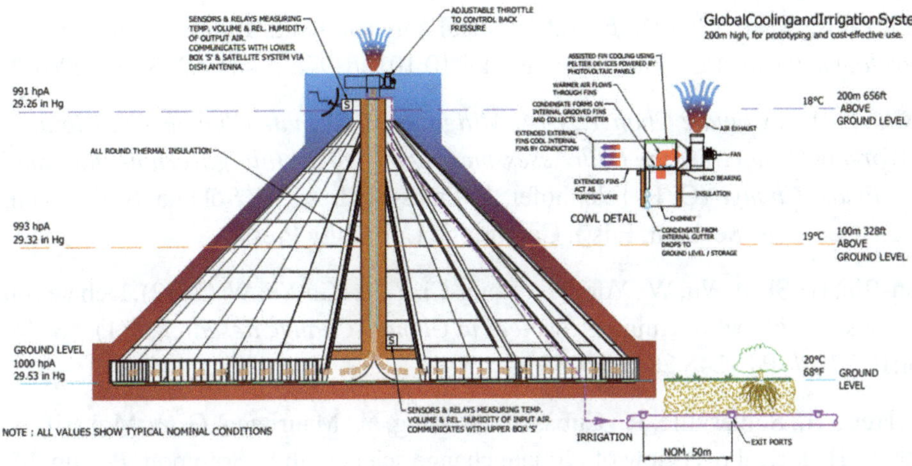

With regard to the origin of Global Warming, it should be noted that many aspects of the view that increased heat in the atmosphere is caused by increased carbon seem self-contradictory. First, from the point of view of chemistry and physics, CO_2 is not a cause of heat, but a result of combustion, and thus a result of the process of the creation of heat. Moreover, the low-carbon strategies suggested to control the increase of heat seem slow, difficult to measure and control, and so far, ineffective. In industry, it is axiomatic that good technologies have measurable results, and always work in the same way if conditions are as foreseen.

REFERENCES

Beenstock, M., Reingewertz, Y., & Paldor, N. (2012). Polynomial cointegration tests of anthropogenic impact on global warming. *Earth System Dynamics: ESD*, *3*(2), 173–188. doi:10.5194/esd-3-173-2012

Cassia, R., Nocioni, M., Correa-Aragunde, N., & Lamattina, L. (2018). Climate change and the impact of greenhouse gasses: CO_2 and NO, friends and foes of plant oxidative stress. *Frontiers in Plant Science*, *9*, 273. doi:10.3389/fpls.2018.00273 PMID:29545820

Cruft, S. (n.d.) The Global Cooling and Irrigation System [Video]. Global Cooling and Irrigation System.https://globalcoolingandirrigationsystem.org/

Estrada, F., Perron, P., & Martínez-López, B. (2013). Statistically derived contributions of diverse human influences to twentieth-century temperature changes. *Nature Geoscience*, *6*(12), 1050–1055. doi:10.1038/ngeo1999

Inglezakis, V. J. (2016). Extraterrestrial environment. In *Environment and Development* (pp. 453–498). Elsevier. doi:10.1016/B978-0-444-62733-9.00007-1

IPCC. (2014). *Climate Change 2014: Mitigation of Climate Change. Contribution of Working Group III to the Fifth Assessment Report of the Intergovernmental Panel on Climate Change* (O. R. Edenhofer, Y. Pichs-Madruga, E. Sokona, S. Farahani, K. Kadner, & A. Seyboth, Eds.). Cambridge University Press.

Jian-Bin, H., Shao-Wu, W., Yong, L., Zong-Ci, Z., & Xin-Yu, W. (2012). Debates on the causes of global warming. *Advances in Climate Change Research*, *3*(1), 38–44. doi:10.3724/SP.J.1248.2012.00038

Le Treut, H., Somerville, R., Cubasch, U., Ding, Y., Mauritzen, C., & Mokssit, A. (2007). Historical overview of climate change science. In S. Solomon, D. Qin, M. Manning, Z. Chen, M. Marquis, & K. B. Averyt, (Eds.), *Climate change 2007: The physical science basis. Contribution of Working Group I to the Fourth Assessment Report of the Intergovernmental Panel on Climate Change*. Cambridge University Press.

Pörtner, H. O., Roberts, D. C., Adams, H., Adler, C., Aldunce, P., Ali, E., & Birkmann, J. (2022). [Impacts, adaptation and vulnerability. *IPCC Sixth Assessment Report.*]. *Climatic Change*, 2022.

Reichle, D. (2019). The global carbon cycle and climate change. Elsevier, 1, 388.

KEY TERMS AND DEFINITIONS

Heat: Is caused by combustion, and amplified by greenhouse gases. Heat which is generated at a low level – e.g. by power stations, rocket launches etc. – cannot immediately disperse into space, but moves to cooler media at the speed dictated by the physics of the surrounding environment. **Heat** is a form of energy which only moves from an energised state (i.e. 'hot') to a less-energised state (i.e. 'cold'). As the upper air is colder (less-energised) than the lower air, heat cannot normally come downwards. The Earth can only lose heat at a certain rate without HUMAN technological assistance. New inventions exist which cool the world by moving heat upwards (in the air) to a level from which it cannot return and harm people or other living things on the ground. We can stop flooding and drought! We can cool the World and feed the World!

Compilation of References

Abdel Latef, A. A. H., Kordrostami, M., Zakir, A., Zaki, H., & Saleh, O. (2019a). Eustress with H_2O_2 facilitates plant growth by improving tolerance to salt stress in two wheat cultivars. *Plants*, *8*(9), 303. doi:10.3390/plants8090303 PMID:31461870

Abdel Latef, A. A. H., Mostofa, M. G., Rahman, M. M., Abdel-Farid, I. B., & Tran, L.-S. P. (2019b). Extracts from yeast and carrot roots enhance maize performance under seawater-induced salt stress by altering physio-biochemical characteristics of stressed plants. *Journal of Plant Growth Regulation*, *38*(3), 966–979. doi:10.100700344-018-9906-8

Adzawla, W., Sawaneh, M., & Yusuf, A. M. (2019). Greenhouse gasses emission and economic growth nexus of sub-Saharan Africa. *Scientific American*, *3*, e00065. doi:10.1016/j.sciaf.2019.e00065

Agnihotri, R., Sahni, S., Sharma, M. P., & Gupta, M. M. (2022). Facets of AM fungi in sequestering soil carbon and improving soil health. In Fungal diversity, ecology, and control management, pp. 327-344. doi:10.1007/978-981-16-8877-5_15

Agnolucci, M., Avio, L., Palla, M., Sbrana, C., Turrini, A., & Giovannetti, M. (2020). Health-promoting properties of plant products: The role of mycorrhizal fungi and associated bacteria. *Agronomy (Basel)*, *10*(12), 1864. doi:10.3390/agronomy10121864

AgriMaroc. (2014). *L' agriculture marocaine en chiffres* 2014. AgriMaroc.ma. https://www.agrimaroc.ma/lagriculture-marocaine-en-chiffres-2014/

Ait-El-Mokhtar, M., Anli, M., Ben Laouane, R., Boutasknit, A., Boutaj, H., Draoui, A., Zarik, L., & Fakhech, A. (2019a). Food security and climate change. In Handbook of Research on Global Environmental Changes and Human Health, pp. 53–73. doi:10.4018/978-1-5225-7775-1.ch004

Ait-El-Mokhtar, M., Boutasknit, A., Ben-Laouane, R., Anli, M., El Amerany, F., Toubali, S., Lahbouki, S., Wahbi, S., & Meddich, A. (2020b). Vulnerability of Oasis Agriculture to Climate Change in Morocco In Impacts of Climate Change on Agriculture and Aquaculture, pp. 76–106. doi:10.4018/978-1-7998-3343-7.ch004

Ait-El-Mokhtar, M., Baslam, M., Ben-Laouane, R., Anli, M., Boutasknit, A., Mitsui, T., Wahbi, S., & Meddich, A. (2020a). Alleviation of detrimental effects of salt stress on date palm (*Phoenix dactylifera* L.) by the application of arbuscular mycorrhizal fungi and/or compost. *Frontiers in Sustainable Food Systems*, *4*, 131. doi:10.3389/fsufs.2020.00131

Ait-El-Mokhtar, M., Ben-Laouane, R., Anli, M., Boutasknit, A., Wahbi, S., & Meddich, A. (2019b). Use of mycorrhizal fungi in improving tolerance of the date palm (*Phoenix dactylifera* L.) seedlings to salt stress. *Scientia Horticulturae*, *253*, 429–438. doi:10.1016/j.scienta.2019.04.066

Ait-El-Mokhtar, M., Fakhech, A., Anli, M., Ben-Laouane, R., Boutasknit, A., Wahbi, S., & Meddich, A. (2020c). Infectivity of the palm groves arbuscular mycorrhizal fungi under arid and semi-arid climate and its edaphic determinants towards efficient ecological restoration. *Rhizosphere*, *15*, 1–5. doi:10.1016/j.rhisph.2020.100220

Ait-Rahou, Y., Ait-El-Mokhtar, M., Anli, M., Boutasknit, A., Ben-Laouane, R., Douira, A., Benkirane, R., El Modafar, C., & Meddich, A. (2020). Use of mycorrhizal fungi and compost for improving the growth and yield of tomato and its resistance to *Verticillium dahliae*. *Archiv für Phytopathologie und Pflanzenschutz*, 1–26. doi:10.1080/03235408.2020.1854938

Ajani, A., & Shams, Z. I. (2016). Comparative Status of Sequestered Carbon Stock of Azadirachta indica and Conocarpus erectus at the University of Karachi Campus, Pakistan. *International Journal of Environment*, *5*(2), 89–97. doi:10.3126/ije.v5i2.15009

Akash, A. R., Rao, A. B., & Chandel, M. K. (2016). Prospects of Implementing CO2 Capture and Sequestration (CCS) in the Proposed Supercritical Coal Power Plants in India. *Energy Procedia*, *90*, 604–612. doi:10.1016/j.egypro.2016.11.229

Al Kharusi, L., Al Yahyai, R., & Yaish, M. W. (2019). Antioxidant response to salinity in salt-tolerant and salt-susceptible cultivars of date palm. *Agriculture*, *9*(1), 1–17. doi:10.3390/agriculture9010008

Alam, T. M., Shaukat, K., Luo, S., Hameed, I. A., Sarwar, M. U., Shabbir, S., Li, J., & Khushi, M. (2020). An Investigation of Credit Card Default Prediction in the Imbalanced Datasets. *IEEE Access: Practical Innovations, Open Solutions*, *8*, 201173–201198. doi:10.1109/ACCESS.2020.3033784

Alaux, P. L., César, V., Naveau, F., Cranenbrouck, S., & Declerck, S. (2018). Impact of Rhizophagus irregularis MUCL 41833 on disease symptoms caused by *Phytophthora infestans* in potato grown under field conditions. *Crop Protection (Guildford, Surrey)*, *107*, 26–33. doi:10.1016/j.cropro.2018.01.003

Allred, B., Martinez, L., Fessehazion, M. K., Rouse, G., Koganti, T., Freeland, R., Eash, N., Wishart, D. B., & Featheringill, R. (2021). Time of day impact on mapping agricultural subsurface drainage systems with UAV thermal infrared imagery. *Agricultural Water Management*, *256*, 107071. doi:10.1016/j.agwat.2021.107071

Amiri, N., Lahlali, R., Amiri, S. E. L., Jarroudi, M., Khebiza, M. Y., & Messouli, M. (2021). Development of an Integrated Model to Assess the Impact of Agricultural Practices and Land Use on Agricultural Production in Morocco under Climate Stress over the Next Twenty Years. *Sustainability*, *13*(21), 11943. doi:10.3390u132111943

Andrino, A., Guggenberger, G., Kernchen, S., Mikutta, R., Sauheitl, L., & Boy, J. (2021). Production of organic acids by arbuscular mycorrhizal fungi and their contribution in the mobilization of phosphorus bound to iron oxides. *Frontiers in Plant Science*, *12*, 661842. doi:10.3389/fpls.2021.661842 PMID:34335645

Anli, M., Baslam, M., Tahiri, A., Raklami, A., Symanczik, S., Boutasknit, A., Ait-El-Mokhtar, M., Ben-Laouane, R., Toubali, S., Ait-Rahou, Y., Ait Chitt, M., Oufdou, K., Mitsui, T., Hafidi, M., & Meddich, A. (2020a). Biofertilizers as strategies to improve photosynthetic apparatus, growth, and drought stress tolerance in the date palm. *Frontiers in Plant Science*, *11*, 1–21. doi:10.3389/fpls.2020.516818 PMID:33193464

Anli, M., El Kaoua, M., Ait-El-Mokhtar, M., Boutasknit, A., Ben-Laouane, R., Toubali, S., Baslam, M., Lyamlouli, K., Hafidi, M., & Meddich, A. (2020b). Seaweed extract application and arbuscular mycorrhizal fungal inoculation: A tool for promoting growth and development of date palm (*Phoenix dactylifera* L.) cv «Boufgous». *South African Journal of Botany*, *132*, 15–21. doi:10.1016/j.sajb.2020.04.004

Anli, M., Symanczik, S., El Abbassi, A., Ait-El-Mokhtar, M., Boutasknit, A., Ben-Laouane, R., Toubali, S., Baslam, M., Mäder, P., Hafidi, M., & Meddich, A. (2020c). Use of arbuscular mycorrhizal fungus *Rhizoglomus irregulare* and compost to improve growth and physiological responses of *Phoenix dactylifera* "Boufgouss.". *Plant Biosystems*, 1–14. doi:10.1080/11263504.2020.1779848

Aridoculture. (n.d.) SAADA. *Climat*. https://sites.google.com/site/aridoculture/home/climat

Arvind, G., Anthra, V. G., Haripriya, H., Akshaya, R. R., & Aravind, S. (2017). Automated irrigation with advanced seed germination and pest control. *IEEE International Conference on Technological Innovations in ICT For Agriculture and Rural Development (TIAR 2017), Tiar*, (pp. 64–67). IEEE. 10.1109/TIAR.2017.8273687

Arvola, A., Malkamäki, A., Penttilä, J., & Toppinen, A. (2019). Mapping the Future Market Potential of Timber from Small-Scale Tree Farmers: Perspectives from the Southern Highlands in Tanzania. *Small-scale Forestry*, *18*(2), 189–212. doi:10.100711842-019-09414-8

Arya, A., Shalini Negi, S., Kathota, J. C., Patel, A. N., Kalubarme, M. H., & Garg, J. K. (2017). Carbon Sequestration Analysis of dominant tree species using Geo-informatics Technology in Gujarat State (INDIA). *International Journal of Environment and Geoinformatics*, *4*(2), 79–93. doi:10.30897/ijegeo.312517

Asseng, S., Ewert, F., Martre, P., Rötter, R. P., Lobell, D. B., Cammarano, D., Kimball, B. A., Ottman, M. J., Wall, G. W., White, J. W., Reynolds, M. P., Alderman, P. D., Prasad, P. V. V., Aggarwal, P. K., Anothai, J., Basso, B., Biernath, C., Challinor, A. J., De Sanctis, G., & Zhu, Y. (2015). Rising temperatures reduce global wheat production. *Nature Climate Change*, *5*(2), 143–147. doi:10.1038/nclimate2470

Ater, M., Barbara, H., & Kassout, J. (2016). Importance des variétés locales, de l'oléastre et des pratiques traditionnelles de l'oléiculture dans la région de Chefchaouen (Nord du Maroc). L 'oléiculture au Maroc de la préhistoire à nos jours: pratiques, diversité, adaptation, usages, commerce et politiques. CIHEAM.

Ayinu, Y. T., Ayal, D. Y., Zeleke, T. T., & Beketie, K. T. (2022). Impact of climate variability on household food security in Godere District, Gambella Region, Ethiopia. *Climate Services*, *27*, 100307. doi:10.1016/j.cliser.2022.100307

Babyak, M. A. (2004). What You See May Not Be What You Get: A Brief, Nontechnical Introduction to Overfitting in Regression-Type Models. *Psychosomatic Medicine*, *66*(3), 411–421. PMID:15184705

Bahadur, A., Batool, A., Nasir, F., Jiang, S., Mingsen, Q., Zhang, Q., Pan, J., Liu, Y., & Feng, H. (2019). Mechanistic insights into arbuscular mycorrhizal fungi-mediated drought stress tolerance in plants. *International Journal of Molecular Sciences*, *20*(17), 1–18. doi:10.3390/ijms20174199 PMID:31461957

Balaghi, R., Tychon, B., Eerens, H., & Jlibene, M. (2008). Empirical regression models using NDVI, rainfall and temperature data for the early prediction of wheat grain yields in Morocco. *International Journal of Applied Earth Observation and Geoinformation*, *10*(4), 438–452. doi:10.1016/j.jag.2006.12.001

Balestrini, R., Brunetti, C., Chitarra, W., & Nerva, L. (2020). Photosynthetic Traits and Nitrogen Uptake in Crops: Which Is the Role of Arbuscular Mycorrhizal Fungi? *Plants*, *9*(9), 1105. doi:10.3390/plants9091105 PMID:32867243

Barkhordarian, A., von Storch, H., & Bhend, J. (2013). The expectation of future precipitation change over the Mediterranean region is different from what we observe. *Climate Dynamics*, *40*(1–2), 225–244. doi:10.100700382-012-1497-7

Bárzana, G., Aroca, R., & Ruiz-Lozano, J. M. (2015). Localized and non-localized effects of arbuscular mycorrhizal symbiosis on accumulation of osmolytes and aquaporins and on antioxidant systems in maize plants subjected to total or partial root drying. *Plant, Cell & Environment*, *38*(8), 1613–1627. doi:10.1111/pce.12507 PMID:25630435

Basu, S., Rabara, R. C., & Negi, S. (2018). AMF: The future prospect for sustainable agriculture. *Physiological and Molecular Plant Pathology*, *102*, 36–45. doi:10.1016/j.pmpp.2017.11.007

Battisti, D. S., & Naylor, R. L. (2009). Historical warnings of future food insecurity with unprecedented seasonal heat. *Science*, *323*(5911), 240–244. doi:10.1126cience.1164363 PMID:19131626

Beck, S., & Mahony, M. (2018). The IPCC and the new map of science and politics. *Wiley Interdisciplinary Reviews: Climate Change*, *9*(6), 1–16. doi:10.1002/wcc.547

Beenstock, M., Reingewertz, Y., & Paldor, N. (2012). Polynomial cointegration tests of anthropogenic impact on global warming. *Earth System Dynamics: ESD*, *3*(2), 173–188. doi:10.5194/esd-3-173-2012

Beghin. (2015). Assessing Food Security in Tanzania in the Next Decade. *Agricultural Policy Review, 201*(3), 3.

Begum, N., Ahanger, M. A., & Zhang, L. (2020). AMF inoculation and phosphorus supplementation alleviates drought induced growth and photosynthetic decline in *Nicotiana tabacum* by up-regulating antioxidant metabolism and osmolyte accumulation. *Environmental and Experimental Botany, 176*, 104088. doi:10.1016/j.envexpbot.2020.104088

Ben-Laouane, R., Ait-El-Mokhtar, M., Anli, M., Boutasknit, A., Ait Rahou, Y., Raklami, A., Oufdou, K., Wahbi, S., & Meddich, A. (2020a). Green compost combined with mycorrhizae and rhizobia: A strategy for improving alfalfa growth and yield under field conditions. *Gesunde Pflanzen, 73*(2), 193–207. doi:10.100710343-020-00537-z

Ben-Laouane, R., Baslam, M., Ait-El-Mokhtar, M., Anli, M., Boutasknit, A., Ait-Rahou, Y., Toubali, S., Mitsui, T., Oufdou, K., Wahbi, S., & Meddich, A. (2020b). Potential of native arbuscular mycorrhizal fungi, rhizobia, and/or green compost as alfalfa (*Medicago sativa*) enhancers under salinity. *Microorganisms, 8*(11), 1–27. doi:10.3390/microorganisms8111695 PMID:33143245

Ben-Laouane, R., Meddich, A., Bechtaoui, N., Oufdou, K., & Wahbi, S. (2019). Effects of arbuscular mycorrhizal fungi and rhizobia symbiosis on the tolerance of *Medicago sativa* to salt stress. *Gesunde Pflanzen, 71*(2), 135–146. doi:10.100710343-019-00461-x

Benoît, É. (2008). Les changements climatiques: vulnérabilité, impacts et adaptation dans le monde de la médecine traditionnelle au Burkina Faso. *VertigO-la revue électronique en sciences de l'environnement, 8*(1).

Bernardo, L., Carletti, P., Badeck, F. W., Rizza, F., Morcia, C., Ghizzoni, R., Rouphael, Y., Colla, G., Terzi, V., & Lucini, L. (2019). Metabolomic responses triggered by arbuscular mycorrhiza enhance tolerance to water stress in wheat cultivars. *Plant Physiology and Biochemistry, 137*, 203–212. doi:10.1016/j.plaphy.2019.02.007 PMID:30802803

Besnard, G., Terral, J. F., & Cornille, A. (2018). On the origins and domestication of the olive: A review and perspectives. *Annals of Botany, 121*(3), 385–403. doi:10.1093/aob/mcx145 PMID:29293871

Betteena, D., Fernando, S., Sabarishwaran, S., Priya, R. R., & Santhoshini, S. (2020, August). Smart agriculture monitoring system using IoT. [UAE]. *IACSIT International Journal of Engineering and Technology, 7*(4), 308–311. doi:10.35940/ijeat.D7916.049420

Bhagat, P. R., Naz, F., & Magda, R. (2022). Artificial intelligence solutions enabling sustainable agriculture : A bibliometric analysis. *PLoS One, 2022*(June), 1–19. doi:10.1371/journal.pone.0268989 PMID:35679287

Bhantana, P., Malla, R., Vista, S. P., Rana, M. S., Mohamed, G., Joshi, B. D., Shah, S., Khadka, D., Timsina, G. P., Poudel, K., Zhihao, D., & Hu, C. X. (2021). Use of arbuscular mycorrhizal fungi (AMF) and zinc fertilizers in an adaptation of plant from drought and heat stress. *Biomedical Journal of Scientific & Technical Research, 38*(3), 30357–30373. doi:10.26717/BJSTR.2021.38.006152

Bhowmik, A., & Ramasubramanian, V. & Chandrahas, & Kumar, A. (2011). Logistic Regression for Classification in Agricultural Ergonomics. *Advances in Applied Research.*, *3*, 163–170.

Bohre, P., & Chaubey, O. P. (2016). Biomass production and carbon sequestration by Azadirachta indica in coal mined lands. *International Journal of Bio-Science and Bio-Technology*, *8*(2), 111–120. doi:10.14257/ijbsbt.2016.8.2.10

Bona, E., Todeschini, V., Cantamessa, S., Cesaro, P., Copetta, A., Lingua, G., Gamalero, E., Bertà, G., & Massa, N. (2018). Combined bacterial and mycorrhizal inocula improve tomato quality at reduced fertilization. *Scientia Horticulturae*, *234*, 160–165. doi:10.1016/j.scienta.2018.02.026

Bongaarts, J. (2009). Human population growth and the demographic transition. *Philosophical Transactions of the Royal Society of London. Series B, Biological Sciences*, *364*(1532), 2985–2990. doi:10.1098/rstb.2009.0137 PMID:19770150

Bonoli, A., Zanni, S., & Awere, E. (2019). Organic waste composting and sustainability in low-income communities in Palestine: Lessons from a pilot project in the village of Al Jalameh, Jenin. *International Journal of Recycling of Organic Waste in Agriculture*, *8*(3), 253–262. doi:10.100740093-019-0264-8

Boretti, A., & Rosa, L. (2019). Reassessing the projections of the World Water Development Report. *Npj Clean Water*, *2*(1), 1–6. doi:10.103841545-019-0039-9

Boselli, V., Ouallali, A., Briak, H., Houssni, M., Kassout, J., El Ouahrani, A., & Michailidi, E. M. (2020). System dynamics applied to terraced agroecosystems: The case study of assaragh (Anti-Atlas Mountains, Morocco). *Water (Basel)*, *12*(6), 1693. doi:10.3390/w12061693

Boucher, D., Elias, P., Faires, J., & Smith, S. (2014). Deforestation Success Stories. In *Union of Concerned Scientists* (June Issue). http://search.ebscohost.com/login.aspx?direct=true&db=8gh&AN=109565479&site=ehost-live

Boutasknit, A., Ait-Rahou, Y., Anli, M., Ait-El-Mokhtar, M., Ben-Laouane, R., & Meddich, A. (2020a). Improvement of garlic growth, physiology, biochemical traits and soil fertility by *rhizophagus irregularis* and compost. *Gesunde Pflanzen*, *73*(2), 149–160. doi:10.100710343-020-00533-3

Boutasknit, A., Baslam, M., Ait-El-Mokhtar, M., Anli, M., Ben-Laouane, R., Ait-Rahou, Y., Mitsui, T., Douira, A., El Modafar, C., Wahbi, S., & Meddich, A. (2021). Assemblage of indigenous arbuscular mycorrhizal fungi and green waste compost enhance drought stress tolerance in carob (*Ceratonia siliqua* L.) trees. *Scientific Reports*, *11*(1), 1–23. doi:10.103841598-021-02018-3 PMID:34819547

Boutasknit, A., Baslam, M., Ait-El-Mokhtar, M., Anli, M., Ben-Laouane, R., Douira, A., El Modafar, C., Mitsui, T., Wahbi, S., & Meddich, A. (2020b). Arbuscular mycorrhizal fungi mediate drought tolerance and recovery in two contrasting carob (*Ceratonia siliqua* L.) ecotypes by regulating stomatal, water relations, and (In)organic adjustments. *Plants*, *9*(1), 80. doi:10.3390/plants9010080 PMID:31936327

Buras, A., Rammig, A. S., & Zang, C. (2020). Quantifying impacts of the 2018 drought on European ecosystems in comparison to 2003. *Biogeosciences*, *17*(6), 1655–1672. doi:10.5194/bg-17-1655-2020

Canadell, J. G., Monteiro, P. M., Costa, M. H., Da Cunha, L. C., Cox, P. M., Alexey, V., et al. (2021). *Global carbon and other biogeochemical cycles and feedbacks*. Cambridge University Press.

Caser, M., Victorino, Í. M. M., Demasi, S., Berruti, A., Donno, D., Lumini, E., Bianciotto, V., & Scariot, V. (2019). Saffron cultivation in marginal alpine environments: How AMF inoculation modulates yield and bioactive compounds. *Agronomy (Basel)*, *9*(1), 12. doi:10.3390/agronomy9010012

Cassia, R., Nocioni, M., Correa-Aragunde, N., & Lamattina, L. (2018). Climate change and the impact of greenhouse gasses: CO2 and NO, friends and foes of plant oxidative stress. *Frontiers in Plant Science*, *9*, 273. doi:10.3389/fpls.2018.00273 PMID:29545820

Chandrasekaran, M. (2020). A meta-analytical approach on arbuscular mycorrhizal fungi inoculation efficiency on plant growth and nutrient uptake. *Agriculture*, *10*(9), 1–12. doi:10.3390/agriculture10090370

Chang, W., Sui, X., Fan, X. X., Jia, T. T., & Song, F. Q. (2018). Arbuscular mycorrhizal symbiosis modulates antioxidant response and ion distribution in salt-stressed *Elaeagnus angustifolia* seedlings. *Frontiers in Microbiology*, *9*, 1–14. doi:10.3389/fmicb.2018.00652 PMID:29675008

Chaturvedi, S., & Khurana, S. M. P. (2017). *Biotechnological Approaches for Urban Solid Waste Management*. 860–866. doi:10.21276/sajb.2017.5.12.7

Chauvy, R., Meunier, N., Thomas, D., & De Weireld, G. (2019). Selecting emerging CO2 utilization products for short- to mid-term deployment. *Applied Energy*, *236*(February), 662–680. doi:10.1016/j.apenergy.2018.11.096

Chen, B., Hu, Y., Li, T., Xu, L., Zhang, X., & Hao, Z. (2018). The role of AM symbiosis in plant adaptation to drought stress. *Journal of Integrated Field Science*, *15*, 22–25.

Chen, S., Zhao, H., Zou, C., Li, Y., Chen, Y., Wang, Z., Jiang, Y., Liu, A., Zhao, P., Wang, M., & Ahammed, G. J. (2017). Combined inoculation with multiple arbuscular mycorrhizal fungi improves growth, nutrient uptake and photosynthesis in cucumber seedlings. *Frontiers in Microbiology*, *8*, 1–11. doi:10.3389/fmicb.2017.02516 PMID:29312217

Chidi, C. L., Zhao, W., Thapa, P., Paudel, B., Chaudhary, S., & Khanal, N. R. (2022). Evaluation of traditional rain-fed agricultural terraces for soil erosion control through UAV observation in the middle mountain of Nepal. *Applied Geography (Sevenoaks, England)*, *148*, 102793. doi:10.1016/j.apgeog.2022.102793

Chun, S. C., Paramasivan, M., & Chandrasekaran, M. (2018). Proline accumulation influenced by osmotic stress in arbuscular mycorrhizal symbiotic plants. *Frontiers in Microbiology*, *9*, 2525. doi:10.3389/fmicb.2018.02525 PMID:30459731

Cline, W. R. (2007). *Global Warming and Agriculture: Impact Estimates by Country*. Peterson Institute.

Coeckelbergh, M. (2021). AI for climate: Freedom, justice, and other ethical and political challenges. *AI and Ethics*, *1*(1), 67–72. doi:10.100743681-020-00007-2

Couliably, S., Kamsu-Foguem, B., Kamissoko, D., & Traore, D. (2022). Deep learning for precision agriculture: A bibliometric analysis. *Intelligent Systems with Applications*, *200102*, 200102. Advance online publication. doi:10.1016/j.iswa.2022.200102

Cruft, S. (n.d.) The Global Cooling and Irrigation System [Video]. Global Cooling and Irrigation System.https://globalcoolingandirrigationsystem.org/

Cruz, J. R. D., Baldovino, R. G., Bandala, A. A., & Dadios, E. P. (2017). Water usage optimization of Smart Farm Automated Irrigation System using artificial neural network. *2017 5th International Conference on Information and Communication Technology*. IEEE. 10.1109/ICoICT.2017.8074668

Culhane, M., Cardona, C., Goldsmith, T. J., St. Charles, K., Suskovic, G., Thompson, B., & Starkey, M. (2018). Building an all-hazards agricultural emergency response system to maintain business continuity and promote the sustainable supply of food and agricultural products. *Cogent Food & Agriculture*, *4*(1), 1550907. doi:10.1080/23311932.2018.1550907

Davidson, D. (2016). Gaps in agricultural climate adaptation research. *Nature Climate Change*, *6*(5), 433–435. doi:10.1038/nclimate3007

de Lima, R. P., da Silva, A. R., & da Silva, Á. P. (2021). An R package for simulation of soil compaction induced by agricultural field traffic. *Soil & Tillage Research*, *206*, 104824. doi:10.1016/j.still.2020.104824

de Wrachien, D. (2014). The Great Potential of Micro-Irrigation Technology for Poor-Rural Communities. *Irrigation & Drainage Systems Engineering*, *3*(2). doi:10.4172/2168-9768.1000e124

Debangshi, U. (2021). Drones - Applications in Agriculture. *Chronicle of Bioresource Management*, *5*(3), 115–120. doi:10.5281/zenodo.5554734

Déqué, M., Somot, S., Sanchez-Gomez, E., Goodess, C. M., Jacob, D., Lenderink, G., & Christensen, O. B. (2012). The spread amongst ENSEMBLES regional scenarios: Regional climate models, driving general circulation models and interannual variability. *Climate Dynamics*, *38*(5), 951–964. doi:10.100700382-011-1053-x

Diagne, N., Ngom, M., Djighaly, P. I., Fall, D., Hocher, V., & Svistoonoff, S. (2020). Roles of arbuscular mycorrhizal fungi on plant growth and performance: Importance in biotic and abiotic stressed regulation. *Diversity (Basel)*, *12*(10), 1–25. doi:10.3390/d12100370

Driouech, F., Déqué, M., & Sánchez-Gómez, E. (2010). Weather regimes—Moroccan precipitation link in a regional climate change simulation. *Global and Planetary Change*, *72*(1-2), 1–10. doi:10.1016/j.gloplacha.2010.03.004

Dyer, G. A., López-Feldman, A., Yúnez-Naude, A., & Taylor, J. E. (2014). Genetic erosion in maize's center of origin. *Proceedings of the National Academy of Sciences of the United States of America, 111*(39), 14094–14099. doi:10.1073/pnas.1407033111

El Ajhar, E. A., Douae, E. K., Ahmed, E. B., Khadija, E. K., & Driss, B. (2015). *Evolution De La Pluviométrie De 1960 À 2015 Au Maroc,* 10. Filière céréalière, Ministère de l'agriculture. https://www.agriculture.gov.ma/fr/filiere/Cerealiere

El Amerany, F., Rhazi, M., Wahbi, S., Taourirte, M., & Meddich, A. (2020). The effect of chitosan, arbuscular mycorrhizal fungi, and compost applied individually or in combination on growth, nutrient uptake, and stem anatomy of tomato. *Scientia Horticulturae, 261*, 109015. doi:10.1016/j.scienta.2019.109015

El Bilali, H., & Allahyari, M. S. (2018). Transition towards sustainability in agriculture and food systems: Role of information and communication technologies. *Information Processing in Agriculture, 5*(4), 456–464. doi:10.1016/j.inpa.2018.06.006

Engel, S., Pagiola, S., Wunder, S., & CIFOR. (2008). Moving ahead with REDD. In Angelson (Ed.), *Issues, options and implications. Bogor,* (*Vol. 65*, Issue 4). CIROR. http://scholar.google.com/scholar?hl=en&btnG=Search&q=intitle:Moving+Ahead+with+REDD#1

Engemann, K. J., & Henderson, D. M. (2012). Business continuity and risk management: essentials of organizational resilience. Rothstein Associates Inc. Food and Agriculture Organization of the United Nations. (n.y). *Q&A: COVID-19 pandemic – impact on food and agriculture.* https://www.fao.org/2019-ncov/q-and-a/impact-on-food-and-agriculture/en/

Estrada, F., Perron, P., & Martínez-López, B. (2013). Statistically derived contributions of diverse human influences to twentieth-century temperature changes. *Nature Geoscience, 6*(12), 1050–1055. doi:10.1038/ngeo1999

Etesami, H., Jeong, B. R., & Glick, B. R. (2021). Contribution of arbuscular mycorrhizal fungi, phosphate–solubilizing bacteria, and silicon to P uptake by plant. *Frontiers in Plant Science, 12*, 699618. doi:10.3389/fpls.2021.699618 PMID:34276750

Evelin, H., Devi, T. S., Gupta, S., & Kapoor, R. (2019). Mitigation of salinity stress in plants by arbuscular mycorrhizal symbiosis: Current understanding and new challenges. *Frontiers in Plant Science, 10*, 1–21. doi:10.3389/fpls.2019.00470 PMID:31031793

FAO. (2010). *Global Forest Resource Assessment.* [Data set]. FAO.

FAO. (2017). Sustainable Woodfuel For Food Security, a Smart Choice: gree, renewable and affordable. In FAO working paper.

FAO. (2018). *The state of world ' s forests: forest pathways to sustainable development.* FAO.

FAO. (2019). *The statistical database.* FAOSTAT.

FAO. (2020). *Global Forest Resources Assessment 2020; Report United Republic of Tanzania*. FAO.

Farooq, M., Hussain, M., & Siddique, K. H. M. (2014). Drought Stress in Wheat during Flowering and Grain-filling Periods. *Critical Reviews in Plant Sciences, 33*(4), 331–349. doi:10.1080/07352689.2014.875291

Farooq, M., Hussain, M., Wakeel, A., & Siddique, K. H. M. (2015). Salt stress in maize: Effects, resistance mechanisms, and management. A review. *Agronomy for Sustainable Development, 35*(2), 461–481. doi:10.100713593-015-0287-0

Faysse, N. (2015). The rationale of the Green Morocco Plan: Missing links between goals and implementation. *The Journal of North African Studies, 20*(4). https://www.tandfonline.com/doi/abs/10.1080/13629387.2015.1053112

Feng, J., Lv, W., Xu, J., Huang, Z., Rui, W., Lei, X., Ju, X., & Li, Z. (2022). Overlapping root architecture and gene expression of nitrogen transporters for nitrogen acquisition of tomato plants colonized with isolates of *Funneliformis mosseae* in hydroponic production. *Plants, 11*(9), 1176. doi:10.3390/plants11091176 PMID:35567176

Fernández-Escobar, R., De la Rosa, R., Leon, L., Gomez, J. A., Testi, L., Orgaz, F., & Masallem, M. (2013). Evolution and sustainability of the olive production systems. *Options Méditerranéennes, 106*, 11–42.

Fernández-Romero, M. L., Parras-Alcántara, L., Lozano-García, B., Clark, J. M., & Collins, C. D. (2016). Soil quality assessment based on carbon stratification index in different olive grove management practices in Mediterranean areas. *Catena, 137*, 449–458. doi:10.1016/j.catena.2015.10.019

Field, K. J., & Pressel, S. (2018). Unity in diversity: Structural and functional insights into the ancient partnerships between plants and fungi. *The New Phytologist, 220*(4), 996–1011. doi:10.1111/nph.15158 PMID:29696662

Fountas, S., Carli, G., Sørensen, C. G., Tsiropoulos, Z., Cavalaris, C., Vatsanidou, A., Liakos, B., Canavari, M., Wiebensohn, J., & Tisserye, B. (2015). Farm management information systems: Current situation and future perspectives. *Computers and Electronics in Agriculture, 115*, 40–50. doi:10.1016/j.compag.2015.05.011

Fróna, D., Szenderák, J., & Harangi-Rákos, M. (2019). The Challenge of Feeding the World. *Sustainability, 11*(20), 5816. doi:10.3390u11205816

Fue, K., Geofrey, A., & Mlozi, P. M. R. S. (2016). Analyzing Usage of Crowdsourcing Platform "Ushaurikilimo" by Pastoral and Agro-Pastoral Communities in Tanzania. *International Journal of Instructional Technology and Distance Learning, 13*(January), 3–23.

Fuss, S., Canadell, J. G., Peters, G. P., Tavoni, M., Andrew, R. M., Ciais, P., Jackson, R. B., Jones, C. D., Kraxner, F., Nakicenovic, N., Le Quéré, C., Raupach, M. R., Sharifi, A., Smith, P., & Yamagata, Y. (2014). Commentary: Betting on negative emissions. *Nature Climate Change, 4*(10), 850–853. doi:10.1038/nclimate2392

Gałązka, A., Gawryjołek, K., Grządziel, J., & Księżak, J. (2017). Effect of different agricultural management practices on soil biological parameters including glomalin fraction. *Plant, Soil and Environment*, *63*(7), 300–306. doi:10.17221/207/2017-PSE

Gałązka, A., Niedźwiecki, J., Grządziel, J., & Gawryjołek, K. (2020). Evaluation of changes in Glomalin-Related Soil Proteins (GRSP) content, microbial diversity and physical properties depending on the type of soil as the important biotic determinants of soil quality. *Agronomy (Basel)*, *10*(9), 1279. doi:10.3390/agronomy10091279

Gao, G., Wang, M., & Huang, H. (2021). (PREPRINT). *Agricultural Irrigation Area Prediction Based on Improved Random Forest Model*. Advance online publication. doi:10.21203/rs.3.rs-156767/v1

Gao, X., Guo, H., Zhang, Q., Guo, H., Zhang, L., Zhang, C., Gou, Z., Liu, Y., Wei, J., Chen, A., Chu, Z., & Zeng, F. (2020). Arbuscular mycorrhizal fungi (AMF) enhanced the growth, yield, fiber quality and phosphorus regulation in upland cotton (*Gossypium hirsutum* L.). *Scientific Reports*, *10*(1), 1–13. doi:10.103841598-020-59180-3 PMID:32034269

Garcia, C. L., Dattamudi, S., Chanda, S., & Jayachandran, K. (2019). Effect of Salinity Stress and Microbial Inoculations on Glomalin Production and Plant Growth Parameters of Snap Bean (*Phaseolus vulgaris*). *Agronomy (Basel)*, *9*(9), 545. doi:10.3390/agronomy9090545

García, L., Parra, L., Jimenez, J. M., Lloret, J., Lorenz, P., García, L., Parra, L., Jimenez, J. M., & Lloret, J., & Irri-, P. L. I. S. (2022). IoT-Based Smart Irrigation Systems : An Overview on the Recent Trends on Sensors and IoT Systems for Irrigation in Precision Agriculture To cite this version : IoT-Based Smart Irrigation Systems : An Overview on. *Sensors (Basel)*, *2020*(20).

Gasmi, I., Aljoumani, B., Sànchez-Espigares, J. A., Mechergui, M., & Moussa, M. (2019). A linear mixed effect (LME) model for soil nutrients and soil salinity changes based on two localized irrigation techniques (drip irrigation and buried diffuser). *Arabian Journal of Geosciences*, *12*(11), 1–12. doi:10.100712517-019-4400-6

Ghanbarzadeh, Z., Mohsenzadeh, S., Rowshan, V., & Moradshahi, A. (2019). Evaluation of the growth, essential oil composition and antioxidant activity of *Dracocephalum moldavica* under water deficit stress and symbiosis with *Claroideoglomus etunicatum* and *Micrococcus yunnanensis*. *Scientia Horticulturae*, *256*, 108652. doi:10.1016/j.scienta.2019.108652

Ghosh, U. K., Islam, M. N., Siddiqui, M. N., Cao, X., & Khan, M. A. R. (2022). Proline, a multifaceted signalling molecule in plant responses to abiotic stress: Understanding the physiological mechanisms. *Plant Biology*, *24*(2), 227–239. doi:10.1111/plb.13363 PMID:34796604

Giacomo, R., & David, G. (2018). E-Argriculture In Action: Drones For Agriculture. In E-Agriculture in Action: Drones for Agriculture.

Gibson, D. (2011). *Managing Risks in Information Systems*. Jones & Bartlett Learning.

Giovannini, L., Palla, M., Agnolucci, M., Avio, L., Sbrana, C., Turrini, A., & Giovannetti, M. (2020). Arbuscular mycorrhizal fungi and associated microbiota as plant biostimulants: Research strategies for the selection of the best performing inocula. *Agronomy (Basel)*, *10*(1), 106. doi:10.3390/agronomy10010106

Goicoechea, N., Baslam, M., Erice, G., & Irigoyen, J. J. (2014). Increased photosynthetic acclimation in alfalfa associated with arbuscular mycorrhizal fungi (AMF) and cultivated in greenhouse under elevated CO_2. *Journal of Plant Physiology*, *171*(18), 1774–1781. doi:10.1016/j.jplph.2014.07.027 PMID:25240322

Golubkina, N., Amagova, Z., Matsadze, V., Zamana, S., Tallarita, A., & Caruso, G. (2020). Effects of arbuscular mycorrhizal fungi on yield, biochemical characteristics, and elemental composition of garlic and onion under selenium supply. *Plants*, *9*(1), 84. doi:10.3390/plants9010084 PMID:31936528

Gornall, J., Betts, R., Burke, E., Clark, R., Camp, J., Willett, K., & Wiltshire, A. (2010). Implications of climate change for agricultural productivity in the early twenty-first century. *Philosophical Transactions of the Royal Society of London. Series B, Biological Sciences*, *365*(1554), 2973–2989. doi:10.1098/rstb.2010.0158

Hájek, P. & Urbancova, Hana. (2013). Using of Business Continuity Standards in Agriculture, Industry and ICT. *AGRIS On-Line Papers in Economics and Informatics*, *5*, 55–67.

Haji, M. (2015). Youth employment in Tanzania: Taking stock of the evidence and knowledge gaps. International Development Research Centre, 1–30.

Hansson, A., Fridahl, M., Haikola, S., Yanda, P., Pauline, N., & Mabhuye, E. (2020). Preconditions for bioenergy with carbon capture and storage (BECCS) in sub-Saharan Africa: The case of Tanzania. *Environment, Development and Sustainability*, *22*(7), 6851–6875. doi:10.100710668-019-00517-y

Harrison, S., Herbohn, J., & Niskanen, A. (2002). Non-industrial, smallholder, small-scale and family forestry: What's in a name? *Small-Scale Forest Economics. Management and Policy*, *1*(1), 1–11. doi:10.100711842-002-0001-3

Harris-Valle, C., Esqueda, M., Gutiérrez, A., Castellanos, A. E., Gardea, A. A., & Berbara, R. (2018). Physiological response of *Cucurbita pepo* var. pepo mycorrhized by *Sonoran desert* native arbuscular fungi to drought and salinity stresses. *Brazilian Journal of Microbiology*, *49*(1), 45–53. doi:10.1016/j.bjm.2017.04.005 PMID:28887008

Hashem, A., Alqarawi, A. A., Radhakrishnan, R., Al-Arjani, A. B. F., Aldehaish, H. A., Egamberdieva, D., & Abd Allah, E. F. (2018). Arbuscular mycorrhizal fungi regulate the oxidative system, hormones and ionic equilibrium to trigger salt stress tolerance in *Cucumis sativus* L. *Saudi Journal of Biological Sciences*, *25*, 1102–1114. doi:10.1016/j.sjbs.2018.03.009 PMID:30174509

HCP. (2022). *Site institutionnel du Haut-Commissariat au Plan du Royaume du Maroc*. Haut-Commissariat. https://www.hcp.ma/

Held, C., Jacovelli, P., Techel, G., Nutto, L., Wathum, G., & Wittmann, N. (2017b). *Tanzanian Wood Product Market Study Report for the Forestry Development Trust*. 139.

Held, C., Jacovelli, P., Techel, G., Nutto, L., Wathum, G., & Wittmann, N. (2017a). *Tanzanian Wood Product Market Study*. Issue November.

Herring, S. C., Christidis, N., Hoell, A., Kossin, J. P., Schreck, C. J. III, & Stott, P. A. (2018). Explaining Extreme Events of 2016 From A Climate Perspective. *Bulletin of the American Meteorological Society*, *99*(1), S1–S157. https://journals.ametsoc.org/view/journals/bams/99/1/bams-explainingextremeevents2016.1.xml. doi:10.1175/BAMS-ExplainingExtremeEvents2016.1

He, Y., Yang, R., Lei, G., Li, M., Li, T., Zhan, F., & Li, Y. (2021). Arbuscular mycorrhizal fungus–induced decrease in phosphorus loss due to leaching in red soils under simulated heavy rainfall. *Journal of Soils and Sediments*, *21*(2), 881–889. doi:10.100711368-020-02849-z

Hilal, M. H., & Hilal, M. M. (2000). Application of magnetic technologies in desert agriculture. I-Seed germination and seedling emergence of some crops in a saline calcareous soil. *Egyptian Journal of Soil Science*, *40*(3), 413–422.

Hmimsa, Y., & Ater, M. (2008). Agrodiversity in the traditional agrosystems of the Rif mountains (North of Morocco). *Biodiversity (Nepean)*, *9*(1-2), 78–81. doi:10.1080/14888386.2008.9712890

Hoogenboom, G. (2020). Agronomic data and the coronavirus epidemic. *The Journal of Agricultural Science*, *158*(3), 161–162. doi:10.1017/S0021859620000623

Hu, J., Lin, X., Wang, J., Dai, J., Cui, X., Chen, R., & Zhang, J. (2009). Arbuscular mycorrhizal fungus enhances crop yield and P-uptake of maize (*Zea mays* L.): A field case study on a sandy loam soil as affected by long-term P-deficiency fertilization. *Soil Biology & Biochemistry*, *41*(12), 2460–2465. doi:10.1016/j.soilbio.2009.09.002

Hu, Y., Xu, X., Wu, F., Sun, Z., Xia, H., Meng, Q., Huang, W., Zhou, H., Gao, J., Li, W., Peng, D., & Xiao, X. (2020). Estimating forest stock volume in Hunan Province, China, by integrating in situ plot data, Sentinel-2 images, and linear and machine learning regression models. *Remote Sensing*, *12*(1), 186. Advance online publication. doi:10.3390/rs12010186

Inglezakis, V. J. (2016). Extraterrestrial environment. In *Environment and Development* (pp. 453–498). Elsevier. doi:10.1016/B978-0-444-62733-9.00007-1

IPCC. (2014). Climate Change 2014 Part A: Global and Sectoral Aspects. In Climate Change 2014: Impacts, Adaptation, and Vulnerability. Part A: Global and Sectoral Aspects. Contribution of Working Group II to the Fifth Assessment Report of the Intergovernmental Panel on Climate Change. IEEE. papers2://publication/uuid/B8BF5043-C873-4AFD-97F9-A630782E590D

IPCC. (2014). *Climate Change 2014: Mitigation of Climate Change. Contribution of Working Group III to the Fifth Assessment Report of the Intergovernmental Panel on Climate Change* (O. R. Edenhofer, Y. Pichs-Madruga, E. Sokona, S. Farahani, K. Kadner, & A. Seyboth, Eds.). Cambridge University Press.

Javaid, M., Haleem, A., Khan, I. H., & Suman, R. (2022). *Understanding the potential applications of Artificial Intelligence in Agriculture Sector*. Advanced Agrochem. doi:10.1016/j.aac.2022.10.001

Jeong, J. H., Resop, J. P., Mueller, N. D., Fleisher, D. H., Yun, K., Butler, E. E., Timlin, D. J., Shim, K. M., Gerber, J. S., Reddy, V. R., & Kim, S. H. (2016). Random Forests for Global and Regional Crop Yield Predictions. *PLoS One*, *11*(6), e0156571. doi:10.1371/journal.pone.0156571 PMID:27257967

Jian-Bin, H., Shao-Wu, W., Yong, L., Zong-Ci, Z., & Xin-Yu, W. (2012). Debates on the causes of global warming. *Advances in Climate Change Research*, *3*(1), 38–44. doi:10.3724/SP.J.1248.2012.00038

Jiang, Y., Zhou, H., Chen, L., Yuan, Y., Fang, H., Luan, L., Chen, Y., Wang, X., Liu, M., Li, H., Peng, X., & Sun, B. (2018). Nematodes and Microorganisms Interactively Stimulate Soil Organic Carbon Turnover in the Macroaggregates. *Frontiers in Microbiology*, *9*, 2803. doi:10.3389/fmicb.2018.02803 PMID:30534114

Jiawei, H., Micheline, K., & Jian, P. (2011). *Data Mining Concepts and Techniques, Third Edition*. Elsevier.

Jones, B. F., & Olken, B. A. (2010). Climate shocks and exports. *The American Economic Review*, *100*(2), 454–459. doi:10.1257/aer.100.2.454

Jung, J., Maeda, M., Chang, A., Bhandari, M., Ashapure, A., & Landivar-Bowles, J. (2021). The potential of remote sensing and artificial intelligence as tools to improve the resilience of agriculture production systems. *Current Opinion in Biotechnology*, *70*, 15–22. doi:10.1016/j.copbio.2020.09.003 PMID:33038780

Kaggle (n.d). Computational Complexity of Machine Learning Models –II. *Kaggle*. https://www.kaggle.com/general/263127

Kamilaris, A., Kartakoullis, A., & Prenafeta-Boldú, F. X. (2017). A review on the practice of big data analysis in agriculture. *Computers and Electronics in Agriculture*, *143*, 23–37. doi:10.1016/j.compag.2017.09.037

Kaniewski, D., Van Campo, E., Boiy, T., Terral, J. F., Khadari, B., & Besnard, G. (2012). Primary domestication and early uses of the emblematic olive tree: Palaeobotanical, historical and molecular evidence from the Middle East. *Biological Reviews of the Cambridge Philosophical Society*, *87*(4), 885–899. doi:10.1111/j.1469-185X.2012.00229.x PMID:22512893

Karmaoui, A. (2022a). Network Analysis on Artificial Intelligence in Agriculture, a Bibliometric Review. In J. Marx Gómez & M. R. Lorini (Eds.), *Digital Transformation for Sustainability. Progress in IS*. Springer. doi:10.1007/978-3-031-15420-1_20

Karmaoui, A. (2022b). Ordovician-Cambrian Palaeontological Heritage of Zagora Province: A Bibliometric Analysis from 1984 to 2020 (Anti-Atlas, Morocco). *Geoheritage*, *14*(2), 55. doi:10.100712371-022-00695-8

Karmaoui, A. (2023). Flood vulnerability in developing countries, international collaboration, and network visualization: A Bibliometric analysis. In V. O. Oladokun, D. Proverbs, O. A. Adebimpe, & T. Adedeji (Eds.), *Handbook of Flood Risk Management in Developing Countries* (p. 384). Routledge. https://www.routledge.com/Handbook-of-Flood-Risk-Management-in-Developing-Countries/Oladokun-Proverbs-Adebimpe-Adedeji/p/book/9780367749743

Karmaoui, A., El Jaafari, S., Chaachouay, H., & Hajj, L. (2022b). The socio-ecological system of the pre-Sahara zone of Morocco: A conceptual framework to analyse the impact of drought and desertification. *GeoJournal*, *87*(6), 4961–4974. doi:10.100710708-021-10546-8

Kassout, J., Barbara, H., Ivorra, S., Terral, J. F., & Ater, M. (2016). Etude préliminaire de la variation de caractères anatomiques du bois d'une forme spontanée et de sept variétés traditionnelles d'olivier (Olea europaeaL.) de la région Nord du Maroc (Chefchaouen et Ouazzane). L'oléiculture au Maroc de la préhistoire à nos jours: pratiques, diversité, adaptation, usages, commerce et politiques, eds M. Ater, L. Essalouh, H. Ilbert, A. Moukhli, and B. Khadari (Montpellier: CIHEAM, Options Méditerranéennes: Série A. Séminaires Méditerranéens, (118), 181-189.

Kassout, J., Ater, M., Ivorra, S., Barbara, H., Limier, B., Ros, J., Girard, V., Paradis, L., & Terral, J.-F. (2021). Resisting aridification: Adaptation of sap conduction performance in Moroccan wild olive subspecies distributed over an aridity gradient. *Frontiers in Plant Science*, *12*, 1288. doi:10.3389/fpls.2021.663721 PMID:34276726

Kaur, S., & Suseela, V. (2020). Unraveling Arbuscular Mycorrhiza-Induced Changes in Plant Primary and Secondary Metabolome. *Metabolites*, *10*(8), 335. doi:10.3390/metabo10080335 PMID:32824704

Khadari, B., Charafi, J., Moukhli, A., & Ater, M. (2008). Substantial genetic diversity in cultivated Moroccan olive despite a single major cultivar: A paradoxical situation evidenced by the use of SSR loci. *Tree Genetics & Genomes*, *4*(2), 213–221. doi:10.100711295-007-0102-4

Khalid, N. A. C. & S. (2018). Reshaping the Terrain Landscape Restoration in Tanzania. *Global Landscape Forum,* 1–4.

Kideghesho. (2015). Realities on Deforestation in Tanzania — Trends, Drivers, Implications and the Way Forward. In *Intech* (*Vol. 32*, pp. 137–144). http://www.trafficj.org/publication/17_Timber-trade-East-Southern-Africa.pdf

Kim, B., Yum, S., Kim, Y.-Y., Yun, N., Shin, S.-Y., & You, S. (2014). An Analysis of Factors Relating to Agricultural Machinery Farm-Work Accidents Using Logistic Regression. *Journal of Biosystems Engineering*, *39*(3), 151–157. doi:10.5307/JBE.2014.39.3.151

Kim, D. G., Thomas, A. D., Pelster, D., Rosenstock, T. S., & Sanz-Cobena, A. (2016). Greenhouse gas emissions from natural ecosystems and agricultural lands in sub-Saharan Africa: Synthesis of available data and suggestions for further research. *Biogeosciences*, *13*(16), 4789–4809. doi:10.5194/bg-13-4789-2016

Koch, M., & Missimer, T. M. (2016). Water resources assessment and management in drylands. *Water (Basel)*, *8*(6), 239. doi:10.3390/w8060239

Kok, Z. H., Shariff, A. R. M., Alfatni, M. S. M., & Khairunniza-Bejo, S. (2021). Support vector machine in precision agriculture: A review. *Computers and Electronics in Agriculture*, *191*, 106546. doi:10.1016/j.compag.2021.106546

Komarek, A. M., De Pinto, A., & Smith, V. H. (2020). A review of types of risks in agriculture: What we know and what we need to know. *Agricultural Systems*, *178*, 102738. doi:10.1016/j.agsy.2019.102738

Kour, D., Rana, K. L., Yadav, A. N., Yadav, N., Kumar, M., Kumar, V., Vyas, P., Dhaliwal, H. S., & Saxena, A. K. (2020). Microbial biofertilizers: Bioresources and eco-friendly technologies for agricultural and environmental sustainability. *Biocatalysis and Agricultural Biotechnology*, *23*, 101487. doi:10.1016/j.bcab.2019.101487

Kumar, V. (1998). An early warning system for agricultural drought in an arid region using limited data. *Journal of Arid Environments*, *40*(2), 199–209. doi:10.1006/jare.1998.0437

Lal, R. (2004). Soil carbon sequestration impacts on global climate change and food security. *science*, *304*(5677), 1623-1627.

Lal, R. (2008). Carbon sequestration. *Philosophical Transactions of the Royal Society of London. Series B, Biological Sciences*, *363*(1492), 815–830. doi:10.1098/rstb.2007.2185 PMID:17761468

Landrigan, P. J., Fuller, R., Acosta, N. J. R., Adeyi, O., Arnold, R., Basu, N., Baldé, A. B., Bertollini, R., Bose-O'Reilly, S., Boufford, J. I., Breysse, P. N., Chiles, T., Mahidol, C., Coll-Seck, A. M., Cropper, M. L., Fobil, J., Fuster, V., Greenstone, M., Haines, A., & Zhong, M. (2018). The Lancet Commission on pollution and health. *Lancet*, *391*(10119), 462–512. doi:10.1016/S0140-6736(17)32345-0 PMID:29056410

Lanfranco, L., Fiorilli, V., Venice, F., & Bonfante, P. (2018). Strigolactones cross the kingdoms: Plants, fungi, and bacteria in the arbuscular mycorrhizal symbiosis. *Journal of Experimental Botany*, *69*(9), 2175–2188. doi:10.1093/jxb/erx432 PMID:29309622

Le Treut, H., Somerville, R., Cubasch, U., Ding, Y., Mauritzen, C., & Mokssit, A. (2007). Historical overview of climate change science. In S. Solomon, D. Qin, M. Manning, Z. Chen, M. Marquis, & K. B. Averyt, (Eds.), *Climate change 2007: The physical science basis. Contribution of Working Group I to the Fourth Assessment Report of the Intergovernmental Panel on Climate Change*. Cambridge University Press.

Leal Filho, W., Totin, E., Franke, J. A., Andrew, S. M., Abubakar, I. R., Azadi, H., & Team, T. G. A. M. I. (2022). Understanding responses to climate-related water scarcity in Africa. *The Science of the Total Environment*, *806*, 150420. doi:10.1016/j.scitotenv.2021.150420 PMID:34571220

Leal Filho, W., Wall, T., Mucova, S. A. R., Nagy, G. J., Balogun, A. L., Luetz, J. M., & Gandhi, O. (2022). Deploying artificial intelligence for climate change adaptation. *Technological Forecasting and Social Change*, *180*, 121662. doi:10.1016/j.techfore.2022.121662

Lee, R. H., Navarro-Navarro, L. A., Ley, A. L., Hartfield, K., Tolleson, D. R., & Scott, C. A. (2021). Spatio-temporal dynamics of climate change, land degradation, and water insecurity in an arid rangeland: The Río San Miguel watershed, Sonora, Mexico. *Journal of Arid Environments*, *193*, 104539. doi:10.1016/j.jaridenv.2021.104539

Leydesdorff, L., & Rafols, I. (2012). Interactive overlays: A new method for generating global journal maps from Web-of-Science data. *Journal of Informetrics*, *6*(2), 318–332. doi:10.1016/j.joi.2011.11.003

Liang, J. F., An, J., Gao, J. Q., Zhang, X. Y., Song, M. H., & Yu, F. H. (2019). Interactive effects of biochar and AMF on plant growth and greenhouse gas emissions from wetland microcosms. *Geoderma*, *346*, 11–17. doi:10.1016/j.geoderma.2019.03.033

Lionello, P., & Scarascia, L. (2018). The relation between climate change in the Mediterranean region and global warming. *Regional Environmental Change*, *18*(5), 1481–1493. doi:10.100710113-018-1290-1

Liu, J. G., Zhou, Q., Guo, Q., Yang, Z. H., Xie, F., & Han, J. T. (2017). Knowledge diffusion of dynamical network in terms of interaction frequency. *Scientific Reports*, *7*(1), 1–7. doi:10.103841598-017-11057-8 PMID:28883456

Liu, X., Ghazali, K. H., Han, F., & Mohamed, I. I. (2021). Automatic Detection of Oil Palm Tree from UAV Images Based on the Deep Learning Method. *Applied Artificial Intelligence*, *35*(1), 13–24. doi:10.1080/08839514.2020.1831226

Liu, Y., Li, G., Zeng, P., Zhang, X., Tian, T., Feng, H., & Che, Y. (2022). Challenge of rainwater harvesting in Shanghai, China: A public psychological perspective. *Journal of Environmental Management*, *318*, 115584. doi:10.1016/j.jenvman.2022.115584 PMID:35753131

Li, X., Liu, L., Xie, J., Wang, Z., Yang, S., Zhang, Z., Qi, S., & Li, Y. (2020). Optimizing the quantity and spatial patterns of farmland shelter forests increases cotton productivity in arid lands. *Agriculture, Ecosystems & Environment*, *292*, 106832. doi:10.1016/j.agee.2020.106832

Lloret, F., de la Riva, E. G., Pérez-Ramos, I. M., Marañón, T., Saura-Mas, S., Díaz-Delgado, R., & Villar, R. (2016). Climatic events inducing die-off in Mediterranean shrublands: Are species' responses related to their functional traits? *Oecologia*, *180*(4), 961–973. doi:10.100700442-016-3550-4 PMID:26801493

Lopez-Bellido, P. J., Lopez-Bellido, L., Fernandez-Garcia, P., Muñoz-Romero, V., & Lopez-Bellido, F. J. (2016). Assessment of carbon sequestration and the carbon footprint in olive groves in Southern Spain. *Carbon Management*, *7*(3-4), 161–170. doi:10.1080/17583004.2016.1213126

Lorencová, E., Frélichová, J., Nelson, E., & Vačkář, D. (2013). Past and future impacts of land use and climate change on agricultural ecosystem services in the Czech Republic. *Land Use Policy*, *33*, 183–194. doi:10.1016/j.landusepol.2012.12.012

Lorite, I. J., Gabaldón-Leal, C., Ruiz-Ramos, M., Belaj, A., De La Rosa, R., León, L., & Santos, C. (2018). Evaluation of olive response and adaptation strategies to climate change under semi-arid conditions. *Agricultural Water Management, 204*, 247–261. doi:10.1016/j.agwat.2018.04.008

Ludwig, F., & Asseng, S. (2006). Climate change impacts on wheat production in a Mediterranean environment in Western Australia. *Agricultural Systems, 90*(1), 159–179. https://doi.org/10.1016/j.agsy.2005.12.002

Lukumbuzya, K., & Sianga, C. (2017). Overview of the Timber Trade in East and Southern Africa: National Perspectives and Regional Trade Linkage. *Traffic Joint Report.* http://www.trafficj.org/publication/17_Timber-trade-East-Southern-Africa.pdf

Lupala, Z. J., Lusambo, L. P., & Ngaga, Y. M. (2014). Management, Growth, and Carbon Storage in Miombo Woodlands of Tanzania. *International Journal of Forestry Research, 2014*, 1–11. doi:10.1155/2014/629317

Lusambo, L. P., Nyanda, S. S., & Mhando, D. G. (2021). Profitability Analysis of Tree Growing in the Southern Highlands of Tanzania. *International Journal of Forestry Research, 2021*, 1–10. doi:10.1155/2021/8872211

Lüthi, C., Panasar, A., Schütze, T., Norström, A., McConville, J., & Saywell, D. (2010). Sustainable Sanitation in Cities: A Framework for Action. *Sustainable Sanitation Alliance.* https://publications.lib.chalmers.se/records/fulltext/192469/local_192469.pdf

Lu, W. C. (2017). Greenhouse gas emissions, energy consumption and economic growth: A panel cointegration analysis for 16 Asian countries. *International Journal of Environmental Research and Public Health, 14*(11), 15. doi:10.3390/ijerph14111436 PMID:29165399

Maazouz, S. (2016). *Le développement de l'irrigation durable au Maroc.* AgriMaroc. https://www.agrimaroc.ma/le-developpement-de-lirrigation-durable-au-maroc/

Madawala, H. M. S. P. (2021). Arbuscular mycorrhizal fungi as biofertilizers: Current trends, challenges, and future prospects. *Biofertilizers*, 83-93. doi:10.1016/B978-0-12-821667-5.00029-4

Mahfudz, M., Saleh, S., Antara, M., Anshary, A., Bachri, S., Made, U., Hasanah, U., & Rauf, R. A. (2019). Adoption and advantages of eco-friendly technology application at the shallot farming system in Indonesia. *Agronomy Research (Tartu), 17*, 1679–1687. doi:10.15159/AR.19.188

Malamiri, H. R. G., Aliabad, F. A., Shojaei, S., Morad, M., & Band, S. S. (2021). A study on the use of UAV images to improve the separation accuracy of agricultural land areas. *Computers and Electronics in Agriculture, 184*, 106079. doi:10.1016/j.compag.2021.106079

Malunguja, G. K., Devi, A., Kilonzo, M., & Rubanza, C. D. K. (2020). Climate change mitigation through carbon dioxide (CO2) sequestration in community reserved forests of northwest Tanzania. *Archives of Agriculture and Environmental Science, 5*(3), 231–240. doi:10.26832/24566632.2020.050301

MAPMDREF. (2019). Filière céréalière. Retrieved 27 December 2021, from https://www.agriculture.gov.ma/fr/filiere/Cerealiere

MAPMDREF. (2021). *Publication d'un rapport sur le bilan et impacts du Plan Maroc Vert.* Ministère de l'agriculture. https://www.agriculture.gov.ma/fr/actualites/publication-dun-rapport-sur-le-bilan-et-impacts-du-plan-maroc-vert

Mariotte, P., Mehrabi, Z., Bezemer, T. M., De Deyn, G. B., Kulmatiski, A., Drigo, B., Veen, G. F. C., van der Heijden, M. G. A., & Kardol, P. (2018). Plant–soil feedback: Bridging natural and agricultural sciences. *Trends in Ecology & Evolution, 33*(2), 129–142. doi:10.1016/j.tree.2017.11.005 PMID:29241940

Martins Silva, J. P., Marques da Silva, M. L., Ferreira da Silva, E., Fernandes da Silva, G., Ribeiro de Mendonça, A., Cabacinha, C. D., Araújo, E. F., Santos, J. S., Vieira, G. C., Felix de Almeida, M. N., & Fernandes, M. R. (2019). Computational techniques applied to volume and biomass estimation of trees in Brazilian savanna. *Journal of Environmental Management, 249*(August), 109368. doi:10.1016/j.jenvman.2019.109368 PMID:31421480

Masanja, N. M., & Mkumbo, H. (2020). The Application of Open Source Artificial Intelligence as an Approach to Frugal Innovation in Tanzania. [IJRIAS]. *International Journal of Research and Innovation in Applied Science, 5*(3), 2454–6194. www.rsisinternational.org

Matesanz, S., & Valladares, F. (2014). Ecological and evolutionary responses of Mediterranean plants to global change. *Environmental and Experimental Botany, 103*, 53–67. doi:10.1016/j.envexpbot.2013.09.004

Mateus, I. D., Masclaux, F. G., Aletti, C., Rojas, E. C., Savary, R., Dupuis, C., & Sanders, I. R. (2019). Dual RNA-seq reveals large-scale non-conserved genotype × genotype-specific genetic reprograming and molecular crosstalk in the mycorrhizal symbiosis. *Multidisciplinary Journal of Microbial Ecology, 13*(5), 1226–1238. doi:10.103841396-018-0342-3 PMID:30647457

Mattingly, D. J., Bokbot, Y., Sterry, M., Cuénod, A., Fenwick, C., Gatto, M. C., & Mugnai, N. (2017). Long-term history in a Moroccan oasis zone: The Middle Draa Project 2015. *Journal of African Archaeology, 15*(2), 141–172. doi:10.1163/21915784-12340009

Meddich, A., Ait Rahou, Y., Boutasknit, A., Ait-El-Mokhtar, M., Fakhech, A., Lahbouki, S., Benaffari, W., Ben-laouanen R., & Wahbi, S. (2021). Role of mycorrhizal fungi in improving the tolerance of melon (*Cucumus melo*) under two water deficit partial root drying and regulated deficit irrigation. *Plant Biosystems-An International Journal Dealing with all Aspects of Plant Biology*, 1-22.

Meddich, A., Oufdou, K., Boutasknit, A., Raklami, A., Tahiri, A., Ben-Laouane, R., Ait-El-Mokhtar, M., Anli, M., Mitsui, T., Wahbi, S., & Baslam, M. (2020). Use of Organic and Biological Fertilizers as Strategies to Improve Crop Biomass, Yields and Physicochemical Parameters of Soil. In Nutrient Dynamics for Sustainable Crop Production, pp. 247–288. Springer. doi:10.1007/978-981-13-8660-2_9

Meddich, A., Ait El Mokhtar, M., Bourzik, W., Mitsui, T., Baslam, M., & Hafidi, M. (2018). Optimizing growth and tolerance of date palm (*Phoenix dactylifera* L.) to drought, salinity, and vascular fusarium-induced wilt (*Fusarium oxysporum*) by application of arbuscular mycorrhizal fungi (AMF). In *Root biology* (pp. 239–258). Springer. doi:10.1007/978-3-319-75910-4_9

Meddich, A., Ait El Mokhtar, M., Wahbi, S., & Boumezzough, A. (2017). Evaluation of the mycorrhizal potential in relation with the physico-chemical properties of soils in Moroccan palm groves (Marrakech and Tafilalet). *Cahiers Agricultures*, *26*, 45012. doi:10.1051/cagri/2017044

Meglouli, H., Fontaine, J., Verdin, A., Magnin-Robert, M., Tisserant, B., Hijri, M., & Sahraoui, A. L. H. (2019). Aided phytoremediation to clean up dioxins/furans-aged contaminated soil: Correlation between microbial communities and pollutant dissipation. *Microorganisms*, *7*(11), 1–18. doi:10.3390/microorganisms7110523 PMID:31684182

Meshram, V., Patil, K., Meshram, V., Hanchate, D., & Ramkteke, S. D. (2021). Machine learning in agriculture domain: A state-of-art survey. *Artificial Intelligence in the Life Sciences*, *1*, 100010. doi:10.1016/j.ailsci.2021.100010

Miranda, C., Urrestarazu, J., & Santesteban, L. G. (2021). An R package for climate adaptation assessment of temperate fruit species. *Computers and Electronics in Agriculture*, *180*, 105879. doi:10.1016/j.compag.2020.105879

Mirzabaev, A., Wu, J., Evans, J., Garcia-Oliva, F., Hussein, I. A. G., Iqbal, M. H., & Tena, F. (2019). Desertification. In Climate Change and Land: an IPCC special report on climate change, desertification, land degradation, sustainable land management, food security, and greenhouse gas fluxes in terrestrial ecosystems.

Mkonda, M. Y., & He, X. (2016). Production Trends of Food Crops: Opportunities, Challenges and Prospects to Improve Tanzanian Rural Livelihoods. *Natural Resources and Conservation*, *4*(4), 51–59. doi:10.13189/nrc.2016.040402

Möhring, N., Kudsk, P., Jørgensen, L. N., Ørum, J. E., & Finger, R. (2021). An R package to calculate potential environmental and human health risks from pesticide applications using the 'Pesticide Load' indicator applied in Denmark. *Computers and Electronics in Agriculture*, *191*, 106498. doi:10.1016/j.compag.2021.106498

Montano, R. A. N. R., Sanquetta, C. R., Wojciechowski, J., Mattar, E., Corte, A. P. D., & Todt, E. (2017). Artificial Intelligence Models to Estimate Biomass of Tropical Forest Trees. *Polibits*, *56*(1), 29–37.

Montgomery, D. C., Peck, E. A., & Vining, G. G. (2013). Introduction to Linear Regression Analysis. In *Wiley Series in Probability and Statistics* (5th ed.). John Wiley & Sons.

Morisset, J., Cunningham, V., & Haji, M. (2014). *Who Wants a Job?* The Magnetic Power of Cities.

Morocco, M. O. S. A. I. C. C. Retrieved 30 March 2022, from http://www.changementclimatique.ma/indexe77d.html?q=en/node/20

Moumane, A.; El Ghazali, F.E.; Al Karkouri, J.; Delorme, J.; Batchi, M.; Chafiki, D.; & Karmaoui, A. (2021). Monitoring spatiotemporal variation of groundwater level and salinity under land use change using integrated field measurements, GIS, geostatistical, and remote-sensing approach: case study of the Feija aquifer, Middle Draa watershed, Moroccan Sahara. *Environ Monit Assess, 193*, 769. doi:10.1007/s10661-021-09581-2

Moumane, A., Al Karkouri, J., Benmansour, A., El Ghazali, F. E., Fico, J., Karmaoui, A., & Batchi, M. (2022). Monitoring long-term land use, land cover change, and desertification in the Ternata oasis, Middle Draa Valley, Morocco. *Remote Sensing Applications: Society and Environment, 100745*, 100745. doi:10.1016/j.rsase.2022.100745

Moussa, S., Kyereh, B., Tougiani, A. A., & Saadou, M. (2018). Carbon stocks of neem tree (Azadirachta indica A. Juss.) in different urban land use and land cover types in Niamey city. *South Asian Journal of Biological Research, 1*(2), 153–165. http://aiipub.com/

Msuya & Kidegesho. (2015). Mainstreaming Agroforestry Policy in Tanzania Legal Framework. *Agroforestry for Biodiversity and Ecosystem Services - Science and Practice, March*, 130–140. doi:10.5772/35859

Msuya, N., Masanja, E., & Temu, A. K. (2011). Environmental Burden of Charcoal Production and Use in Dar es Salaam, Tanzania. *Journal of Environmental Protection, 02*(10), 1364–1369. https://doi.org/10.4236/jep.2011.210158

Mueller T.G., Cetin H., Fleming R.A., Dillon C.R., Karathanasis A.D, Shearer S.A.. Erosion probability maps: Calibrating precision agriculture data with soil surveys using logistic regression. (2005), *Journal of Soil and Water Conservation November, 60* (6), 462-468.

Muthuraj, R., & Mekonnen, T. (2018). Recent progress in carbon dioxide (CO2) as feedstock for sustainable materials development: Co-polymers and polymer blends. *Polymer, 145*, 348–373. https://doi.org/10.1016/j.polymer.2018.04.078

Mwaseba, D. L., Erkkilä, A., Friis-Hansen, E., Maro, A. H., & Maziku, J. D. (2020). Responsibilization in governance of non-industrial private forestry: Experiences from the Southern Highlands of Tanzania. *Forest Policy and Economics, 118*(April 2019), 102243. doi:10.1016/j.forpol.2020.102243

Nanang, D. M., Day, R. J., & Amaligo, J. N. (1997). Growth and yield of neem (Azadirachta indica A. Juss.) plantations in Northern Ghana. *Commonwealth Forestry Review, 76*(245), 103–106.

Nasir, F., Shi, S., Tian, L., Chang, C., Ma, L., Li, X., Gao, Y., & Tian, C. (2019). Strigolactones shape the rhizomicrobiome in rice (*Oryza sativa*). *Plant Science, 286*, 118–133. doi:10.1016/j.plantsci.2019.05.016 PMID:31300137

Nassar, Y. F., Abdunnabi, M. J., Sbeta, M. N., Hafez, A. A., Amer, K. A., Ahmed, A. Y., & Belgasim, B. (2021). Dynamic analysis and sizing optimization of a pumped hydroelectric storage-integrated hybrid PV/Wind system: A case study. *Energy Conversion and Management, 229*, 113744. doi:10.1016/j.enconman.2020.113744

Newman, J. A. (2004). Climate change and cereal aphids: The relative effects of increasing CO2 and temperature on aphid population dynamics. *Global Change Biology*, *10*(1), 5–15. doi:10.1111/j.1365-2486.2003.00709.x

Ngaga, Y. M. (2011). A Platform for Stakeholders in African Forestry. *Forest Pl Antations and Wo Odlots in Tanzania. 1*(16), 80.

Ngo, H. C., Hashim, U. R., Sek, Y. W., Kumar, Y. J., & Ke, W. S. (2019). Weeds detection in agricultural fields using convolutional neural network. *International Journal of Innovative Technology and Exploring Engineering*, *8*(11), 292–296. https://doi.org/10.35940/ijitee.K1327.0981119

Nogia, P., Sidhu, G. K., Mehrotra, R., & Mehrotra, S. (2016). Capturing atmospheric carbon: Biological and nonbiological methods. *The International Journal of Low Carbon Technologies*, *11*(2), 266–274. https://doi.org/10.1093/ijlct/ctt077

Noumi, V. N., Djongmo, V. A., Nyeck, B., Mbobda, R. B. T., & Zapfack, L. (2018). Vegetation structure, carbon sequestration potential and species conservation in four agroforestry systems in Cameroon (Tropical africa). *Acta Botanica Brasílica*, *32*(2), 212–221. https://doi.org/10.1590/0102-33062017abb0279

Odenberger, M., Kjärstad, J., & Johnsson, F. (2013). Prospects for CCS in the EU energy roadmap to 2050. *Energy Procedia*, *37*, 7573–7581. https://doi.org/10.1016/j.egypro.2013.06.701

Ok, A. O., Özlem, A., & Oguz, G. (2011). Evaluation of random forest method for agricultural crop classification. *European Journal of Remote Sensing*, *45*(1), 421–432. doi:10.5721/EuJRS20124535

Oladosu, Y., Rafii, M. Y., Abdullah, N., Magaji, U., Hussin, G., Ramli, A., & Miah, G. (2016). Fermentation Quality and Additives: A Case of Rice Straw Silage. *BioMed Research International*, *2016*. doi:10.1155/2016/7985167

Olivares, O. E. A., Torres, S. S., Jiménez, B. S. I., Enríquez, C. J. O., Zignol, F., Reygadas, Y., & Tiefenbacher, J. P. (2019). Climate change, land use/land cover change, and population growth as drivers of groundwater depletion in the Central Valleys, Oaxaca, Mexico. *Remote Sensing*, *11*(11), 1290. doi:10.3390/rs11111290

Olivier & Peters. (2020). Trends In Global Co 2 And Total Greenhouse Gas 2020 Report. *PBL Netherlands Environmental Assessment Agency, PBL public*(December), 85.

ORMVA (2018). *Office National de Mise en Valeur Agricole.* Ministère de l'Agriculture, de la Pèche Maritime, du Développement Rural et Eaux et Forêts du Maroc.

Pahandle Regional Services (n.d.). *The PRPC 1.* PRS. http://webcache.googleusercontent.com/search?q=cache:583KGJS00EwJ:theprpc.org/Programs/RegionalServices/Cattle-Feedyard-Industry/PRPC_CattleFeed/bcdoc/Texas%20Cattle%20Feedyard%20BCG%20Supplement.docx+&cd=3&hl=en&ct=clnk&gl=cz

Pahandle Regional Services (n.d.). *The PRPC 2.* PRS. http://webcache.googleusercontent.com/search?q=cache:OayhnqhEWRcJ:www.theprpc.org/Programs/RegionalServices/Dairy-Industry/prpc_dairy/bcdoc/Texas%20Dairy%20Farm%20BCG%20Template.docx+&cd=2&hl=en&ct=clnk&gl=cz

Pan, J., Huang, C., Peng, F., Zhang, W., Luo, J., Ma, S., & Xue, X. (2020). Effect of arbuscular mycorrhizal fungi (AMF) and plant growth-promoting bacteria (PGPR) inoculations on *Elaeagnus angustifolia* L. in saline Soil. *Applied Sciences (Basel, Switzerland)*, *10*(3), 945. doi:10.3390/app10030945

Parras-Alcántara, L., Martín-Carrillo, M., & Lozano-García, B. (2013). Impacts of land use change in soil carbon and nitrogen in a Mediterranean agricultural area (Southern Spain). *Solid Earth*, *4*(1), 167–177. doi:10.5194e-4-167-2013

Partel, V., Charan Kakarla, S., & Ampatzidis, Y. (2019). Development and evaluation of a low-cost and smart technology for precision weed management utilizing artificial intelligence. *Computers and Electronics in Agriculture*, *157*(January), 339–350. https://doi.org/10.1016/j.compag.2018.12.048

Paustian, K., Lehmann, J., Ogle, S., Reay, D., Robertson, G. P., & Smith, P. (2016). Climate-smart soils. *Nature*, *532*(7597), 49–57. doi:10.1038/nature17174 PMID:27078564

Pedersen, R. H. (2017). The Political Economy of Private Forestry in Tanzania : a Review. In *Danish Institute for International Studies (DIIS)*. www.diis.dk

Peters, G. P. (2010). Carbon footprints and embodied carbon at multiple scales. *Current Opinion in Environmental Sustainability*, *2*(4), 245–250. doi:10.1016/j.cosust.2010.05.004

PFP. (2016). United Republic of Tanzania. In *Private Forestry Program*. doi:10.5089/9781513547442.002

Podaras, A. (2021): Multi-Approach Classification for Cattle and Dairy Business Functions for Business Continuity Policy Formulation. figshare. [Dataset]. doi:10.6084/m9.figshare.17429579.v1

Podaras, A. (2022). BCM_AGRO_DATA. figshare. [Dataset]. doi:10.6084/m9.figshare.19500164

Podaras, A. (2018). Measuring the Accuracy Levels Regarding the Dual Business Function Criticality Classifier. *IEEE Access: Practical Innovations, Open Solutions*, *6*, 41598–41606. doi:10.1109/ACCESS.2018.2860976

Podaras, A., Antlová, K., & Motejlek, J. (2016). Information management tools for implementing an effective enterprise business continuity strategy. *E+M. Ekonomie a Management*, *19*(1), 165–182. doi:10.15240/tul/001/2016-1-012

Podaras, A., Moirogiorgou, K., & Zervakis, M. (2021). Regression-Based Recovery Time Predictions in Business Continuity Management: A Public College Case Study. In *Handbook of Research on Global Challenges for Improving Public Services and Government Operations*. IGI Global. doi:10.4018/978-1-7998-4978-0.ch020

Portail national des collectivités territoriales. (2022). *La Région*. Portail national des collectivités territoriales. https://www.collectivites-territoriales.gov.ma/fr/la-region

Pörtner, H. O., Roberts, D. C., Adams, H., Adler, C., Aldunce, P., Ali, E., & Birkmann, J. (2022). [Impacts, adaptation and vulnerability. *IPCC Sixth Assessment Report*.]. *Climatic Change*.

Pošík, P., & Pošík, P. (n.y.). Czech Technical University in Prague, Faculty of Electrical Engineering Department of Cybernetics. *Bias-variance trade-off. Crossvalidation. Regularization.* https://cw.fel.cvut.cz/old/_media/courses/ae3m33ui/lectures/biasvariance-slides.pdf

Pourmohamad, Y., Alizadeh, A., Baygi, M. M., Gebremichael, M., Ziaei, A. N., & Bannayan, M. (2019). Optimizing cropping area by proposing a combined water-energy productivity function for Neyshabur Basin, Iran. *Agricultural Water Management*, *217*, 131–140. doi:10.1016/j.agwat.2019.02.040

Proietti, P., Sdringola, P., Brunori, A., Ilarioni, L., Nasini, L., Regni, L., Pelleri, F., Desideri, U., & Proietti, S. (2016). Assessment of carbon balance in intensive and extensive tree cultivation systems for oak, olive, poplar and walnut plantation. *Journal of Cleaner Production*, *112*, 2613–2624. doi:10.1016/j.jclepro.2015.10.009

Proietti, S., Sdringola, P., Desideri, U., Zepparelli, F., Brunori, A., Ilarioni, L., Nasini, L., Regni, L., & Proietti, P. (2014). Carbon footprint of an olive tree grove. *Applied Energy*, *127*, 115–124. doi:10.1016/j.apenergy.2014.04.019

Pye, S., Watkiss, P., Savage, M., & Goodwin, J. (2010). *Opportunities for Low Carbon Investment in Tanzania: An Assessment of Future Emissions Growth and Low Carbon Reduction Potential* (Issue December). SEI, Dew Point, Global Climate Adoption.http://www.economics-of-cc-in-tanzania.org/images/Tanzania_low_carbon_growth_assessment_v7.pdf

Pye-Smith, C. (2008). Farming Trees, Banishing Hunger How an agroforestry programme is helping smallholders in. *America*.

Pylak, M., Oszust, K., & Frąc, M. (2019). Review report on the role of bioproducts, biopreparations, biostimulants and microbial inoculants in organic production of fruit. *Reviews in Environmental Science and Biotechnology*, *18*(3), 597–616. doi:10.100711157-019-09500-5

Qin, M., Zhang, Q., Pan, J., Jiang, S., Liu, Y., Bahadur, A., Peng, Z., Yang, Y., & Feng, H. (2020). Effect of arbuscular mycorrhizal fungi on soil enzyme activity is coupled with increased plant biomass. *European Journal of Soil Science*, *71*, 84–92. doi:10.1111/ejss.12815

Quiroga, G., Erice, G., Aroca, R., & Ruiz-Lozano, J. M. (2020). Elucidating the possible involvement of maize aquaporins in the plant boron transport and homeostasis mediated by *Rhizophagus irregularis* under drought stress conditions. *International Journal of Molecular Sciences*, *21*(5), 148. doi:10.3390/ijms21051748 PMID:32143345

Rahman, M. F. F., Fan, S., Zhang, Y., & Chen, L. (2021). A comparative study on application of unmanned aerial vehicle systems in agriculture. *Agriculture (Switzerland)*, *11*(1), 1–26. doi:10.3390/agriculture11010022

Raparelli, E., & Bajocco, S. (2019). A bibliometric analysis on the use of unmanned aerial vehicles in agricultural and forestry studies. *International Journal of Remote Sensing*, *40*(24), 9070–9083. doi:10.1080/01431161.2019.1569793

Razouk, R., Hssaini, L., Alghoum, M., Adiba, A., & Hamdani, A. (2022). Phenotyping Olive Cultivars for Drought Tolerance Using Leaf Macro-Characteristics. *Horticulturae*, *8*(10), 939. doi:10.3390/horticulturae8100939

Reichle, D. (2019). The global carbon cycle and climate change. Elsevier, 1, 388.

REIT. (2016). Tanzania Climate Action Report for 2016. In Resilience and Economic Inclusion Team.

Renaut, S., Daoud, R., Masse, J., Vialle, A., & Hijri, M. (2020). Inoculation with *Rhizophagus irregularis* does not alter arbuscular mycorrhizal fungal community structure within the roots of corn, wheat, and soybean crops. *Microorganisms*, *8*(1), 83. doi:10.3390/microorganisms8010083 PMID:31936180

Riaz, M., Kamran, M., Fang, Y., Wang, Q., Cao, H., Yang, G., Deng, L., Wang, Y., Zhou, Y., Anastopoulos, I., & Wang, X. (2021). Arbuscular mycorrhizal fungi-induced mitigation of heavy metal phytotoxicity in metal contaminated soils: A critical review. *Journal of Hazardous Materials*, *402*, 123919. doi:10.1016/j.jhazmat.2020.123919 PMID:33254825

Ribeiro, A. F. S., Russo, A., Gouveia, C. M., & Páscoa, P. (2019). Modelling drought-related yield losses in Iberia using remote sensing and multiscalar indices. *Theoretical and Applied Climatology*, *136*(1), 203–220. https://doi.org/10.1007/s00704-018-2478-5

Rimington, W. R., Pressel, S., Duckett, J. G., Field, K. J., Read, D. J., & Bidartondo, M. I. (2018). Ancient plants with ancient fungi: Liverworts associate with early-diverging arbuscular mycorrhizal fungi. *Proceedings. Biological Sciences*, *285*(1888), 20181600. doi:10.1098/rspb.2018.1600 PMID:30305437

Rittinghouse, J. W., & Ransome, J. F. (2005). Business Impact Analysis (BIA) - Prioritization and Classification of Critical Business Functions. In: Business Continuity and Disaster Recovery for Infosec Managers (pp. 83–86). Elsevier Press.

Rivera-Becerril, F., van Tuinen, D., Chatagnier, O., Rouard, N., Béguet, J., Kuszala, C., Soulas, G., Gianinazzi-Pearson, V., & Martin-Laurent, F. (2017). Impact of a pesticide cocktail (fenhexamid, folpel, deltamethrin) on the abundance of Glomeromycota in two agricultural soils. *The Science of the Total Environment*, *577*, 84–93. doi:10.1016/j.scitotenv.2016.10.098 PMID:27817923

Rizos, B., & Drabik, R. T. (2018). The Role of Business in the Circular Economy: Markets, Processes and Enabling Policies. In *The Centre for European Policy Studies* (Issue March). https://www.scopus.com/inward/record.url?eid=2-s2.0-84949807776&partnerID=40&md5=736fc93fe5bc8635ff79c068db9c0e1b

Rojas-Downing, M. M., Nejadhashemi, A. P., Harrigan, T., & Woznicki, S. A. (2017). Climate change and livestock: Impacts, adaptation, and mitigation. *Climate Risk Management*, *16*, 145–163. https://doi.org/10.1016/j.crm.2017.02.001

Rosle, R., Che'ya, N. N., Ang, Y., Rahmat, F., Wayayok, A., Berahim, Z., Ilahi, W. F. F., Ismail, M. R., & Omar, M. H. (2021). Weed detection in rice fields using remote sensing technique: A review. *Applied Sciences (Switzerland)*, *11*(22), 25. doi:10.3390/app112210701

Rumyantsev, A., Sidorin, A., Volkov, A., Al Shanaa, O., Sambuk, E., & Padkina, M. (2021). Transcriptome analysis unveils the effects of proline on gene expression in the Yeast *Komagataella phaffii*. *Microorganisms*, *10*(1), 67. doi:10.3390/microorganisms10010067 PMID:35056516

Russell, S., & Norvig, P. P. (2020). Artificial Intelligence: A Modern Approach (4th ed.). Pearson.

Sahebjamnia, N., Torabi, S. A., & Mansouri, A. (2014). Integrated Business Continuity and Disaster Recovery Planning: Towards Organizational Resilience. *European Journal of Operational Research*, *242*(1), 261–273. doi:10.1016/j.ejor.2014.09.055

Saka, M.G., Osho, J. S. A., & Iso, H. D. (2020). Assessment of biomass contents and carbon stock of plantation grown neem specie in Yola, Nigeria. *Journal of Research in Forestry, Wildlife & Environment*, *12*(1), 284–289.

Sangwan, S., & Prasanna, R. (2021). Mycorrhizae helper bacteria: Unlocking their potential as bioenhancers of plant–arbuscular mycorrhizal fungal associations. *Microbial Ecology*, *84*(1), 1–10. doi:10.100700248-021-01831-7 PMID:34417849

Schilling, J., Freier, K. P., Hertig, E., & Scheffran, J. (2012). Climate change, vulnerability and adaptation in North Africa with focus on Morocco. *Agriculture, Ecosystems & Environment*, *156*, 12–26. doi:10.1016/j.agee.2012.04.021

Schilling, J., Hertig, E., Tramblay, Y., & Scheffran, J. (2020). Climate change vulnerability, water resources and social implications in North Africa. *Regional Environmental Change*, *20*(1), 15. https://doi.org/10.1007/s10113-020-01597-7

Schotten, M., El Aisati, M., Meester, W. J., Steiginga, S., & Ross, C. A. (2017). A brief history of Scopus: The world's largest abstract and citation database of scientific literature. Research analytics: Boosting university productivity and competitiveness through. *Scientometrics*, *31*. https://www.taylorfrancis.com/chapters/brief-history-scopus-world-largest-abstract-citation-database-scientific-literature-michiel-schotten-hamed-el-aisati-wim-meester-susanne-steiginga-cameron-ross/e/10.1201/9781315155890-3

Seekings, J. (2020). International actors and social protection in Africa, 2000-2020. *University of Capetown*, *2*(March), 1–17.

Sereno, D., Oury, B., Geiger, A., Vela, A., Karmaoui, A., & Desquesnes, M. (2022). Isothermal Nucleic Acid Amplification to Detect Infection Caused by Parasites of the Trypanosomatidae Family: A Literature Review and Opinion on the Laboratory to Field Applicability. *International Journal of Molecular Sciences*, *23*(14), 7543. doi:10.3390/ijms23147543 PMID:35886895

Shafi, A., Zahoor, I., & Mushtaq, U. (2019). Proline accumulation and oxidative stress: Diverse roles and mechanism of tolerance and adaptation under salinity stress. In *Salt Stress, Microbes, and Plant Interactions: Mechanisms and Molecular Approaches* (pp. 269–300). Springer. doi:10.1007/978-981-13-8805-7_13

Shaikh, T. A., Rasool, T., & Lone, F. R. (2022). Towards leveraging the role of machine learning and artificial intelligence in precision agriculture and smart farming. *Computers and Electronics in Agriculture*, *198*, 107119. doi:10.1016/j.compag.2022.107119

Shannak, S. (2022). Optimizing dynamics of water-energy-food nexus in a desert climate. *Energy Policy*, *164*(C), 112884. doi:10.1016/j.enpol.2022.112884

Shapa, M., Trojer, L., & Machuve, D. (2021). Mobile-based Decision Support System for Poultry Farmers: A Case of Tanzania. *International Journal of Advanced Computer Science and Applications*, *12*(2), 584–590. https://doi.org/10.14569/IJACSA.2021.0120273

Sharma, K., Gupta, S., Thokchom, S. D., & Kapoor, R. (2021). The Potential Roles of arbuscular mycorrhizal fungi in soil health and conservation. *International journal of plant and environment*, *7*, 39-48.

Sharma, A., Shahzad, B., Kumar, V., Kohli, S. K., Sidhu, G. P. S., Bali, A. S., Handa, N., Kapoor, D., Bhardwaj, R., & Zheng, B. (2019). Phytohormones regulate accumulation of osmolytes under abiotic stress. *Biomolecules*, *9*(7), 285. doi:10.3390/biom9070285 PMID:31319576

Sharma, M. P., Grover, M., Chourasiya, D., Bharti, A., Agnihotri, R., Maheshwari, H. S., Pareek, A., Buyer, J. S., Sharma, S. K., Schütz, L., Mathimaran, N., Singla-Pareek, S. L., Grossman, J. M., & Bagyaraj, D. J. (2020). Deciphering the role of trehalose in tripartite symbiosis among rhizobia, arbuscular mycorrhizal fungi, and legumes for enhancing abiotic stress tolerance in crop plants. *Frontiers in Microbiology*, *11*, 509919. doi:10.3389/fmicb.2020.509919 PMID:33042042

Shaukat, K., Luo, S., Chen, S., & Liu, D. (2020). Cyber Threat Detection Using Machine Learning Techniques: A Performance Evaluation Perspective," In: *Proc. 2020 International Conference on Cyber Warfare and Security (ICCWS)*, 1-6. 10.1109/ICCWS48432.2020.9292388

Shaukat, K., Luo, S., Varadharajan, V., Hameed, I. A., Chen, S., Liu, D., & Li, J. (2020). Performance Comparison and Current Challenges of Using Machine Learning Techniques in Cybersecurity. *Energies*, *13*(10), 2509. doi:10.3390/en13102509

Shaukat, K., Luo, S., Varadharajan, V., Hameed, I. A., & Xu, M. (2020). A Survey on Machine Learning Techniques for Cyber Security in the Last Decade. *IEEE Access: Practical Innovations, Open Solutions*, *8*, 222310–222354. doi:10.1109/ACCESS.2020.3041951

Shi, Y., Wang, S., Guo, J., Xu, Z., Wang, S., & Sang, Y. (2022). Effects of arbuscular mycorrhizal inoculation on the phytoremediation of PAH-contaminated soil: A meta-analysis. *Chemosphere*, *136033*, 136033. doi:10.1016/j.chemosphere.2022.136033 PMID:35981621

Singh, B., Strømman, A. H., & Hertwich, E. G. (2011). Comparative impact assessment of CCS portfolio: Life cycle perspective. *Energy Procedia*, *4*, 2486–2493. https://doi.org/10.1016/j.egypro.2011.02.144

Singh, N., & Singh, A. N. (2020). Odysseys of agriculture sensors: Current challenges and forthcoming prospects. *Computers and Electronics in Agriculture*, *171*, 105328. doi:10.1016/j.compag.2020.105328

Sivanappan, R. K. (1995). Soil and water management in the dry lands of India. *Land Use Policy*, *12*(2), 165–175. doi:10.1016/0264-8377(94)00001-G

Smith, P. (2004). Carbon sequestration in croplands: The potential in Europe and the global context. *European Journal of Agronomy*, *20*(3), 229–236. doi:10.1016/j.eja.2003.08.002

Snedaker, S. (2007).Business Impact Analysis. In: *Business Continuity and Disaster Recovery for IT Professionals*. Elsevier.

Sofo, A., Nuzzo, V., Palese, A. M., Xiloyannis, C., Celano, G., Zukowskyj, P., & Dichio, B. (2005). Net CO2 storage in Mediterranean olive and peach orchards. *Scientia Horticulturae*, *107*(1), 17–24. doi:10.1016/j.scienta.2005.06.001

Soh, M. C., Mitchell, N. J., Ridley, A. R., Butler, C. W., Puan, C. L., & Peh, K. S. H. (2019). Impacts of habitat degradation on tropical montane biodiversity and ecosystem services: A systematic map for identifying future research priorities. *Frontiers in Forests and Global Change*, *2*, 83. doi:10.3389/ffgc.2019.00083

Soylu, S., & Çarman, K. (2021). Fuzzy logic based automatic slip control system for agricultural tractors. *Journal of Terramechanics*, *95*, 25–32. https://doi.org/10.1016/j.jterra.2021.03.001

Stackexchange Network. (n.d). Cross Validated. *Stackexchange*. https://stats.stackexchange.com/questions/450322/what-is-the-time-complexity-of-k-fold-cross-validation

Subhashini, D. V. (2016). Effect of NPK fertilizers and co-inoculation with phosphate-solubilizing arbuscular mycorrhizal fungus and potassium-mobilizing bacteria on growth, yield, nutrient acquisition, and quality of tobacco (*Nicotiana tabacum* L.). *Communications in Soil Science and Plant Analysis*, *47*(3), 328–337. doi:10.1080/00103624.2015.1123724

Supriadi, L. S. R., & Sui Pheng, L. (2018). Business Continuity Management (BCM). In *Business Continuity Management in Construction. Management in the Built Environment* (pp. 41–73). Springer.

Surendran, U., Sandeep, O., & Joseph, E. J. (2016). impacts of magnetic treatment of irrigation water on plant, water and soil characteristics. *Agricultural Water Management*, *178*, 21–29. doi:10.1016/j.agwat.2016.08.016

Suryawanshi, M. N., Patel, A. R., Kale, T. S., & Patil, P. R. (2014). Carbon sequestration potential of tree species in the environment of North Maharashtra University Campus, Jalgaon (MS) India. *Bioscience Discovery*, *5*(2), 175–179.

Swamila, M., Philip, D., Akyoo, A. M., Sieber, S., Bekunda, M., & Kimaro, A. A. (2020). Gliricidia agroforestry technology adoption potential in selected dryland areas of Dodoma Region, Tanzania. *Agriculture (Switzerland), 10*(7), 1–17. doi:10.3390/agriculture10070306

Symanczik, S., Krützmann, J., Nehls, U., Boller, T., & Courty, P. E. (2020). Expression of major intrinsic protein genes in Sorghum bicolor roots under water deficit depends on arbuscular mycorrhizal fungal species. *Soil Biology & Biochemistry, 140*, 107643. doi:10.1016/j.soilbio.2019.107643

Symanczik, S., Lehmann, M. F., Wiemken, A., Boller, T., & Courty, P. E. (2018). Effects of two contrasted arbuscular mycorrhizal fungal isolates on nutrient uptake by *Sorghum bicolor* under drought. *Mycorrhiza, 28*(8), 779–785. doi:10.100700572-018-0853-9 PMID:30006910

Taimourya, H., Oussible, M., Bourarach, E., El Harif, A., Hassanain, N., Masmoudi, L., & BAAMAL, L. (2015). Évaluation de la productivité de la pomme de terre (Solanum Tuberosum L.), sous l'effet de l'irrigation avec une eau traitée magnétiquement dans la région de Chaouia (Maroc). *Revue Marocaine des Sciences Agronomiques et Vétérinaires, 3*(3), 30-36. https://agromaroc.com/index.php/Actes_IAVH2/article/view/406

Tak, A. A., & Kakde, U. B. (2020). Analysis of carbon sequestration by dominant trees in urban areas of Thane city. *International Journal of Global Warming, 20*(1), 1–11. https://doi.org/10.1504/IJGW.2020.104615

Talaviya, T., Shah, D., Patel, N., Yagnik, H., & Shah, M. (2020). Implementation of artificial intelligence in agriculture for optimisation of irrigation and application of pesticides and herbicides. *Artificial Intelligence in Agriculture., 4*, 58–73. doi:10.1016/j.aiia.2020.04.002

Tcvetkov, P., Cherepovitsyn, A., & Fedoseev, S. (2019). The changing role of CO2 in the transition to a circular economy: Review of carbon sequestration projects. *Sustainability (Switzerland), 11*(20), 1–19. doi:10.3390/su11205834

Teutscherova, N., Vazquez, E., Arango, J., Arevalo, A., Benito, M., & Pulleman, M. (2019). Native arbuscular mycorrhizal fungi increase the abundance of ammonia-oxidizing bacteria, but suppress nitrous oxide emissions shortly after urea application. *Geoderma, 338*, 493–501. doi:10.1016/j.geoderma.2018.09.023

Tigkas, D., & Tsakiris, G. (2015). Early Estimation of Drought Impacts on Rainfed Wheat Yield in Mediterranean Climate. *Environmental Processes, 2*(1), 97–114. doi:10.1007/s40710-014-0052-4

Tol, R. S. J. (2010). The Economic Impact of Climate Change. *Perspektiven der Wirtschaftspolitik, 11*(Supplement), 13–37. https://doi.org/10.1111/1468-2516.0326

Torabi, S. A., Rezaei, H., & Sahebjamnia, N. (2014). A New Framework for Business Impact Analysis in Business Continuity Management (with a case study). *Safety Science, 68*, 309–323. doi:10.1016/j.ssci.2014.04.017

Toubali, S., Tahiri, A., Anli, M., Symanczik, S., Boutasknit, A., Ait-El-Mokhtar, M., Ben-Laouane, R., Oufdou, K., Ait-Rahou, Y., Ben-Ahmed, H., Jemo, M., Hafidi, M., & Meddich, A. (2020). Physiological and biochemical behaviors of date palm vitroplants treated with microbial consortia and compost in response to salt stress. *Applied Sciences (Basel, Switzerland)*, *10*(23), 8665. doi:10.3390/app10238665

Umar, A. (2017). Phytohyperaccumulator-AMF (arbuscular mycorrhizal fungi) interaction in heavy metals detoxification of soil. *Acta Biológica Paranaense*, *46*, 123–148. doi:10.5380/abpr.v46i0.57147

UNCTAD. (2019). Commodity Dependence, Climate Change and the Paris Agreement. In *Commodities and Development Report*. https://unctad.org/en/PublicationsLibrary/ditccom2019d3_en.pdf

UNDP. (2009). Climate Change Adaptation. Morocco. Retrieved 30 March 2022, from https://www.adaptation-undp.org/explore/arab-states/morocco

United Nations. (2017). World population projected to reach 9.8 billion in 2050, and 11.2 billion in 2100. Retrieved 20 March 2022, from https://www.un.org/en/desa/world-population-projected-reach-98-billion-2050-and-112-billion-2100

United Republic of Tanzania (URT). (2021a). *2020 Tanzania in Figures*. Issue June.

United Republic of Tanzania (URT). (2021b). *National Environment Policy*. URT.

University of California San Francisco IT Service Management. (2016). *IT Service Continuity Management, Business Impact Analysis Process Activity*. UCSF. https://itsm.ucsf.edu/sites/itsm.ucsf.edu/files/ITSM%20BIA%20Process%20v1.2-1.pdf

Urbanec, J., & Urbancova, H. (2014). Adoption of business continuity management standards in Czech organizations. *Scientia Agriculturae Bohemica*, *45*(1), 66–74. doi:10.7160ab.2014.450109

URT. (2011). *Tanzania Agriculture and Food Security Investment Plan (TAFSIP)* (Issue October 2011). URT.

URT. (2014). *Tanzania: Second national communication to the United Nations framework convention climate change*. URT .

URT. (2015a). National Forest Resources Monitoring and Assessment of Tanzania Mainland (NAFORMA). Main Results Report. pp.33.

URT. (2015b). *Tanzania Intended Nationally Determined Contributions (INDC)* (Issue 2007). URT. http://www4.unfccc.int/submissions/INDC/Published Documents/Turkey/1/The_INDC_of_TURKEY_v.15.19.30.pdf

URT. (2018). *Formal Sector Employment and Earnings Survey*. URT.

URT. (2019a). Health and Pollution Action Plan. In United Nations Industrial Development Organization, 2019 (Issue February).

URT. (2019b). *Tanzania National Nutrition Survey 2018.* Issue June.

URT. (2020). *National Climate Change Statistics Report 2019.* NBS.

Valverde, P., de Carvalho, M., Serralheiro, R., Maia, R., Ramos, V., & Oliveira, B. (2015). Climate change impacts on rainfed agriculture in the Guadiana river basin (Portugal). *Agricultural Water Management, 150*, 35–45. doi:10.1016/j.agwat.2014.11.008

Van der Werf, G. R., Morton, D. C., DeFries, R. S., Olivier, J. G., Kasibhatla, P. S., Jackson, R. B., Collatz, G. J., & Randerson, J. T. (2009). CO2 emissions from forest loss. *Nature Geoscience, 2*(11), 737–738. doi:10.1038/ngeo671

van der Zwaan, B., & Gerlagh, R. (2016). Offshore CCS and ocean acidification: A global long-term probabilistic cost-benefit analysis of climate change mitigation. *Climatic Change, 137*(1–2), 157–170. https://doi.org/10.1007/s10584-016-1674-5

van Minnen, J. G., Strengers, B. J., Eickhout, B., Swart, R. J., & Leemans, R. (2008). Quantifying the effectiveness of climate change mitigation through forest plantations and carbon sequestration with an integrated land-use model. *Carbon Balance and Management, 3*, 1–20. https://doi.org/10.1186/1750-0680-3-3

Ventures, S. (2019). Artificial Intelligence in Tanzania, What's Happening . *The Medium.*

Verma, R., Lobos-Ossandón, V., Merigó, J. M., Cancino, C., & Sienz, J. (2021). Forty years of applied mathematical modelling: A bibliometric study. *Applied Mathematical Modelling, 89*, 1177–1197. doi:10.1016/j.apm.2020.07.004

Viebahn, P., Vallentin, D., & Höller, S. (2015). Prospects of carbon capture and storage (CCS) in China's power sector - An integrated assessment. *Applied Energy, 157*, 229–244. https://doi.org/10.1016/j.apenergy.2015.07.023

Wang, F., Sun, Y., & Shi, Z. (2019). Arbuscular mycorrhiza enhances biomass production and salt tolerance of sweet sorghum. *Microorganisms, 7*(9), 289. doi:10.3390/microorganisms7090289 PMID:31450847

Wang, J., & Liu, Z. (2014). A bibliometric analysis on rural studies in human geography and related disciplines. *Scientometrics, 101*(1), 39–59. doi:10.100711192-014-1388-2

WB. (2018). Agricultural Pollution. Field Burning. *World Bank Group*, 1–7. https://documents.worldbank.org/curated/en/989351521207797690/pdf/124342-repl-WB-Knowledge-Burning.pdf%0Ahttps://openknowledge.worldbank.org/bitstream/handle/10986/29504/124342-BRI-p153343-PUBLIC-march-22-9-pm-WB-Knowledge-Burning.pdf?sequence=1&isAllowed=

Western Australian Government. (2009). *Business Continuity Management Guidelines (2nd Edition).* WAG. https://www.complianceonline.com/articlefiles/Western%20Australia_BCM_Guidelines.pdf

WHO. (2016). *Ambient air pollution: A global assessment of exposure and burden of disease*. WHO.

WHO. (2018). Air pollution and child health. *Book, WHO/CED/PH*(NC-SA 3.0 IGO), 18.

Wipf, D., Krajinski, F., van Tuinen, D., Recorbet, G., & Courty, P. E. (2019). Trading on the arbuscular mycorrhiza market: From arbuscules to common mycorrhizal networks. *The New Phytologist*, *223*(3), 1127–1142. doi:10.1111/nph.15775 PMID:30843207

Wongchai, A., Jenjeti, D., Priyadarsini, A. I., Deb, N., Bhardwaj, A., & Tomar, P. (2022b). Farm monitoring and disease prediction by classification based on deep learning architectures in sustainable agriculture. *Ecological Modelling*, *474*, 110167. doi:10.1016/j.ecolmodel.2022.110167

Wongchai, A., Shukla, S. K., Ahmed, M. A., Sakthi, U., Jagdish, M., & kumar, R. (2022a). Artificial intelligence-enabled soft sensor and internet of things for sustainable agriculture using ensemble deep learning architecture. *Computers & Electrical Engineering*, *102*, 108128. doi:10.1016/j.compeleceng.2022.108128

World Bank Climate Change Knowledge Portal. (2021). Morocco. Retrieved December 2021, from https://climateknowledgeportal.worldbank.org/country/morocco

World Bank. (2012). *Carbon sequestration in agricultural soils*. World Bank.

World Bank. (2021). Morocco. Total Population. Retrieved 7 January 2022, from https://data.worldbank.org/indicator/SP.POP.TOTL?locations=MA

Wrachien, D. D. (2013). Food Production and Irrigation and Drainage Systems Development Perspective and Challenges. *Irrigation & Drainage Systems Engineering*, *2*(3). doi:10.4172/2168-9768.1000e122

WRI-CAIT. (2021). *HISTORICAL GHG EMISSION BY SECTOR IN TANZANIA (MTCO2e)*. WHO.

Xing, K., Sun, D., Zhang, J., & Zhao, F. (2021). Wide Diurnal Temperature Amplitude and High Population Density Can Positively Affect the Life History of Sitobion avenae (Hemiptera: Aphididae). *Journal of Insect Science*, *21*(2), 6. doi:10.1093/jisesa/ieab011

Yang, F., Watson, P., Koukoula, M., & Anagnostou, E. N. (2020). Enhancing Weather-Related Power Outage Prediction by Event Severity Classification. *IEEE Access: Practical Innovations, Open Solutions*, *8*, 60029–60042. doi:10.1109/ACCESS.2020.2983159

Yin, X., Feng, Q., Li, Y., Liu, W., Zhu, M., Xu, G., Zheng, X., & Sindikubwabo, C. (2021). Induced soil degradation risks and plant responses by salinity and sodicity in intensive irrigated agroecosystems of seasonally-frozen arid regions. *Journal of Hydrology (Amsterdam)*, *603*, 127036. doi:10.1016/j.jhydrol.2021.127036

Zafar, M., & Salahuddin, K. (2009). On the Use of K-Fold Cross-Validation to Choose Cutoff Values and Assess the Performance of Predictive Models in Stepwise Regression. *The International Journal of Biostatistics*, *5*(1).

Zemni, N., Slama, F., Bouksila, F., & Bouhlila, R. (2022). Simulating and monitoring water flow, salinity distribution and yield production under buried diffuser irrigation for date palm tree in Saharan Jemna oasis (North Africa). *Agriculture, Ecosystems & Environment, 325*, 107772. doi:10.1016/j.agee.2021.107772

Zhang, F., Zou, Y. N., & Wu, Q. S. (2018). Quantitative estimation of water uptake by mycorrhizal extraradical hyphae in citrus under drought stress. *Scientia Horticulturae, 229*, 132–136. doi:10.1016/j.scienta.2017.10.038

Zhang, F., Zou, Y. N., Wu, Q. S., & Kuča, K. (2020). Arbuscular mycorrhizas modulate root polyamine metabolism to enhance drought tolerance of trifoliate orange. *Environmental and Experimental Botany, 171*, 103926. doi:10.1016/j.envexpbot.2019.103926

Zhang, L., Xu, M., Liu, Y., Zhang, F., Hodge, A., & Feng, G. (2016). Carbon and phosphorus exchange may enable cooperation between an arbuscular mycorrhizal fungus and a phosphate-solubilizing bacterium. *The New Phytologist, 210*(3), 1022–1032. doi:10.1111/nph.13838 PMID:27074400

Zhang, Z., Mallik, A., Zhang, J., Huang, Y., & Zhou, L. (2019). Effects of arbuscular mycorrhizal fungi on inoculated seedling growth and rhizosphere soil aggregates. *Soil & Tillage Research, 194*, 104340. doi:10.1016/j.still.2019.104340

Zheng, H., Zhou, X., Cheng, T., Yao, X., Tian, Y., Cao, W., & Zhu, Y. (2016). Comparative analysis of vegetation indices, non-parametric and physical retrieval methods for monitoring nitrogen in wheat using UAV-based multispectral imagery, 7350–7353. IEEE.

Zhongming, Z., Linong, L., Xiaona, Y., Wangqiang, Z., & Wei, L. (2018). 'Soil probiotics' promise bigger, healthier crops, but there's a downside. *Global S&T Development Trend Analysis Platform of Resources and Environment*. http://resp.llas.ac.cn/C666/handle/2XK7JSWQ/124877

Zhou, B., Yang, L., Chen, X., Ye, S., Peng, Y., & Liang, C. (2021). Effect of magnetic water irrigation on the improvement of salinized soil and cotton growth in Xinjiang. *Agricultural Water Management, 248*, 106784. doi:10.1016/j.agwat.2021.106784

Zhou, X., Harrington, R., Woiwod, I. P., Perry, J. N., Bale, J. S., & Clark, S. J. (1995). Effects of temperature on aphid phenology. *Global Change Biology, 1*(4), 303–313. doi:10.1111/j.1365-2486.1995.tb00029.x

Zhu, H., Zhou, Y., Zhai, H., He, S., Zhao, N., & Liu, Q. (2020). A novel sweetpotato *WRKY* transcription factor, *IBWRKY2*, positively regulates drought and salt tolerance in transgenic *Arabidopsis*. *Biomolecules, 10*(4), 506. doi:10.3390/biom10040506 PMID:32230780

Related References

To continue our tradition of advancing academic research, we have compiled a list of recommended IGI Global readings. These references will provide additional information and guidance to further enrich your knowledge and assist you with your own research and future publications.

Ajiboye, O. E., & Yusuff, O. S. (2017). Foreign Land Acquisition: Food Security and Food Chains – The Nigerian Experience. In I. Management Association (Ed.), Natural Resources Management: Concepts, Methodologies, Tools, and Applications (pp. 1524-1545). Hershey, PA: IGI Global. https://doi.org/ doi:10.4018/978-1-5225-0803-8.ch072

Alapiki, H. E., & Amadi, L. A. (2018). Sustainable Food Consumption in the Neoliberal Order: Challenges and Policy Implications. In A. Obayelu (Ed.), *Food Systems Sustainability and Environmental Policies in Modern Economies* (pp. 90–123). Hershey, PA: IGI Global. doi:10.4018/978-1-5225-3631-4.ch005

Altaş, A. (2018). Geographical Information System Applications Utilized in Museums in Turkey Within the Scope of the Cultural Heritage Tourism: A Case Study of Mobile Application of Müze Asist. In S. Chaudhuri & N. Ray (Eds.), *GIS Applications in the Tourism and Hospitality Industry* (pp. 42–60). Hershey, PA: IGI Global. doi:10.4018/978-1-5225-5088-4.ch002

Andreea, I. R. (2018). Beyond Macroeconomics of Food and Nutrition Security. *International Journal of Sustainable Economies Management*, *7*(1), 13–22. doi:10.4018/IJSEM.2018010102

Anwar, J. (2017). Reproductive and Mental Health during Natural Disaster: Implications and Issues for Women in Developing Nations – A Case Example. In I. Management Association (Ed.), Gaming and Technology Addiction: Breakthroughs in Research and Practice (pp. 446-472). Hershey, PA: IGI Global. https://doi.org/ doi:10.4018/978-1-5225-0778-9.ch021

Awadh, H., Aksissou, M., Benhardouze, W., Darasi, F., & Snaiki, J. (2018). Socioeconomic Status of Artisanal Fishers in the West Part of Moroccan Mediterranean. *International Journal of Social Ecology and Sustainable Development*, *9*(1), 40–52. doi:10.4018/IJSESD.2018010104

Aye, G. C., & Haruna, R. F. (2018). Effect of Climate Change on Crop Productivity and Prices in Benue State, Nigeria: Implications for Food Security. In V. Erokhin (Ed.), *Establishing Food Security and Alternatives to International Trade in Emerging Economies* (pp. 244–268). Hershey, PA: IGI Global. doi:10.4018/978-1-5225-2733-6.ch012

Azizan, S. A., & Suki, N. M. (2017). Consumers' Intentions to Purchase Organic Food Products. In T. Esakki (Ed.), *Green Marketing and Environmental Responsibility in Modern Corporations* (pp. 68–82). Hershey, PA: IGI Global. doi:10.4018/978-1-5225-2331-4.ch005

Barakabitze, A. A., Fue, K. G., Kitindi, E. J., & Sanga, C. A. (2017). Developing a Framework for Next Generation Integrated Agro Food-Advisory Systems in Developing Countries. In I. Management Association (Ed.), Agri-Food Supply Chain Management: Breakthroughs in Research and Practice (pp. 47-67). Hershey, PA: IGI Global. doi:10.4018/978-1-5225-1629-3.ch004

Beachcroft-Shaw, H., & Ellis, D. (2018). Using Successful Cases to Promote Environmental Sustainability: A Social Marketing Approach. In I. Management Association (Ed.), Sustainable Development: Concepts, Methodologies, Tools, and Applications (pp. 936-953). Hershey, PA: IGI Global. doi:10.4018/978-1-5225-3817-2.ch042

Behnassi, M., Kahime, K., Boussaa, S., Boumezzough, A., & Messouli, M. (2017). Infectious Diseases and Climate Vulnerability in Morocco: Governance and Adaptation Options. In I. Management Association (Ed.), Public Health and Welfare: Concepts, Methodologies, Tools, and Applications (pp. 91-109). Hershey, PA: IGI Global. https://doi.org/ doi:10.4018/978-1-5225-1674-3.ch005

Bekele, F., & Bekele, I. (2017). Social and Environmental Impacts on Agricultural Development. In W. Ganpat, R. Dyer, & W. Isaac (Eds.), *Agricultural Development and Food Security in Developing Nations* (pp. 21–56). Hershey, PA: IGI Global. doi:10.4018/978-1-5225-0942-4.ch002

Benaouda, A., & García-Peñalvo, F. J. (2018). Towards an Intelligent System for the Territorial Planning: Agricultural Case. In F. García-Peñalvo (Ed.), *Global Implications of Emerging Technology Trends* (pp. 158–178). Hershey, PA: IGI Global. doi:10.4018/978-1-5225-4944-4.ch010

Bhaskar, A., Rao, G. B., & Vencatesan, J. (2017). Characterization and Management Concerns of Water Resources around Pallikaranai Marsh, South Chennai. In P. Rao & Y. Patil (Eds.), *Reconsidering the Impact of Climate Change on Global Water Supply, Use, and Management* (pp. 102–121). Hershey, PA: IGI Global. doi:10.4018/978-1-5225-1046-8.ch007

Bhyan, P., Shrivastava, B., & Kumar, N. (2022). Requisite Sustainable Development Contemplating Buildings: Economic and Environmental Sustainability. In A. Hussain, K. Tiwari, & A. Gupta (Eds.), *Addressing Environmental Challenges Through Spatial Planning* (pp. 269–288). IGI Global. https://doi.org/10.4018/978-1-7998-8331-9.ch014

Bogataj, D., & Drobne, D. (2017). Control of Perishable Goods in Cold Logistic Chains by Bionanosensors. In S. Joo (Ed.), *Applying Nanotechnology for Environmental Sustainability* (pp. 376–402). Hershey, PA: IGI Global. doi:10.4018/978-1-5225-0585-3.ch016

Bogataj, D., & Drobne, D. (2017). Control of Perishable Goods in Cold Logistic Chains by Bionanosensors. In I. Management Association (Ed.), Materials Science and Engineering: Concepts, Methodologies, Tools, and Applications (pp. 471-497). Hershey, PA: IGI Global. https://doi.org/ doi:10.4018/978-1-5225-1798-6.ch019

Bogueva, D., & Marinova, D. (2018). What Is More Important: Perception of Masculinity or Personal Health and the Environment? In D. Bogueva, D. Marinova, & T. Raphaely (Eds.), *Handbook of Research on Social Marketing and Its Influence on Animal Origin Food Product Consumption* (pp. 148–162). Hershey, PA: IGI Global. doi:10.4018/978-1-5225-4757-0.ch010

Bouzid, M. (2017). Waterborne Diseases and Climate Change: Impact and Implications. In M. Bouzid (Ed.), *Examining the Role of Environmental Change on Emerging Infectious Diseases and Pandemics* (pp. 89–108). Hershey, PA: IGI Global. doi:10.4018/978-1-5225-0553-2.ch004

Bowles, D. C. (2017). Climate Change-Associated Conflict and Infectious Disease. In M. Bouzid (Ed.), *Examining the Role of Environmental Change on Emerging Infectious Diseases and Pandemics* (pp. 68–88). Hershey, PA: IGI Global. doi:10.4018/978-1-5225-0553-2.ch003

Buck, J. J., & Lowry, R. K. (2017). Oceanographic Data Management: Quills and Free Text to the Digital Age and "Big Data". In P. Diviacco, A. Leadbetter, & H. Glaves (Eds.), *Oceanographic and Marine Cross-Domain Data Management for Sustainable Development* (pp. 1–22). Hershey, PA: IGI Global. doi:10.4018/978-1-5225-0700-0.ch001

Buse, C. G. (2017). Are Climate Change Adaptation Policies a Game Changer?: A Case Study of Perspectives from Public Health Officials in Ontario, Canada. In M. Bouzid (Ed.), *Examining the Role of Environmental Change on Emerging Infectious Diseases and Pandemics* (pp. 230–257). Hershey, PA: IGI Global. doi:10.4018/978-1-5225-0553-2.ch010

Calderon, F. A., Giolo, E. G., Frau, C. D., Rengel, M. G., Rodriguez, H., Tornello, M., ... Gallucci, R. (2018). Seismic Microzonation and Site Effects Detection Through Microtremors Measures: A Review. In N. Ceryan (Ed.), *Handbook of Research on Trends and Digital Advances in Engineering Geology* (pp. 326–349). Hershey, PA: IGI Global. doi:10.4018/978-1-5225-2709-1.ch009

Carfì, D., Donato, A., & Panuccio, D. (2018). A Game Theory Coopetitive Perspective for Sustainability of Global Feeding: Agreements Among Vegan and Non-Vegan Food Firms. In I. Management Association (Ed.), Game Theory: Breakthroughs in Research and Practice (pp. 71-104). Hershey, PA: IGI Global. https://doi.org/doi:10.4018/978-1-5225-2594-3.ch004

Castagnolo, V. (2018). Analyzing, Classifying, Safeguarding: Drawing for the Borgo Murattiano Neighbourhood of Bari. In G. Carlone, N. Martinelli, & F. Rotondo (Eds.), *Designing Grid Cities for Optimized Urban Development and Planning* (pp. 93–108). Hershey, PA: IGI Global. doi:10.4018/978-1-5225-3613-0.ch006

Chekima, B. (2018). The Dilemma of Purchase Intention: A Conceptual Framework for Understanding Actual Consumption of Organic Food. *International Journal of Sustainable Economies Management*, *7*(2), 1–13. doi:10.4018/IJSEM.2018040101

Chen, Y. (2017). Sustainable Supply Chains and International Soft Landings: A Case of Wetland Entrepreneurship. In B. Christiansen & F. Kasarcı (Eds.), *Corporate Espionage, Geopolitics, and Diplomacy Issues in International Business* (pp. 232–247). Hershey, PA: IGI Global. doi:10.4018/978-1-5225-1031-4.ch013

Çıtak, L., Akel, V., & Ersoy, E. (2018). Investors' Reactions to the Announcement of New Constituents of BIST Sustainability Index: An Analysis by Event Study and Mean-Median Tests. In M. Risso & S. Testarmata (Eds.), *Value Sharing for Sustainable and Inclusive Development* (pp. 270–289). Hershey, PA: IGI Global. doi:10.4018/978-1-5225-3147-0.ch012

D'Aleo, V., D'Aleo, F., & Bonanno, R. (2018). New Food Industries Toward a New Level of Sustainable Supply: Success Stories, Business Models, and Strategies. In V. Erokhin (Ed.), *Establishing Food Security and Alternatives to International Trade in Emerging Economies* (pp. 74–97). Hershey, PA: IGI Global. doi:10.4018/978-1-5225-2733-6.ch004

Dagevos, H., & Reinders, M. J. (2018). Flexitarianism and Social Marketing: Reflections on Eating Meat in Moderation. In D. Bogueva, D. Marinova, & T. Raphaely (Eds.), *Handbook of Research on Social Marketing and Its Influence on Animal Origin Food Product Consumption* (pp. 105–120). Hershey, PA: IGI Global. doi:10.4018/978-1-5225-4757-0.ch007

Danisman, G. O. (2022). What Drives Eco-Design Innovations in European SMEs? In U. Akkucuk (Ed.), *Disruptive Technologies and Eco-Innovation for Sustainable Development* (pp. 191–206). IGI Global. https://doi.org/10.4018/978-1-7998-8900-7.ch011

Deenapanray, P. N., & Ramma, I. (2017). Adaptations to Climate Change and Climate Variability in the Agriculture Sector in Mauritius: Lessons from a Technical Needs Assessment. In I. Management Association (Ed.), Natural Resources Management: Concepts, Methodologies, Tools, and Applications (pp. 655-680). Hershey, PA: IGI Global. https://doi.org/ doi:10.4018/978-1-5225-0803-8.ch030

Deenapanray, P. N., & Ramma, I. (2017). Adaptations to Climate Change and Climate Variability in the Agriculture Sector in Mauritius: Lessons from a Technical Needs Assessment. In I. Management Association (Ed.), Natural Resources Management: Concepts, Methodologies, Tools, and Applications (pp. 655-680). Hershey, PA: IGI Global. https://doi.org/ doi:10.4018/978-1-5225-0803-8.ch030

Deshpande, S., Basu, S. K., Li, X., & Chen, X. (2017). Smart, Innovative and Intelligent Technologies Used in Drug Designing. In I. Management Association (Ed.), Pharmaceutical Sciences: Breakthroughs in Research and Practice (pp. 1175-1191). Hershey, PA: IGI Global. https://doi.org/ doi:10.4018/978-1-5225-1762-7.ch045

Dlamini, P. N. (2017). Use of Information and Communication Technologies Tools to Capture, Store, and Disseminate Indigenous Knowledge: A Literature Review. In P. Ngulube (Ed.), *Handbook of Research on Theoretical Perspectives on Indigenous Knowledge Systems in Developing Countries* (pp. 225–247). Hershey, PA: IGI Global. doi:10.4018/978-1-5225-0833-5.ch010

Dolejsova, M., & Kera, D. (2017). The Fermentation GutHub Project and the Internet of Microbes. In S. Konomi & G. Roussos (Eds.), *Enriching Urban Spaces with Ambient Computing, the Internet of Things, and Smart City Design* (pp. 25–46). Hershey, PA: IGI Global. doi:10.4018/978-1-5225-0827-4.ch002

Dolunay, O. (2018). A Paradigm Shift: Empowering Farmers to Eliminate the Waste in the Form of Fresh Water and Energy Through the Implementation of 4R+T. In I. Management Association (Ed.), Sustainable Development: Concepts, Methodologies, Tools, and Applications (pp. 882-892). Hershey, PA: IGI Global. https://doi.org/doi:10.4018/978-1-5225-3817-2.ch039

Dube, P., Heijman, W. J., Ihle, R., & Ochieng, J. (2018). The Potential of Traditional Leafy Vegetables for Improving Food Security in Africa. In V. Erokhin (Ed.), *Establishing Food Security and Alternatives to International Trade in Emerging Economies* (pp. 220–243). Hershey, PA: IGI Global. doi:10.4018/978-1-5225-2733-6.ch011

Duruji, M. M., & Urenma, D. F. (2017). The Environmentalism and Politics of Climate Change: A Study of the Process of Global Convergence through UNFCCC Conferences. In I. Management Association (Ed.), *Natural Resources Management: Concepts, Methodologies, Tools, and Applications* (pp. 77-108). Hershey, PA: IGI Global. 10.4018/978-1-5225-0803-8.ch004

Dutta, U. (2017). Agro-Geoinformatics, Potato Cultivation, and Climate Change. In S. Londhe (Ed.), *Sustainable Potato Production and the Impact of Climate Change* (pp. 247–271). Hershey, PA: IGI Global. doi:10.4018/978-1-5225-1715-3.ch012

Edirisinghe, R., Stranieri, A., & Wickramasinghe, N. (2017). A Taxonomy for mHealth. In N. Wickramasinghe (Ed.), *Handbook of Research on Healthcare Administration and Management* (pp. 596–615). Hershey, PA: IGI Global. doi:10.4018/978-1-5225-0920-2.ch036

Ekpeni, N. M., & Ayeni, A. O. (2018). Global Natural Hazard and Disaster Vulnerability Management. In A. Eneanya (Ed.), *Handbook of Research on Environmental Policies for Emergency Management and Public Safety* (pp. 83–104). Hershey, PA: IGI Global. doi:10.4018/978-1-5225-3194-4.ch005

Ene, C., Voica, M. C., & Panait, M. (2017). Green Investments and Food Security: Opportunities and Future Directions in the Context of Sustainable Development. In M. Mieila (Ed.), *Measuring Sustainable Development and Green Investments in Contemporary Economies* (pp. 163–200). Hershey, PA: IGI Global. doi:10.4018/978-1-5225-2081-8.ch007

Escamilla, I., Ruíz, M. T., Ibarra, M. M., Soto, V. L., Quintero, R., & Guzmán, G. (2018). Geocoding Tweets Based on Semantic Web and Ontologies. In M. Lytras, N. Aljohani, E. Damiani, & K. Chui (Eds.), *Innovations, Developments, and Applications of Semantic Web and Information Systems* (pp. 372–392). Hershey, PA: IGI Global. doi:10.4018/978-1-5225-5042-6.ch014

Escribano, A. J. (2018). Marketing Strategies for Trendy Animal Products: Sustainability as a Core. In F. Quoquab, R. Thurasamy, & J. Mohammad (Eds.), *Driving Green Consumerism Through Strategic Sustainability Marketing* (pp. 169–203). Hershey, PA: IGI Global. doi:10.4018/978-1-5225-2912-5.ch010

Eudoxie, G., & Roopnarine, R. (2017). Climate Change Adaptation and Disaster Risk Management in the Caribbean. In W. Ganpat & W. Isaac (Eds.), *Environmental Sustainability and Climate Change Adaptation Strategies* (pp. 97–125). Hershey, PA: IGI Global. doi:10.4018/978-1-5225-1607-1.ch004

Farmer, L. S. (2017). Data Analytics for Strategic Management: Getting the Right Data. In V. Wang (Ed.), *Encyclopedia of Strategic Leadership and Management* (pp. 810–822). Hershey, PA: IGI Global. doi:10.4018/978-1-5225-1049-9.ch056

Fattal, L. R. (2017). Catastrophe: An Uncanny Catalyst for Creativity. In R. Shin (Ed.), *Convergence of Contemporary Art, Visual Culture, and Global Civic Engagement* (pp. 244–262). Hershey, PA: IGI Global. doi:10.4018/978-1-5225-1665-1.ch014

Forti, I. (2017). A Cross Reading of Landscape through Digital Landscape Models: The Case of Southern Garda. In A. Ippolito (Ed.), *Handbook of Research on Emerging Technologies for Architectural and Archaeological Heritage* (pp. 532–561). Hershey, PA: IGI Global. doi:10.4018/978-1-5225-0675-1.ch018

Gharbi, A., De Runz, C., & Akdag, H. (2017). Urban Development Modelling: A Survey. In S. Faiz & K. Mahmoudi (Eds.), *Handbook of Research on Geographic Information Systems Applications and Advancements* (pp. 96–124). Hershey, PA: IGI Global. doi:10.4018/978-1-5225-0937-0.ch004

Ghosh, I., & Ghoshal, I. (2018). Implications of Trade Liberalization for Food Security Under the ASEAN-India Strategic Partnership: A Gravity Model Approach. In V. Erokhin (Ed.), *Establishing Food Security and Alternatives to International Trade in Emerging Economies* (pp. 98–118). Hershey, PA: IGI Global. doi:10.4018/978-1-5225-2733-6.ch005

Glaves, H. M. (2017). Developing a Common Global Framework for Marine Data Management. In P. Diviacco, A. Leadbetter, & H. Glaves (Eds.), *Oceanographic and Marine Cross-Domain Data Management for Sustainable Development* (pp. 47–68). Hershey, PA: IGI Global. doi:10.4018/978-1-5225-0700-0.ch003

Godulla, A., & Wolf, C. (2018). Future of Food: Transmedia Strategies of National Geographic. In R. Gambarato & G. Alzamora (Eds.), *Exploring Transmedia Journalism in the Digital Age* (pp. 162–182). Hershey, PA: IGI Global. doi:10.4018/978-1-5225-3781-6.ch010

Gomes, P. P. (2018). Food and Environment: A Review on the Sustainability of Six Different Dietary Patterns. In A. Obayelu (Ed.), *Food Systems Sustainability and Environmental Policies in Modern Economies* (pp. 15–31). Hershey, PA: IGI Global. doi:10.4018/978-1-5225-3631-4.ch002

Gonzalez-Feliu, J. (2018). Sustainability Evaluation of Green Urban Logistics Systems: Literature Overview and Proposed Framework. In A. Paul, D. Bhattacharyya, & S. Anand (Eds.), *Green Initiatives for Business Sustainability and Value Creation* (pp. 103–134). Hershey, PA: IGI Global. doi:10.4018/978-1-5225-2662-9.ch005

Goodland, R. (2017). A Fresh Look at Livestock Greenhouse Gas Emissions and Mitigation. In I. Management Association (Ed.), Natural Resources Management: Concepts, Methodologies, Tools, and Applications (pp. 124-139). Hershey, PA: IGI Global. doi:10.4018/978-1-5225-0803-8.ch006

Goundar, S., & Appana, S. (2018). Mainstreaming Development Policies for Climate Change in Fiji: A Policy Gap Analysis and the Role of ICTs. In I. Management Association (Ed.), Sustainable Development: Concepts, Methodologies, Tools, and Applications (pp. 402-432). Hershey, PA: IGI Global. doi:10.4018/978-1-5225-3817-2.ch020

Granell-Canut, C., & Aguilar-Moreno, E. (2018). Geospatial Influence in Science Mapping. In M. Khosrow-Pour, D.B.A. (Ed.), Encyclopedia of Information Science and Technology, Fourth Edition (pp. 3473-3483). Hershey, PA: IGI Global. doi:10.4018/978-1-5225-2255-3.ch302

Grigelis, A., Blažauskas, N., Gelumbauskaitė, L. Ž., Gulbinskas, S., Suzdalev, S., & Ferrarin, C. (2017). Marine Environment Data Management Related to the Human Activity in the South-Eastern Baltic Sea (The Lithuanian Sector). In P. Diviacco, A. Leadbetter, & H. Glaves (Eds.), *Oceanographic and Marine Cross-Domain Data Management for Sustainable Development* (pp. 282–302). Hershey, PA: IGI Global. doi:10.4018/978-1-5225-0700-0.ch012

Guma, I. P., Rwashana, A. S., & Oyo, B. (2018). Food Security Policy Analysis Using System Dynamics: The Case of Uganda. *International Journal of Information Technologies and Systems Approach, 11*(1), 72–90. doi:10.4018/IJITSA.2018010104

Guma, I. P., Rwashana, A. S., & Oyo, B. (2018). Food Security Indicators for Subsistence Farmers Sustainability: A System Dynamics Approach. *International Journal of System Dynamics Applications, 7*(1), 45–64. doi:10.4018/IJSDA.2018010103

Gupta, P., & Goyal, S. (2017). Wildlife Habitat Evaluation. In A. Santra & S. Mitra (Eds.), *Remote Sensing Techniques and GIS Applications in Earth and Environmental Studies* (pp. 258–264). Hershey, PA: IGI Global. doi:10.4018/978-1-5225-1814-3.ch013

Hanson, T., & Hildebrand, E. (2018). GPS Travel Diaries in Rural Transportation Research: A Focus on Older Drivers. In I. Management Association (Ed.), Intelligent Transportation and Planning: Breakthroughs in Research and Practice (pp. 609-625). Hershey, PA: IGI Global. doi:10.4018/978-1-5225-5210-9.ch027

Hartman, M. B. (2017). Research-Based Climate Change Public Education Programs. In I. Management Association (Ed.), Natural Resources Management: Concepts, Methodologies, Tools, and Applications (pp. 992-1003). Hershey, PA: IGI Global. doi:10.4018/978-1-5225-0803-8.ch046

Hashim, N. (2017). Zanzibari Seaweed: Global Climate Change and the Promise of Adaptation. In I. Management Association (Ed.), Natural Resources Management: Concepts, Methodologies, Tools, and Applications (pp. 365-391). Hershey, PA: IGI Global. https://doi.org/ doi:10.4018/978-1-5225-0803-8.ch019

Herrera, J. E., Argüello, L. V., Gonzalez-Feliu, J., & Jaimes, W. A. (2017). Decision Support System Design Requirements, Information Management, and Urban Logistics Efficiency: Case Study of Bogotá, Colombia. In G. Jamil, A. Soares, & C. Pessoa (Eds.), *Handbook of Research on Information Management for Effective Logistics and Supply Chains* (pp. 223–238). Hershey, PA: IGI Global. doi:10.4018/978-1-5225-0973-8.ch012

Huizinga, T., Ayanso, A., Smoor, M., & Wronski, T. (2017). Exploring Insurance and Natural Disaster Tweets Using Text Analytics. *International Journal of Business Analytics*, *4*(1), 1–17. doi:10.4018/IJBAN.2017010101

Hung, K., Kalantari, M., & Rajabifard, A. (2017). An Integrated Method for Assessing the Text Content Quality of Volunteered Geographic Information in Disaster Management. *International Journal of Information Systems for Crisis Response and Management*, *9*(2), 1–17. doi:10.4018/IJISCRAM.2017040101

Husnain, A., & Avdic, A. (2018). Identifying the Contemporary Status of E-Service Sustainability Research. In I. Management Association (Ed.), Sustainable Development: Concepts, Methodologies, Tools, and Applications (pp. 467-485). Hershey, PA: IGI Global. https://doi.org/ doi:10.4018/978-1-5225-3817-2.ch022

Iarossi, M. P., & Ferro, L. (2017). "The Past is Never Dead. It's Not Even Past": Virtual Archaeological Promenade. In A. Ippolito & M. Cigola (Eds.), *Handbook of Research on Emerging Technologies for Digital Preservation and Information Modeling* (pp. 228–255). Hershey, PA: IGI Global. doi:10.4018/978-1-5225-0680-5.ch010

Ignjatijević, S., & Cvijanović, D. (2018). Analysis of Serbian Honey Production and Exports. In *Exploring the Global Competitiveness of Agri-Food Sectors and Serbia's Dominant Presence: Emerging Research and Opportunities* (pp. 109–139). Hershey, PA: IGI Global. doi:10.4018/978-1-5225-2762-6.ch005

Jana, S. K., & Karmakar, A. K. (2017). Globalization, Governance, and Food Security: The Case of BRICS. In I. Management Association (Ed.), Natural Resources Management: Concepts, Methodologies, Tools, and Applications (pp. 692-712). Hershey, PA: IGI Global. https://doi.org/ doi:10.4018/978-1-5225-0803-8.ch032

Jana, S. K., & Karmakar, A. K. (2017). Food Security in Asia: Is There Convergence? In I. Management Association (Ed.), Natural Resources Management: Concepts, Methodologies, Tools, and Applications (pp. 109-123). Hershey, PA: IGI Global. https://doi.org/ doi:10.4018/978-1-5225-0803-8.ch005

John, J., & Kumar, S. (2018). A Locational Decision Making Framework for Shipbreaking Under Multiple Criteria. In I. Management Association (Ed.), Operations and Service Management: Concepts, Methodologies, Tools, and Applications (pp. 504-527). Hershey, PA: IGI Global. doi:10.4018/978-1-5225-3909-4.ch024

John, J., & Srivastava, R. K. (2018). Decision Insights for Shipbreaking using Environmental Impact Assessment: Review and Perspectives. *International Journal of Strategic Decision Sciences*, *9*(1), 45–62. doi:10.4018/IJSDS.2018010104

Joshi, Y., & Rahman, Z. (2018). Determinants of Sustainable Consumption Behaviour: Review and Conceptual Framework. In A. Paul, D. Bhattacharyya, & S. Anand (Eds.), *Green Initiatives for Business Sustainability and Value Creation* (pp. 239–262). Hershey, PA: IGI Global. doi:10.4018/978-1-5225-2662-9.ch011

Juma, D. W., Reuben, M., Wang, H., & Li, F. (2018). Adaptive Coevolution: Realigning the Water Governance Regime to the Changing Climate. In I. Management Association (Ed.), Hydrology and Water Resource Management: Breakthroughs in Research and Practice (pp. 346-357). Hershey, PA: IGI Global. https://doi.org/ doi:10.4018/978-1-5225-3427-3.ch014

K., S., & Tripathy, B. K. (2018). Neighborhood Rough-Sets-Based Spatial Data Analytics. In M. Khosrow-Pour, D.B.A. (Ed.), *Encyclopedia of Information Science and Technology, Fourth Edition* (pp. 1835-1844). Hershey, PA: IGI Global. https://doi.org/ doi:10.4018/978-1-5225-2255-3.ch160

Kabir, F. (2018). Towards a More Gender-Inclusive Climate Change Policy. In N. Mahtab, T. Haque, I. Khan, M. Islam, & I. Wahid (Eds.), *Handbook of Research on Women's Issues and Rights in the Developing World* (pp. 354–369). Hershey, PA: IGI Global. doi:10.4018/978-1-5225-3018-3.ch022

Kabir, F. (2018). Towards a More Gender-Inclusive Climate Change Policy. In I. Management Association (Ed.), Climate Change and Environmental Concerns: Breakthroughs in Research and Practice (pp. 525-540). Hershey, PA: IGI Global. https://doi.org/ doi:10.4018/978-1-5225-5487-5.ch027

Kanyamuka, J. S., Jumbe, C. B., & Ricker-Gilbert, J. (2018). Making Agricultural Input Subsidies More Effective and Profitable in Africa: The Role of Complementary Interventions. In A. Obayelu (Ed.), *Food Systems Sustainability and Environmental Policies in Modern Economies* (pp. 172–187). Hershey, PA: IGI Global. doi:10.4018/978-1-5225-3631-4.ch008

Karimi, H., & Gholamrezafahimi, F. (2017). Study of Integrated Coastal Zone Management and Its Environmental Effects: A Case of Iran. In R. Singh, A. Singh, & V. Srivastava (Eds.), *Environmental Issues Surrounding Human Overpopulation* (pp. 64–88). Hershey, PA: IGI Global. doi:10.4018/978-1-5225-1683-5.ch004

Kaya, I. R., Hutabarat, J., & Bambang, A. N. (2018). "Sasi": A New Path to Sustain Seaweed Farming From Up-Stream to Down-Stream in Kotania Bay, Molucass. *International Journal of Social Ecology and Sustainable Development, 9*(2), 28–36. doi:10.4018/IJSESD.2018040103

Khader, V. (2018). Technologies for Food, Health, Livelihood, and Nutrition Security. In I. Management Association (Ed.), Food Science and Nutrition: Breakthroughs in Research and Practice (pp. 94-112). Hershey, PA: IGI Global. https://doi.org/ doi:10.4018/978-1-5225-5207-9.ch005

Kocadağlı, A. Y. (2017). The Temporal and Spatial Development of Organic Agriculture in Turkey. In W. Ganpat, R. Dyer, & W. Isaac (Eds.), *Agricultural Development and Food Security in Developing Nations* (pp. 130–156). Hershey, PA: IGI Global. doi:10.4018/978-1-5225-0942-4.ch006

Koundouri, P., Giannouli, A., & Souliotis, I. (2017). An Integrated Approach for Sustainable Environmental and Socio-Economic Development Using Offshore Infrastructure. In I. Management Association (Ed.), Renewable and Alternative Energy: Concepts, Methodologies, Tools, and Applications (pp. 1581-1601). Hershey, PA: IGI Global. doi:10.4018/978-1-5225-1671-2.ch056

Kumar, A., & Dash, M. K. (2017). Sustainability and Future Generation Infrastructure on Digital Platform: A Study of Generation Y. In N. Ray (Ed.), *Business Infrastructure for Sustainability in Developing Economies* (pp. 124–142). Hershey, PA: IGI Global. doi:10.4018/978-1-5225-2041-2.ch007

Kumar, A., Mukherjee, A. B., & Krishna, A. P. (2017). Application of Conventional Data Mining Techniques and Web Mining to Aid Disaster Management. In A. Kumar (Ed.), *Web Usage Mining Techniques and Applications Across Industries* (pp. 138–167). Hershey, PA: IGI Global. doi:10.4018/978-1-5225-0613-3.ch006

Kumar, C. P. (2017). Impact of Climate Change on Groundwater Resources. In I. Management Association (Ed.), Natural Resources Management: Concepts, Methodologies, Tools, and Applications (pp. 1094-1120). Hershey, PA: IGI Global. doi:10.4018/978-1-5225-0803-8.ch052

Kumari, S., & Patil, Y. (2017). Achieving Climate Smart Agriculture with a Sustainable Use of Water: A Conceptual Framework for Sustaining the Use of Water for Agriculture in the Era of Climate Change. In P. Rao & Y. Patil (Eds.), *Reconsidering the Impact of Climate Change on Global Water Supply, Use, and Management* (pp. 122–143). Hershey, PA: IGI Global. doi:10.4018/978-1-5225-1046-8.ch008

Kumari, S., & Patil, Y. (2018). Achieving Climate Smart Agriculture With a Sustainable Use of Water: A Conceptual Framework for Sustaining the Use of Water for Agriculture in the Era of Climate Change. In I. Management Association (Ed.), Climate Change and Environmental Concerns: Breakthroughs in Research and Practice (pp. 111-133). Hershey, PA: IGI Global. doi:10.4018/978-1-5225-5487-5.ch006

Kursah, M. B. (2017). Least-Cost Pipeline using Geographic Information System: The Limit to Technicalities. *International Journal of Applied Geospatial Research, 8*(3), 1–15. doi:10.4018/ijagr.2017070101

Lahiri, S., Ghosh, D., & Bhakta, J. N. (2017). Role of Microbes in Eco-Remediation of Perturbed Aquatic Ecosystem. In J. Bhakta (Ed.), *Handbook of Research on Inventive Bioremediation Techniques* (pp. 70–107). Hershey, PA: IGI Global. doi:10.4018/978-1-5225-2325-3.ch004

Lallo, C. H., Smalling, S., Facey, A., & Hughes, M. (2018). The Impact of Climate Change on Small Ruminant Performance in Caribbean Communities. In I. Management Association (Ed.), Climate Change and Environmental Concerns: Breakthroughs in Research and Practice (pp. 193-218). Hershey, PA: IGI Global. doi:10.4018/978-1-5225-5487-5.ch010

Laurini, R. (2017). Nature of Geographic Knowledge Bases. In S. Faiz & K. Mahmoudi (Eds.), *Handbook of Research on Geographic Information Systems Applications and Advancements* (pp. 29–60). Hershey, PA: IGI Global. doi:10.4018/978-1-5225-0937-0.ch002

Lawrence, J., Simpson, L., & Piggott, A. (2017). Protected Agriculture: A Climate Change Adaptation for Food and Nutrition Security. In I. Management Association (Ed.), Natural Resources Management: Concepts, Methodologies, Tools, and Applications (pp. 140-158). Hershey, PA: IGI Global. doi:10.4018/978-1-5225-0803-8.ch007

Leadbetter, A., Cheatham, M., Shepherd, A., & Thomas, R. (2017). Linked Ocean Data 2.0. In P. Diviacco, A. Leadbetter, & H. Glaves (Eds.), *Oceanographic and Marine Cross-Domain Data Management for Sustainable Development* (pp. 69–99). Hershey, PA: IGI Global. doi:10.4018/978-1-5225-0700-0.ch004

Lucas, M. R., Rego, C., Vieira, C., & Vieira, I. (2017). Proximity and Cooperation for Innovative Regional Development: The Case of the Science and Technology Park of Alentejo. In L. Carvalho (Ed.), *Handbook of Research on Entrepreneurial Development and Innovation Within Smart Cities* (pp. 199–228). Hershey, PA: IGI Global. doi:10.4018/978-1-5225-1978-2.ch010

Ma, X., Beaulieu, S. E., Fu, L., Fox, P., Di Stefano, M., & West, P. (2017). Documenting Provenance for Reproducible Marine Ecosystem Assessment in Open Science. In P. Diviacco, A. Leadbetter, & H. Glaves (Eds.), *Oceanographic and Marine Cross-Domain Data Management for Sustainable Development* (pp. 100–126). Hershey, PA: IGI Global. doi:10.4018/978-1-5225-0700-0.ch005

Mabe, L. K., & Oladele, O. I. (2017). Application of Information Communication Technologies for Agricultural Development through Extension Services: A Review. In T. Tossy (Ed.), *Information Technology Integration for Socio-Economic Development* (pp. 52–101). Hershey, PA: IGI Global. doi:10.4018/978-1-5225-0539-6.ch003

Malomo, B. I. (2018). A Review of Psychological Resilience as a Response to Natural Hazards in Nigeria. In A. Eneanya (Ed.), *Handbook of Research on Environmental Policies for Emergency Management and Public Safety* (pp. 147–165). Hershey, PA: IGI Global. doi:10.4018/978-1-5225-3194-4.ch008

Manchiraju, S. (2018). Predicting Behavioral Intentions Toward Sustainable Fashion Consumption: A Comparison of Attitude-Behavior and Value-Behavior Consistency Models. In I. Management Association (Ed.), Fashion and Textiles: Breakthroughs in Research and Practice (pp. 1-21). Hershey, PA: IGI Global. doi:10.4018/978-1-5225-3432-7.ch001

Manzella, G. M., Bartolini, R., Bustaffa, F., D'Angelo, P., De Mattei, M., Frontini, F., ... Spada, A. (2017). Semantic Search Engine for Data Management and Sustainable Development: Marine Planning Service Platform. In P. Diviacco, A. Leadbetter, & H. Glaves (Eds.), *Oceanographic and Marine Cross-Domain Data Management for Sustainable Development* (pp. 127–154). Hershey, PA: IGI Global. doi:10.4018/978-1-5225-0700-0.ch006

Mbonigaba, J. (2018). Comparing the Effects of Unsustainable Production and Consumption of Food on Health and Policy Across Developed and Less Developed Countries. In A. Obayelu (Ed.), *Food Systems Sustainability and Environmental Policies in Modern Economies* (pp. 124–158). Hershey, PA: IGI Global. doi:10.4018/978-1-5225-3631-4.ch006

McKeown, A. E. (2017). Nurses, Healthcare, and Environmental Pollution and Solutions: Breaking the Cycle of Harm. In I. Management Association (Ed.), Natural Resources Management: Concepts, Methodologies, Tools, and Applications (pp. 392-415). Hershey, PA: IGI Global. https://doi.org/ doi:10.4018/978-1-5225-0803-8.ch020

Mili, B., Barua, A., & Katyaini, S. (2017). Climate Change and Adaptation through the Lens of Capability Approach: A Case Study from Darjeeling, Eastern Himalaya. In I. Management Association (Ed.), Natural Resources Management: Concepts, Methodologies, Tools, and Applications (pp. 1351-1365). Hershey, PA: IGI Global. https://doi.org/ doi:10.4018/978-1-5225-0803-8.ch064

Mir, S. A., Shah, M. A., Mir, M. M., & Iqbal, U. (2017). New Horizons of Nanotechnology in Agriculture and Food Processing Industry. In B. Nayak, A. Nanda, & M. Bhat (Eds.), *Integrating Biologically-Inspired Nanotechnology into Medical Practice* (pp. 230–258). Hershey, PA: IGI Global. doi:10.4018/978-1-5225-0610-2.ch009

Moallem, M., Sterrett, W. L., Gordon, C. R., Sukhera, S. M., Mahmood, A., & Bashir, A. (2021). An Investigation of the Effects of Integrating Computing and Project- or Problem-Based Learning in the Context of Environmental Sciences: A Case of Pakistani STEM Teachers. In S. Schroth & J. Daniels (Eds.), *Building STEM Skills Through Environmental Education* (pp. 49–89). IGI Global. https://doi.org/10.4018/978-1-7998-2711-5.ch003

Mujere, N., & Moyce, W. (2017). Climate Change Impacts on Surface Water Quality. In W. Ganpat & W. Isaac (Eds.), *Environmental Sustainability and Climate Change Adaptation Strategies* (pp. 322–340). Hershey, PA: IGI Global. doi:10.4018/978-1-5225-1607-1.ch012

Mukherjee, A. B., Krishna, A. P., & Patel, N. (2018). Geospatial Technology for Urban Sciences. In *Geospatial Technologies in Urban System Development: Emerging Research and Opportunities* (pp. 99–120). Hershey, PA: IGI Global. doi:10.4018/978-1-5225-3683-3.ch005

Nagarajan, S. K., & Sangaiah, A. K. (2017). Vegetation Index: Ideas, Methods, Influences, and Trends. In N. Kumar, A. Sangaiah, M. Arun, & S. Anand (Eds.), *Advanced Image Processing Techniques and Applications* (pp. 347–386). Hershey, PA: IGI Global. doi:10.4018/978-1-5225-2053-5.ch016

Naraine, L., & Meehan, K. (2017). Strengthening Food Security with Sustainable Practices by Smallholder Farmers in Lesser Developed Economies. In W. Ganpat, R. Dyer, & W. Isaac (Eds.), *Agricultural Development and Food Security in Developing Nations* (pp. 57–81). Hershey, PA: IGI Global. doi:10.4018/978-1-5225-0942-4.ch003

Naraine, L., & Meehan, K. (2017). Strengthening Food Security with Sustainable Practices by Smallholder Farmers in Lesser Developed Economies. In W. Ganpat, R. Dyer, & W. Isaac (Eds.), *Agricultural Development and Food Security in Developing Nations* (pp. 57–81). Hershey, PA: IGI Global. doi:10.4018/978-1-5225-0942-4.ch003

Naraine, L., & Meehan, K. (2017). Strengthening Food Security with Sustainable Practices by Smallholder Farmers in Lesser Developed Economies. In W. Ganpat, R. Dyer, & W. Isaac (Eds.), *Agricultural Development and Food Security in Developing Nations* (pp. 57–81). Hershey, PA: IGI Global. doi:10.4018/978-1-5225-0942-4.ch003

Nikolaou, K., Tsakiridou, E., Anastasiadis, F., & Mattas, K. (2017). Exploring Alternative Distribution Channels of Agricultural Products. *International Journal of Food and Beverage Manufacturing and Business Models*, *2*(2), 36–66. doi:10.4018/IJFBMBM.2017070103

Nishat, K. J., & Rahman, M. S. (2018). Disaster, Vulnerability, and Violence Against Women: Global Findings and a Research Agenda for Bangladesh. In N. Mahtab, T. Haque, I. Khan, M. Islam, & I. Wahid (Eds.), *Handbook of Research on Women's Issues and Rights in the Developing World* (pp. 235–250). Hershey, PA: IGI Global. doi:10.4018/978-1-5225-3018-3.ch014

O'Hara, S., Jones, D., & Trobman, H. B. (2018). Building an Urban Food System Through UDC Food Hubs. In A. Burtin, J. Fleming, & P. Hampton-Garland (Eds.), *Changing Urban Landscapes Through Public Higher Education* (pp. 116–143). Hershey, PA: IGI Global. doi:10.4018/978-1-5225-3454-9.ch006

Obayelu, A. E. (2018). Integrating Environment, Food Systems, and Sustainability in Feeding the Growing Population in Developing Countries. In A. Obayelu (Ed.), *Food Systems Sustainability and Environmental Policies in Modern Economies* (pp. 1–14). Hershey, PA: IGI Global. doi:10.4018/978-1-5225-3631-4.ch001

Othman, R., Nath, N., & Laswad, F. (2018). Environmental Reporting and Accounting: Sustainability Hybridisation. In G. Azevedo, J. da Silva Oliveira, R. Marques, & A. Ferreira (Eds.), *Handbook of Research on Modernization and Accountability in Public Sector Management* (pp. 130–158). Hershey, PA: IGI Global. doi:10.4018/978-1-5225-3731-1.ch007

Ouadi, A., & Zitouni, A. (2021). Phasor Measurement Improvement Using Digital Filter in a Smart Grid. In A. Recioui & H. Bentarzi (Eds.), *Optimizing and Measuring Smart Grid Operation and Control* (pp. 100–117). IGI Global. https://doi.org/10.4018/978-1-7998-4027-5.ch005

Padigala, B. S. (2018). Traditional Water Management System for Climate Change Adaptation in Mountain Ecosystems. In I. Management Association (Ed.), Climate Change and Environmental Concerns: Breakthroughs in Research and Practice (pp. 630-655). Hershey, PA: IGI Global. doi:10.4018/978-1-5225-5487-5.ch033

Panda, C. K. (2018). Mobile Phone Usage in Agricultural Extension in India: The Current and Future Perspective. In F. Mtenzi, G. Oreku, D. Lupiana, & J. Yonazi (Eds.), *Mobile Technologies and Socio-Economic Development in Emerging Nations* (pp. 1–21). Hershey, PA: IGI Global. doi:10.4018/978-1-5225-4029-8.ch001

Pandian, S. L., Yarrakula, K., & Chaudhury, P. (2018). GIS-Based Decision Support System for Village Level: A Case Study in Andhra Pradesh. In S. Chaudhuri & N. Ray (Eds.), *GIS Applications in the Tourism and Hospitality Industry* (pp. 275–295). Hershey, PA: IGI Global. doi:10.4018/978-1-5225-5088-4.ch012

Quaranta, G., & Salvia, R. (2017). Social-Based Product Innovation and Governance in The Milk Sector: The Case of Carciocacio and Innonatura. In T. Tarnanidis, M. Vlachopoulou, & J. Papathanasiou (Eds.), *Driving Agribusiness With Technology Innovations* (pp. 293–310). Hershey, PA: IGI Global. doi:10.4018/978-1-5225-2107-5.ch015

Rahman, M. K., Schmidlin, T. W., Munro-Stasiuk, M. J., & Curtis, A. (2017). Geospatial Analysis of Land Loss, Land Cover Change, and Landuse Patterns of Kutubdia Island, Bangladesh. *International Journal of Applied Geospatial Research*, *8*(2), 45–60. doi:10.4018/IJAGR.2017040104

Rajack-Talley, T. A. (2017). Agriculture, Trade Liberalization and Poverty in the ACP Countries. In W. Ganpat, R. Dyer, & W. Isaac (Eds.), *Agricultural Development and Food Security in Developing Nations* (pp. 1–20). Hershey, PA: IGI Global. doi:10.4018/978-1-5225-0942-4.ch001

Rajamanickam, S. (2018). Exploring Landscapes in Regional Convergence: Environment and Sustainable Development in South Asia. In I. Management Association (Ed.), Sustainable Development: Concepts, Methodologies, Tools, and Applications (pp. 1051-1087). Hershey, PA: IGI Global. https://doi.org/ doi:10.4018/978-1-5225-3817-2.ch047

Rizvi, S. M., & Dearden, A. (2018). KHETI: ICT Solution for Agriculture Extension and Its Replication in Open and Distance Learning. In U. Pandey & V. Indrakanti (Eds.), *Open and Distance Learning Initiatives for Sustainable Development* (pp. 163–174). Hershey, PA: IGI Global. doi:10.4018/978-1-5225-2621-6.ch008

Roşu, L., & Macarov, L. I. (2017). Management of Drought and Floods in the Dobrogea Region. In I. Management Association (Ed.), Agri-Food Supply Chain Management: Breakthroughs in Research and Practice (pp. 372-403). Hershey, PA: IGI Global. https://doi.org/ doi:10.4018/978-1-5225-1629-3.ch016

Rouzbehani, K., & Rouzbehani, S. (2018). Mapping Women's World: GIS and the Case of Breast Cancer in the US. *International Journal of Public Health Management and Ethics*, *3*(1), 14–25. doi:10.4018/IJPHME.2018010102

Roy, D. (2018). Success Factors of Adoption of Mobile Applications in Rural India: Effect of Service Characteristics on Conceptual Model. In M. Khosrow-Pour, D.B.A. (Ed.), Green Computing Strategies for Competitive Advantage and Business Sustainability (pp. 211-238). Hershey, PA: IGI Global. https://doi.org/ doi:10.4018/978-1-5225-5017-4.ch010

Sajeva, M., Lemon, M., & Sahota, P. S. (2017). Governance for Food Security: A Framework for Social Learning and Scenario Building. *International Journal of Food and Beverage Manufacturing and Business Models*, *2*(2), 67–84. doi:10.4018/IJFBMBM.2017070104

Sambhanthan, A., & Potdar, V. (2017). A Study of the Parameters Impacting Sustainability in Information Technology Organizations. *International Journal of Knowledge-Based Organizations*, *7*(3), 27–39. doi:10.4018/IJKBO.2017070103

Sanga, C., Kalungwizi, V. J., & Msuya, C. P. (2017). Bridging Gender Gaps in Provision of Agricultural Extension Service Using ICT: Experiences from Sokoine University of Agriculture (SUA) Farmer Voice Radio (FVR) Project in Tanzania. In I. Management Association (Ed.), Discrimination and Diversity: Concepts, Methodologies, Tools, and Applications (pp. 682-697). Hershey, PA: IGI Global. https://doi.org/ doi:10.4018/978-1-5225-1933-1.ch031

Santra, A., & Mitra, D. (2017). Role of Remote Sensing in Potential Fishing Zone Forecast. In A. Santra & S. Mitra (Eds.), *Remote Sensing Techniques and GIS Applications in Earth and Environmental Studies* (pp. 243–257). Hershey, PA: IGI Global. doi:10.4018/978-1-5225-1814-3.ch012

Schaap, D. (2017). SeaDataNet: Towards a Pan-European Infrastructure for Marine and Ocean Data Management. In P. Diviacco, A. Leadbetter, & H. Glaves (Eds.), *Oceanographic and Marine Cross-Domain Data Management for Sustainable Development* (pp. 155–177). Hershey, PA: IGI Global. doi:10.4018/978-1-5225-0700-0.ch007

Seckin-Celik, T. (2017). Sustainability Reporting and Sustainability in the Turkish Business Context. In U. Akkucuk (Ed.), *Ethics and Sustainability in Global Supply Chain Management* (pp. 115–132). Hershey, PA: IGI Global. doi:10.4018/978-1-5225-2036-8.ch006

Segbefia, A. Y., Barnes, V. R., Akpalu, L. A., & Mensah, M. (2018). Environmental Location Assessment for Seaweed Cultivation in Ghana: A Spatial Multi-Criteria Approach. *International Journal of Applied Geospatial Research*, *9*(1), 51–64. doi:10.4018/IJAGR.2018010104

Sen, Y. (2018). How to Manage Sustainability: A Framework for Corporate Sustainability Tools. In I. Management Association (Ed.), Sustainable Development: Concepts, Methodologies, Tools, and Applications (pp. 568-589). Hershey, PA: IGI Global. doi:10.4018/978-1-5225-3817-2.ch026

Shamshiry, E., Abdulai, A. M., Mokhtar, M. B., & Komoo, I. (2015). Regional Landfill Site Selection with GIS and Analytical Hierarchy Process Techniques: A Case Study of Langkawi Island, Malaysia. In P. Thomas, M. Srihari, & S. Kaur (Eds.), *Handbook of Research on Cultural and Economic Impacts of the Information Society* (pp. 248–282). Hershey, PA: IGI Global. doi:10.4018/978-1-4666-8598-7.ch011

Sharma, Y. K., Mangla, S. K., Patil, P. P., & Uniyal, S. (2018). Analyzing Sustainable Food Supply Chain Management Challenges in India. In M. Ram & J. Davim (Eds.), *Soft Computing Techniques and Applications in Mechanical Engineering* (pp. 162–180). Hershey, PA: IGI Global. doi:10.4018/978-1-5225-3035-0.ch008

Silvestrelli, P. (2018). The Impact of Events: To Which Extent Are They Sustainable for Tourist Destinations? Some Evidences From Expo Milano 2015. In M. Risso & S. Testarmata (Eds.), *Value Sharing for Sustainable and Inclusive Development* (pp. 185–204). Hershey, PA: IGI Global. doi:10.4018/978-1-5225-3147-0.ch008

Silvius, G. (2018). Sustainability Evaluation of IT/IS Projects. In I. Management Association (Ed.), Sustainable Development: Concepts, Methodologies, Tools, and Applications (pp. 26-40). Hershey, PA: IGI Global. https://doi.org/ doi:10.4018/978-1-5225-3817-2.ch002

Singh, R., Srivastava, P., Singh, P., Upadhyay, S., & Raghubanshi, A. S. (2017). Human Overpopulation and Food Security: Challenges for the Agriculture Sustainability. In R. Singh, A. Singh, & V. Srivastava (Eds.), *Environmental Issues Surrounding Human Overpopulation* (pp. 12–39). Hershey, PA: IGI Global. doi:10.4018/978-1-5225-1683-5.ch002

Singh, R., Srivastava, P., Singh, P., Upadhyay, S., & Raghubanshi, A. S. (2017). Human Overpopulation and Food Security: Challenges for the Agriculture Sustainability. In R. Singh, A. Singh, & V. Srivastava (Eds.), *Environmental Issues Surrounding Human Overpopulation* (pp. 12–39). Hershey, PA: IGI Global. doi:10.4018/978-1-5225-1683-5.ch002

Srivastava, N. (2017). Climate Change Mitigation: Collective Efforts and Responsibly. In I. Management Association (Ed.), Natural Resources Management: Concepts, Methodologies, Tools, and Applications (pp. 64-76). Hershey, PA: IGI Global. https://doi.org/ doi:10.4018/978-1-5225-0803-8.ch003

Stanganelli, M., & Gerundo, C. (2017). Understanding the Role of Urban Morphology and Green Areas Configuration During Heat Waves. *International Journal of Agricultural and Environmental Information Systems*, *8*(2), 50–64. doi:10.4018/IJAEIS.2017040104

Stewart, M. K., Hagood, D., & Ching, C. C. (2017). Virtual Games and Real-World Communities: Environments that Constrain and Enable Physical Activity in Games for Health. *International Journal of Game-Based Learning*, *7*(1), 1–19. doi:10.4018/IJGBL.2017010101

Stone, R. J. (2017). Modelling the Frequency of Tropical Cyclones in the Lower Caribbean Region. In W. Ganpat & W. Isaac (Eds.), *Environmental Sustainability and Climate Change Adaptation Strategies* (pp. 341–349). Hershey, PA: IGI Global. doi:10.4018/978-1-5225-1607-1.ch013

Syed, A., & Jabeen, U. A. (2018). Climate Change Impact on Agriculture and Food Security. In A. Eneanya (Ed.), *Handbook of Research on Environmental Policies for Emergency Management and Public Safety* (pp. 223–237). Hershey, PA: IGI Global. doi:10.4018/978-1-5225-3194-4.ch012

Tam, G. C. (2017). The Global View of Sustainability. In *Managerial Strategies and Green Solutions for Project Sustainability* (pp. 1–24). Hershey, PA: IGI Global. doi:10.4018/978-1-5225-2371-0.ch001

Tam, G. C. (2017). Understanding Project Sustainability. In *Managerial Strategies and Green Solutions for Project Sustainability* (pp. 110–139). Hershey, PA: IGI Global. doi:10.4018/978-1-5225-2371-0.ch005

Tam, G. C. (2017). Perspectives on Sustainability. In *Managerial Strategies and Green Solutions for Project Sustainability* (pp. 53–76). Hershey, PA: IGI Global. doi:10.4018/978-1-5225-2371-0.ch003

Tang, M., & Karunanithi, A. T. (2018). Visual Logic Maps (vLms). In *Advanced Concept Maps in STEM Education: Emerging Research and Opportunities* (pp. 108–149). Hershey, PA: IGI Global. doi:10.4018/978-1-5225-2184-6.ch005

Taşçıoğlu, M., & Yener, D. (2018). The Value and Scope of GIS in Marketing and Tourism Management. In S. Chaudhuri & N. Ray (Eds.), *GIS Applications in the Tourism and Hospitality Industry* (pp. 189–211). Hershey, PA: IGI Global. doi:10.4018/978-1-5225-5088-4.ch009

Thanh Tung, B. (2021). Pharmacology and Therapeutic Applications of Resveratrol. In A. Hussain & S. Behl (Eds.), *Treating Endocrine and Metabolic Disorders With Herbal Medicines* (pp. 321–333). IGI Global. https://doi.org/10.4018/978-1-7998-4808-0.ch014

Tianming, G. (2018). Food Security and Rural Development on Emerging Markets of Northeast Asia: Cases of Chinese North and Russian Far East. In V. Erokhin (Ed.), *Establishing Food Security and Alternatives to International Trade in Emerging Economies* (pp. 155–176). Hershey, PA: IGI Global. doi:10.4018/978-1-5225-2733-6.ch008

Tiftikçigil, B. Y., Yaşgül, Y. S., & Güriş, B. (2017). Sustainability of Foreign Trade Deficit in Energy: The Case of Turkey. In N. Ray (Ed.), *Business Infrastructure for Sustainability in Developing Economies* (pp. 94–109). Hershey, PA: IGI Global. doi:10.4018/978-1-5225-2041-2.ch005

Tiwari, S., Vaish, B., & Singh, P. (2018). Population and Global Food Security: Issues Related to Climate Change. In I. Management Association (Ed.), Climate Change and Environmental Concerns: Breakthroughs in Research and Practice (pp. 41-64). Hershey, PA: IGI Global. https://doi.org/ doi:10.4018/978-1-5225-5487-5.ch003

Toujani, A., & Achour, H. (2018). A Data Mining Framework for Forest Fire Mapping. In I. Management Association (Ed.), Information Retrieval and Management: Concepts, Methodologies, Tools, and Applications (pp. 771-794). Hershey, PA: IGI Global. doi:10.4018/978-1-5225-5191-1.ch033

Tripathy, B., & K., S. B. (2017). Rough Fuzzy Set Theory and Neighbourhood Approximation Based Modelling for Spatial Epidemiology. In I. Management Association (Ed.), *Public Health and Welfare: Concepts, Methodologies, Tools, and Applications* (pp. 1257-1268). Hershey, PA: IGI Global. https://doi.org/ doi:10.4018/978-1-5225-1674-3.ch058

Trukhachev, A. (2017). New Approaches to Regional Branding through Green Production and Utilization of Existing Natural Advantages. In I. Management Association (Ed.), Advertising and Branding: Concepts, Methodologies, Tools, and Applications (pp. 1758-1778). Hershey, PA: IGI Global. doi:10.4018/978-1-5225-1793-1.ch081

Tsobanoglou, G. O., & Vlachopoulou, E. I. (2017). Social-Ecological Systems in Local Fisheries Communities. In G. Korres, E. Kourliouros, & M. Michailidis (Eds.), *Handbook of Research on Policies and Practices for Sustainable Economic Growth and Regional Development* (pp. 306–316). Hershey, PA: IGI Global. doi:10.4018/978-1-5225-2458-8.ch026

Tuydes-Yaman, H., & Karatas, P. (2018). Evaluation of Walkability and Pedestrian Level of Service. In I. Management Association (Ed.), Intelligent Transportation and Planning: Breakthroughs in Research and Practice (pp. 264-291). Hershey, PA: IGI Global. doi:10.4018/978-1-5225-5210-9.ch012

Uddin, S., Chakravorty, S., Ray, A., & Sherpa, K. S. (2018). Optimal Location of Sub-Station Using Q-GIS and Multi-Criteria Decision Making Approach. *International Journal of Decision Support System Technology*, *10*(2), 65–79. doi:10.4018/IJDSST.2018040104

Uddin, S., Chakravorty, S., Sherpa, K. S., & Ray, A. (2018). Power Distribution System Planning Using Q-GIS. *International Journal of Energy Optimization and Engineering*, *7*(2), 61–75. doi:10.4018/IJEOE.2018040103

Uzun, F. V. (2018). Natural Resources Management. In A. Eneanya (Ed.), *Handbook of Research on Environmental Policies for Emergency Management and Public Safety* (pp. 1–21). Hershey, PA: IGI Global. doi:10.4018/978-1-5225-3194-4.ch001

V, M., Agrawal, R., Sharma, V., & T.N., K. (2018). Supply Chain Social Sustainability and Manufacturing. In I. Management Association (Ed.), *Technology Adoption and Social Issues: Concepts, Methodologies, Tools, and Applications* (pp. 226-252). Hershey, PA: IGI Global. https://doi.org/ doi:10.4018/978-1-5225-5201-7.ch011

van der Vliet-Bakker, J. M. (2017). Environmentally Forced Migration and Human Rights. In C. Akrivopoulou (Ed.), *Defending Human Rights and Democracy in the Era of Globalization* (pp. 146–180). Hershey, PA: IGI Global. doi:10.4018/978-1-5225-0723-9.ch007

Vargas-Hernández, J. G. (2022). A Comprehensive Entrepreneurship Model for the Internationalization of Green Innovation Businesses. In U. Akkucuk (Ed.), *Disruptive Technologies and Eco-Innovation for Sustainable Development* (pp. 131–149). IGI Global. https://doi.org/10.4018/978-1-7998-8900-7.ch008

Vaskov, A. G., Lin, Z. Y., Tyagunov, M. G., Shestopalova, T. A., & Deryugina, G. V. (2018). Design of Renewable Sources GIS for ASEAN Countries. In V. Kharchenko & P. Vasant (Eds.), *Handbook of Research on Renewable Energy and Electric Resources for Sustainable Rural Development* (pp. 1–25). Hershey, PA: IGI Global. doi:10.4018/978-1-5225-3867-7.ch001

Vázquez, D. G., & Gil, M. T. (2017). Sustainability in Smart Cities: The Case of Vitoria-Gasteiz (Spain) – A Commitment to a New Urban Paradigm. In L. Carvalho (Ed.), *Handbook of Research on Entrepreneurial Development and Innovation Within Smart Cities* (pp. 248–268). Hershey, PA: IGI Global. doi:10.4018/978-1-5225-1978-2.ch012

Wahab, I. N., & Soonthodu, S. (2018). Geographical Information System in Eco-Tourism. In S. Chaudhuri & N. Ray (Eds.), *GIS Applications in the Tourism and Hospitality Industry* (pp. 61–75). Hershey, PA: IGI Global. doi:10.4018/978-1-5225-5088-4.ch003

Weiss-Randall, D. (2018). Cultivating Environmental Justice. In *Utilizing Innovative Technologies to Address the Public Health Impact of Climate Change: Emerging Research and Opportunities* (pp. 110–143). Hershey, PA: IGI Global. doi:10.4018/978-1-5225-3414-3.ch004

Weiss-Randall, D. (2018). Cultivating Resilience. In *Utilizing Innovative Technologies to Address the Public Health Impact of Climate Change: Emerging Research and Opportunities* (pp. 204–235). Hershey, PA: IGI Global. doi:10.4018/978-1-5225-3414-3.ch007

Weiss-Randall, D. (2018). Climate Change Solutions: Where Do We Go From Here? In *Utilizing Innovative Technologies to Address the Public Health Impact of Climate Change: Emerging Research and Opportunities* (pp. 236–268). Hershey, PA: IGI Global. doi:10.4018/978-1-5225-3414-3.ch008

Whyte, K. P., List, M., Stone, J. V., Grooms, D., Gasteyer, S., Thompson, P. B., . . . Bouri, H. (2018). Uberveillance, Standards, and Anticipation: A Case Study on Nanobiosensors in U.S. Cattle. In I. Management Association (Ed.), Biomedical Engineering: Concepts, Methodologies, Tools, and Applications (pp. 577-596). Hershey, PA: IGI Global. https://doi.org/ doi:10.4018/978-1-5225-3158-6.ch025

Wulff, E. (2017). Data and Operational Oceanography: A Review in Support of Responsible Fisheries and Aquaculture. In P. Diviacco, A. Leadbetter, & H. Glaves (Eds.), *Oceanographic and Marine Cross-Domain Data Management for Sustainable Development* (pp. 303–324). Hershey, PA: IGI Global. doi:10.4018/978-1-5225-0700-0.ch013

Yener, D. (2017). Geographic Information Systems and Its Applications in Marketing Literature. In S. Faiz & K. Mahmoudi (Eds.), *Handbook of Research on Geographic Information Systems Applications and Advancements* (pp. 158–172). Hershey, PA: IGI Global. doi:10.4018/978-1-5225-0937-0.ch006

Yu, K., Liu, Y., & Sharma, A. (2021). Analyze the Effectiveness of the Algorithm for Agricultural Product Delivery Vehicle Routing Problem Based on Mathematical Model. *International Journal of Agricultural and Environmental Information Systems, 12*(3), 26–38. https://doi.org/10.4018/IJAEIS.2021070103

Zhou, X., Sharma, A., & Mohindru, V. (2021). Research on Linear Programming Algorithm for Mathematical Model of Agricultural Machinery Allocation. *International Journal of Agricultural and Environmental Information Systems, 12*(3), 1–12. https://doi.org/10.4018/IJAEIS.2021070101

About the Contributors

Karmaoui Ahmed started his academic career as Research Fellow at Cadi Ayyad University and awarded Ph.D. from the Faculty of Semlalia (UCA); graduated from Publons Academy, and awarded Top 1% of international Reviewers (2018) by Publons. He has published various peer-reviewed international journal articles, book chapters, and books on a range of environmental topics. He is currently the president of the Southern Center for Culture and Science, Morocco, and an assistant professor at FSTE (UMI). Broad areas of his research interests include ecosystem services assessment.

***

Boutasknit Abderrahim is a doctor specialized in biotechnology and plant physiology.

Mohamed Anli has a PhD in Biotechnology and Plant Physiology, Master's degree in Ecosystem Management and Conservation Biology, Bachelor's degree in Biology Applied to Environmental Management, and is Faculty of Science Semlalia, Cadi Ayyad University, Marrakesh, Morocco.

Ben Laouane Raja works at the Laboratory of Agro-Food, Biotechnologies and Valorization of Plant Bioresources (AGROBIOVAL) Plant Biotechnology and Physiology Unit Faculty of Science Semlalia, Cadi Ayyad University Marrakesh, Morocco.

Nabil El Khatri is a PhD candidate in the Department of Agricultural Innovation and Technology Transfer Center AITTC at UM6P. Khatri has an Engineering degree in agronomy and ecology from the Hassan II Institute of Agronomy and Veterinary IAV, Rabat. Khatri's current PhD research is concerned with the Quantification of carbon sequestration in the rehabitated old mining sites in Morocco. Khatri has worked on the carbon balance assessment of olive oil in the region of Marrakech Safi with the National Institute of Agronomic Research of Marrakech and the Cherifian Office of Phosphate OCP. Khatri's research interests include: Agronomy, Forestry, Climate Change, Carbon footprint, and Remote Sensing.

Mohamed Ait El Mokhtar was born in 9 September 1985 in Charleville-Mézières, Ardennes state (France). He finished his high school studies in Marrakesh, Morocco, to obtain his Baccalaureate in "Experimental Sciences" in 2004. He continued his university studies at Cadi Ayyad University in Marrakesh to graduate with bachelor degree in Biology. In 2009, he was graduated from the same university with Master's degree in "Life and Health Sciences". Mr. Ait El Mokhtar started his professional carrier as professor of Biology in the high school since 2010. Four years later, he began his PhD in Plant biotechnology and physiology at Cadi Ayyad University. He's investigating the effect of organic and biological fertilizers on the tolerance of plants to biotic and abiotic stresses and on arbuscular mycorrhizal fungi infectivity of soils from Moroccan palm groves. He's also working on the impact of climate change on ecosystems and especially on the oasis ecosystem.

Manomeet Gupta is a sesearcher-PhD Student in Business Informatics.

Lhoussain Hajji is a full professor in Agro-Food and Food Safety, Food and hive product quality control, Apitherapy, and Global changes.

Proscovia Paschal Kamugisha is lecturer at Mzumbe University - Mbeya Campus College. Dr. Kamugisha attained a PhD in Agricultural Economics in 2015 of Sokoine University of Agriculture (SUA). Dr. Kamugisha is the Head of Department of Administrative Studies (HoD) of Mzumbe University – Mbeya Campus College – Tanzania since 2020. Since then she has been entrusted to serve as Acting Principal in the office of Campus Principal's office at various time periods. She has served a Board member of Mji Mpya Secondary School from 2017 – 2020 and ELU Nursery and Primary School from 2018 – 2021. Initially, Dr. Kamugisha worked as Acting Head of Department of Economics between 2015 and 2017 at various time periods. Currently, she is heading the research project titled "Economic Analysis of Broiler Production using Black Soldier Flies Larvae Meals in Tanzania" as Principal Investigator (PI) since June 2022. Moreover, Dr. Kamugisha has authored number of journal articles and book chapters in Agricultural Economics, environmental Economics and value chain management at both local and international levels. Her research interests include: management of natural resources, development economics, aquaculture economics, Livestock Economics, Crops Economics and development economics.

Sebastian Mhanga is a Tanzanian with a professional background in Agricultural Economics and Natural Resources Management for Sustainable Agriculture. He has worked as a development practitioner for supporting private sector with one, local and three international development organizations covering a period of 14 years. The local organization is PASS Trust, and the international organizations are Technoserve, SNV and MEDA. Currently he is a co- founder and the Managing Director for EXPAND (T) LIMITED a business consulting firm based in Morogoro, Tanzania with a mission to unblock technical and information barriers to entrepreneurs; focusing on markets, investment, technology and management. Based on his career background Sebastian has developed interest in Scientific writing for knowledge development, learning and discovery of new business opportunities. So far he has published four papers and one book chapter. His main areas of research interest are Economic Development and Natural Resources Management linked to Sustainable Agriculture.

Dana Nejedlova is an assistant professor of Informatics.

Athanasios Podaras is an assistant professor of Business Informatics.

Index

A

B

C

D

E

F

G

H

I

L

M

O

P

R

S

T

V

W

Y

www.ingramcontent.com/pod-product-compliance
Lightning Source LLC
LaVergne TN
LVHW080856300926
849162LV00003B/432